From Christoffel Words to Markoff Numbers

From Christoffel Words to Markoff Numbers

Second Edition

CHRISTOPHE REUTENAUER

OXFORD
UNIVERSITY PRESS

OXFORD
UNIVERSITY PRESS

Great Clarendon Street, Oxford, OX2 6DP,
United Kingdom

Oxford University Press is a department of the University of Oxford. It furthers the University's objective of excellence in research, scholarship, and education by publishing worldwide. Oxford is a registered trade mark of Oxford University Press in the UK and in certain other countries.

Published in the United States of America by Oxford University Press
198 Madison Avenue, New York, NY 10016, United States of America

British Library Cataloguing in Publication Data
Data available

Library of Congress Control Number: 2026930345

ISBN 9780197907559

DOI: 10.1093/9780197907535.001.0001

Printed and bound by
CPI Group (UK) Ltd., Croydon, CR0 4YY

MIX
Paper | Supporting responsible forestry
FSC® C013604
FSC www.fsc.org

The manufacturer's authorised representative in the EU for product safety is Oxford University Press España S.A. of El Parque Empresarial San Fernando de Henares, Avenida de Castilla, 2 – 28830 Madrid (www.oup.es/en or product.safety@oup.com). OUP España S.A. also acts as importer into Spain of products made by the manufacturer.

Foreword to the second edition

This book is a revised edition of my book, with the same title, published in 2019. Three chapters have been added: on quadratic forms, on quadratic numbers and their approximations by rationals, on standard words, and a final chapter on the commutator subgroup of $SL_2(\mathbb{Z})$. The book has been somewhat reorganized due to the logical dependence of chapters.

Among the results that have been added are the construction of the minimal automaton of the set of suffixes of a central word by Sciortino and Zamboni; their characterization of prefixes of central words; the characterization of standard words through factorization into squares due to Whiteland, Peltomäki, and Saarela; the freeness of the commutator subgroup of $SL_2(\mathbb{Z})$; the existence of a normal form for Sturmian words and the enumeration formula for Sturmian words by Lipatov and Mignosi; and the results on 3-matrices by Schmutz-Schaller.

Many proofs in the first edition have been improved: the proof of the existence of infinitely many good convergents for Markoff irrationalities and the result by Perrine on Markoff triples arising from bases of the commutator subgroup of $SL_2(\mathbb{Z})$.

Exercises have been added, many of them with sketches of solutions at the end of the book.

Acknowledgements. For many discussions, help, and email exchanges, I want to thank Hugh Thomas, Frédéric Patras, Mélodie Lapointe, James Currie, Yann Bugeaud, Sébastien Labbé, Paul Schmutz Schaller, David Clampitt, Dominique Perrin, Maxime Crochemore, Nathan Bonin, Francis Clavette, Antoine Abram, Thomas Haettel, Valérie Berthé, Gabriele Fici, Jarkko Peltomäki, Mélodie Andrieu, Markus Whiteland, Aleksi Saarela, Shuo Li, Antonio Restivo, Yan Lanciault, Julien Cassaigne, Ian Banfield, Jean-Paul Allouche, Jeffrey Shallit, Franco Saliola, Srečko Brlek, and François Bergeron.

Christophe Reutenauer
Université du Québec à Montréal
May 2025

Contents

Introduction to the first edition

In a short article written in Latin in 1875 [C1], Elwin Bruno Christoffel introduced a class of words on a binary alphabet, now called Christoffel words. They were followed in the 20th century by the theory of Sturmian sequences, introduced by Morse and Hedlund in 1940 [MH] in the framework of symbolic dynamics and developed more recently within combinatorics on words, with connections to discrete geometry. Independently of Christoffel, Andrei Andreyevich Markoff (the same as the one of Markov processes, but writing his name in the French way in his two articles written in French) published in 1879 and 1880 his famous theory [M1, M2], now called Markoff theory. It characterizes certain quadratic forms and certain real numbers by extremal inequalities. Both classes are constructed using certain words, essentially the Christoffel words. Markoff's theory has visibly fascinated many great mathematicians, who reproved one of the two aspects of the theory: Frobenius [F] (1913), Remak [R] (1924), Perron [P1, P2] (1921), Dickson [D1] (1930), Cassels [Ca1, Ca2] (1949, 1957), Harvey Cohn [Co1, Co2] (1955, 1972), and Bombieri [B] (2007). The link between Christoffel words and Markoff theory was noted by Frobenius (1913) and was known to Caroline Series [Se2] (1985).

The present book is motivated by this link. In Part I, we provide the classical theory of Markoff in its two aspects, based on the theory of Christoffel words. In Part II, we give more advanced results of the theory of Christoffel words.

I. Part I presents the classical theory of Markoff. This theory has two facets, both of which are considered. The first culminates in an approximation theorem for irrational real numbers: it gives a sequence of approximations whose error term is progressively smaller, which holds for all irrationals except for an exceptional sequence of quadratic numbers given in the theorem.

Let us be more precise. The first level of approximation is a theorem of Hurwitz: for each irrational real number x, there exist infinitely many rational numbers $\frac{p}{q}$ such that $|x - \frac{p}{q}| < \frac{1}{\sqrt{5}q^2}$.[1] The second level is stated as follows: if one excludes $x = \frac{1+\sqrt{5}}{2}$, the golden ratio, and also all x whose expansions into continued fractions have only 1s after some rank,[2] then there exist infinitely many approximations satisfying $\left|x - \frac{p}{q}\right| < \frac{1}{\sqrt{8}q^2}$. The third level is as follows: if one excludes the previous numbers, and also $x = 1 + \sqrt{2}$, and moreover all x whose expansions into continued fractions have only 2s after some rank,[3] then there exist infinitely many approximations satisfying $|x - \frac{p}{q}| < \frac{1}{\sqrt{221/25}q^2}$. The theorem continues: there is an infinite sequence of exceptions

[1] The theorem of Hurwitz strengthens Dirichlet's theorem: there exist infinitely many rationals such that $|x - \frac{p}{q}| < \frac{1}{q^2}$.

[2] The expansion of the golden ratio has only 1s.

[3] The expansion of $1 + \sqrt{2}$ has only 2s.

and better approximations; Markoff's constants $\sqrt{5}, \sqrt{8}, \sqrt{221/25}, \cdots$ form an infinite increasing sequence of quadratic numbers, whose limit is 3.

Markoff's constants are parametrized by a sequence of natural numbers, called the Markoff numbers: the Markoff number m defines the constant $\sqrt{9 - \frac{4}{m^2}}$; for example, $m = 1$, $m = 2$, and $m = 5$ give the three constants above. The exceptions mentioned above are also quadratic numbers, also parametrized by the Markoff numbers. The Markoff numbers have the remarkable property of being the components of solutions of a Diophantine equation, known as the Markoff equation: $x^2 + y^2 + z^2 = 3xyz$; for example, the triple $(1, 2, 5)$ is a solution.

Christoffel words naturally come into this theory. Indeed, returning to the exceptional numbers above, these are numbers whose continued fraction expansions, seen as infinite words, are of the form p^∞, for some finite words p; these p are exactly the Christoffel words, written on the alphabet $\{22, 11\}$. For example, the word 2211 will give the next exception, namely the number x whose continued fraction expansion is $[\overline{2211}]$, that is to say $x = \frac{9+\sqrt{221}}{10}$.[4]

The approximation theorem is not in the work of Markoff, although it is commonly attributed to him. Markoff's work actually concerns real indefinite binary quadratic forms (that is, forms in two variables, with positive discriminant, and real coefficients). For such a quadratic form $f(x, y) = \alpha x^2 + \beta xy + \gamma y^2$, we define $L(f) = \inf\{|f(x, y)| : x, y \in \mathbb{Z}, (x, y) \neq (0, 0)\}$, that is, the infimum of its images when the variables take integer values, not both 0; define also its discriminant $d(f) = \beta^2 - 4\alpha\gamma$. Markoff's theorem then expresses a result on inequalities (relating d and L) with exceptions that are progressively better, if one excludes some special quadratic forms. The first inequality $L(f) \leq \frac{1}{\sqrt{5}}\sqrt{d(f)}$, due to Korkine and Zolotareff, is true for all quadratic forms. Excluding the form $x^2 - xy - y^2$ and those equivalent to some multiple of it,[5] one has the inequality $L(f) \leq \frac{1}{\sqrt{8}}\sqrt{d(f)}$, also due to Korkine and Zolotareff. The next level is the fact that if a quadratic form is not equivalent to a multiple of $x^2 - xy - y^2$ nor of $2x^2 - 4xy - 2y^2$, then $L(f) \leq \left(1/\frac{\sqrt{221}}{5}\right)\sqrt{d(f)}$. The theorem continues with an infinite sequence of exceptions (the Markoff forms) and constants. The latter are the same as in the approximation theorem for irrational numbers.

The first part of the book begins with the aspects of the theory of Christoffel words that are needed to prove both facets of Markoff theory, which are the subject of the remainder of this part. In particular, the solutions of the Markoff equation (Markoff triples) are characterized: they are in bijection with Christoffel words.

This part of the book has some similarities with the recent book by Martin Aigner [A]. One difference is that we present here, in the most concise way, Markoff theory in its two facets: approximations of real numbers and infima of quadratic forms. The second facet is not treated in [A]. Aigner's book, however, gives many results about the Frobenius uniqueness conjecture, which is only stated here. Another difference is that we chose to work with sequences of integers (continued fractions are indeed

4 In order to obtain this value, one may follow the classical Proposition 5.4.1.

5 Two forms are equivalent if each is obtained from the other by some integral change of variables.

naturally sequences of integers), rather than with bi-infinite sequences, as Markoff himself does, as well as Dickson, Bombieri, and Aigner.

Note that most articles or books that present Markoff theory do so under one of its aspects: approximations of real numbers or minima of quadratic forms. To my knowledge, the only place where the two aspects are discussed is in the article [C1] by Cassels and his book [C2]; Cassels first proves the theorem on quadratic forms and then deduces the theorem on approximations.[6] This deduction is not trivial, and I know of no simple method that allows one to pass from one theorem to the other. However, the proofs of both theorems in the present book are deduced from several results on Christoffel words.

II. The second part of the book presents the theory of Christoffel words. We adopt the perspective of the precursors ([B0, B1, dL0]), Jean Berstel and Aldo de Luca, who turned this theory into a very rich field of combinatorics on words, related to arithmetic and discrete geometry. We investigate the particular palindromes (the central words) involved in this theory, the periods of these palindromes and the characterizations that follow, and the many links between Lyndon words and Christoffel words. Continued fractions play a special role for these words; we therefore study the Stern–Brocot tree, its connection with continued fractions, and also the geometric interpretation of continued fractions due to Felix Klein. The concept of conjugation of words is important for Christoffel words and leads to various characterizations of them; we discuss some of these, in particular, the generation of Christoffel words through Cayley graphs of cyclic groups. We also study their factors, obtaining the bound $n+1$ for the number of circular factors of length n, as well as the palindromic factors. We also study balanced words and (finite) Sturmian words. Sturmian morphisms, already introduced in the first part, are studied in relation to the automorphisms of the free group with two generators. We completely establish the results on the automorphisms of the free group, following Jakob Nielsen, and in particular focus on his characterization of inner automorphisms and an invertibility criterion. We deduce the characterizations of Sturmian morphisms, following results of Filippo Mignosi and Patrice Séébold. A very simple algorithm for the bases of this free group is given.

In this book, we do not address the theory of Sturmian sequences (often called Sturmian words, which are actually infinite words). The adopted perspective is finitary. We also do not address the various presentations of monoids and groups involved in the theory (see [Sb1], Proposition 1; [Lo], Section 2.3.3; [KR]).

A reader who is interested only in the theory of Christoffel words may proceed to Part II of the book after having read Chapter 1, Chapter 2, and Theorem 8.2.3 and its proof.

Acknowledgements: Many thanks to Srečko Brlek, Christian Kassel, Mélodie Lapointe, Valérie Berthé, Vladimir Shpilrain, Ilya Kapovich, Paul Schupp, Enrico Bombieri, Sergey Fomin, Florent Hivert, Dominique Perrin, Antonio Restivo, Amy Glen, Jean Berstel, François Bergeron, Hugh Thomas, David Clampitt, Damien Jamet, Caroline Series, Nancy Wallace, Herman Goulet-Ouellet, Serge Perrine, Fanny Desjardins, for discussions, help, and email exchanges.

[6] His proofs are totally different from those presented in the present book.

Aldo de Luca gave me several useful comments and suggestions: *ti ringrazio molto, Aldo.* The thorough reading by Patrice Séébold and Gwénaël Richomme of the manuscript saved me from many typos, instances of sloppiness, and some mistakes: *un très grand merci, Patrice et Gwénaël.* I am grateful to Aaron Lauve and Neha Siddiqui, who carefully read the whole manuscript, for their mathematical and stylistic suggestions: thanks a lot, Aaron and Neha.

Christophe Reutenauer
Montréal 2018

PART I
THE THEORY OF MARKOFF

1
Basics

We denote by $\mathbb{R}$ (resp. $\mathbb{Z}$, $\mathbb{N}$, $\mathbb{P}$) the set of real numbers (resp. integers, natural integers, positive natural integers). Note that $\mathbb{N}$ contains 0.

Let A be a finite set, called the *alphabet*. A *letter* is an element of the alphabet. A *word* is a finite sequence of letters, including the empty sequence, called the *empty word*; the *length*, or *degree*, of the word w is the length of the sequence, denoted by $|w|$. For a given letter $a \in A$, $|w|_a$ denotes the number of occurrences of the letter a in the word w; it is called the *a-degree* of w. The *free monoid* on A is the set of all words on A and is denoted A^*.

The set A^* is a monoid, the multiplication being the *concatenation* of words: the concatenation of the words u and v is obtained by writing u followed by v. The neutral element of this monoid is the empty word, which is therefore denoted 1; when there is ambiguity (for example, when $A = \mathbb{P}$, so that 1 is a letter), we denote it ϵ. This monoid is *free* on A, which means that each function from A into a monoid has a unique extension as a monoid homomorphism from A^* into this monoid.

For words w, u, and v, if $w = uv$, we say that it is a *factorization* of w; it is *non-trivial* if u and v are both non-empty.

For words m, u, v, and w such that $w = umv$, we say that u is a *prefix* of w, v is a *suffix* of w, and m is *factor* of w; it is *proper* if it is distinct from w.

If m is a factor of w, an *occurrence of m in w* is the data of m and of a position where m starts (equivalently a factorization $w = umv$).

Two words w and w' are *conjugate* if for some words u and v, one has $w = uv$ and $w' = vu$. The *conjugation class* of a word is the set of its conjugates.

Denote the *conjugator* by C: it is the mapping sending each word $a_1 \cdots a_n$ of length n onto $a_2 \cdots a_n a_1$. This defines an action of the cyclic group $\mathbb{Z}/n\mathbb{Z}$ on the set of words of length n.

The conjugates of a word w of length n are the words $\mathrm{C}^i(w)$, $i = 0, \ldots, n-1$. We say that a word is *radical*[1] if it is not a power of another word; for example, $aabb$ is radical and $aabbaabb$ is not. A word w is radical if and only if the words $\mathrm{C}^i(w)$, $i = 0, \ldots, n-1$ are distinct (see Exercise 1.0.2).

We say that a word m is a *circular factor* of the word w, if m is a factor of some conjugate of w. Equivalently, m is a factor of length $\leq |w|$ of the word ww; equivalently, it is a prefix of some conjugate of w.

The *reversal* of the word $w = a_1 \cdots a_n$, $a_i \in A$, is $\tilde{w} = a_n \cdots a_1$. We sometimes need a functional notation: $\mathcal{R}$ denotes the mapping defined by $\mathcal{R}(w) = \tilde{w}$; this mapping is an anti-automorphism of the free monoid, called the *reversal mapping*. A word w is

[1] The usual terminology in combinatorics on words is *primitive*, but here this is misleading because we shall give, in the free group, another meaning to the word "primitive".

From Christoffel Words to Markoff Numbers. Second Edition. Christophe Reutenauer, Oxford University Press.
 DOI: 10.1093/9780197907535.003.0001

a *palindrome* if $w = \tilde{w}$. If two words are conjugate, then their reversals are conjugate: this follows from the equalities $\widetilde{uv} = \tilde{v}\tilde{u}$ and $\widetilde{vu} = \tilde{u}\tilde{v}$.

The *commutative image* of a word w is the A-tuple $(|w|_a)_{a\in A}$; equivalently, it is the image of w under the canonical monoid homomorphism from the free monoid A^* onto the free commutative monoid $\mathbb{N}^A$.

A sequence over A is also called an *infinite word*.[2] Such a sequence $(a_n)_{n\in\mathbb{N}}$ is called *ultimately periodic* if for some $p \in \mathbb{P}$ and some $i \in \mathbb{N}$ one has $a_n = a_{n+p}$ for any $n \geq i$; it is *periodic* if moreover $i = 0$. If s is ultimately periodic as above, each finite factor of s of the form $a_n \cdots a_{n+p-1}$, with $n \geq i$ is called a *periodic pattern* of s.

One may multiply a finite word and an infinite one: $s = wt$ is the infinite word obtained by concatenating the finite word w and the infinite word t. Prefixes and factors of infinite words are finite words and are defined in the obvious way.

If $w = a_0 \cdots a_{k-1}$ is a non-empty word, with $a_i \in A$, we denote by w^∞ the infinite word $(b_n)_{n\in\mathbb{N}}$ such that for any $n \in \mathbb{N}$, $b_n = a_r$, where r is the remainder of the Euclidean division of n by k; w^∞ may also be written $www \cdots w \cdots$. Likewise, ${}^\infty w$ denotes a periodic left infinite word (a *left infinite word* is a sequence indexed by the integers $n \leq 0$).

A *bi-infinite word* is a bi-infinite sequence $(a_n)_{n\in\mathbb{Z}}$. It is called *periodic* if for some $p \geq 1$ and any n one has $a_n = a_{n+p}$. For a non-empty finite word $w = a_0 \cdots a_{k-1}$, we denote by ${}^\infty w^\infty$ the periodic bi-infinite word $(b_n)_{n\in\mathbb{Z}}$ defined for any $n \in \mathbb{Z}$ by $b_n = a_r$, where $r = n \bmod k$.

A homomorphism (we also say simply *morphism*) from a free monoid to another is called *substitution*; this is because applying the morphism to a word amounts to replacing each letter by the corresponding word. If the morphism is *non-erasing*, that is, sending each non-empty word onto a non-empty word, then one may apply it to infinite (resp. bi-infinite) words: it sends such a word onto another such word.

We denote by $\mathrm{F}(a, b)$ the *free group* on two free generators a, b. Each element in the free group may be represented by a *reduced word*, which is a product of the generators and their inverses, without factor xx^{-1} or $x^{-1}x$ with $x = a$ or b. The free monoid $\{a, b\}^*$ is a submonoid of $\mathrm{F}(a, b)$, since each word in $\{a, b\}^*$ is necessarily reduced. There is a canonical surjective group homomorphism from $\mathrm{F}(a, b)$ onto $\mathbb{Z}^2 = \mathbb{Z}^{\{a,b\}}$; the image of an element of $\mathrm{F}(a, b)$ is called its *commutative image*; this is compatible with the terminology for the free monoid.

It is useful to represent an endomorphism of the free monoid $\{a, b\}^*$, or of the free group $\mathrm{F}(a, b)$, where $a < b$, by the pair (u, v): this endomorphism maps a onto u and b onto v. This representation is compatible with the canonical injection of the monoid of endomorphisms of the free monoid $\{a, b\}^*$ into the monoid of endomorphisms of the free group $\mathrm{F}(a, b)$. The *abelianization* of such an endomorphism φ is the 2×2 matrix $Ab(\varphi) = \begin{pmatrix} p & q \\ r & s \end{pmatrix}$, where (p, r) (resp. (q, s)) is the commutative image of u (resp. v). If $h = \varphi(g)$, the commutative images (i, j) and (k, l) of g and h are related by

$$\begin{pmatrix} k \\ l \end{pmatrix} = \begin{pmatrix} p & q \\ r & s \end{pmatrix} \begin{pmatrix} i \\ j \end{pmatrix}$$

[2] We sometimes say "finite word" in order to clearly distinguish between finite and infinite words.

The abelianization map Ab is a monoid homomorphism from the monoid $\mathrm{End}(\mathrm{F}(a,b))$ of endomorphisms of the free group $\mathrm{F}(a,b)$ onto the multiplicative monoid of matrices $\mathbb{Z}^{2\times 2}$. It follows in particular that if such an endomorphism is invertible (that is, an automorphism of $\mathrm{F}(a,b)$), then the corresponding matrix is invertible and hence belongs to $\mathrm{GL}_2(\mathbb{Z})$.

The image of the restriction of Ab to the group $\mathrm{Aut}(\mathrm{F}(a,b))$ of automorphisms of $\mathrm{F}(a,b)$ is the group $\mathrm{GL}_2(\mathbb{Z})$. This is proved as follows: the matrices

$$R = \begin{pmatrix} 1 & 1 \\ 0 & 1 \end{pmatrix}, \quad L = \begin{pmatrix} 1 & 0 \\ 1 & 1 \end{pmatrix} \tag{1.1}$$

generate the group $\mathrm{SL}_2(\mathbb{Z})$; see pp. 312–313 in [KT]. Since the matrix $\begin{pmatrix} 0 & 1 \\ 1 & 0 \end{pmatrix}$ is of determinant -1 and since $\mathrm{SL}_2(\mathbb{Z})$ is a subgroup of index 2 of $\mathrm{GL}_2(\mathbb{Z})$, the latter group is generated by these three matrices. Since they are the abelianizations of the respective automorphisms (a, ab), (ab, a), and (b, a) of $\mathrm{F}(a,b)$, it follows that the abelianization mapping from $\mathrm{Aut}(\mathrm{F}(a,b))$ into $\mathrm{GL}_2(\mathbb{Z})$ is surjective.

We use several times the *Euler phi function*: for $n \geq 1$, $\varphi(n)$ is equal to the number of integers in $\{0, \dots, n-1\}$ which are relatively prime to n.

Note that throughout the book, unless otherwise specified, the alphabet will be $A = \{a, b\}$. If A must be ordered, it will be the order $a < b$.

We need to consider *the lexicographical order* on the set of finite or infinite words on an ordered alphabet: $u <_{lex} v$ if either u is a proper prefix of v, or if $u = mau'$, $v = mbv'$ for words m, u', and v', and letters a and b such that $a < b$. The lexicographical order is the order used in dictionaries, and we also call it the *alphabetical order*.

1.0.1 Exercises

Exercise 1.0.1 *Show that if $w = uv$, then each conjugate of w, distinct from uv and vu, is of the form u_2vu_1 for some non-trivial factorization $u = u_1u_2$ or of the form v_2uv_1 for some non-trivial factorization $v = v_1v_2$.*

Exercise 1.0.2 *Show that the following conditions are equivalent for a word w of length $n \geq 1$:*

(i) w is radical;
(ii) $\{d \in \mathbb{Z}, \mathrm{C}^d(w) = w\} = n\mathbb{Z}$;
(iii) the words $\mathrm{C}^i(w), i = 0, \dots, n-1$ are all distinct;
(iv) w has n conjugates.

Hint: the mapping C induces an action of the group $\mathbb{Z}$ on the set of conjugation classes of w; the subgroup fixing w is of the form $d\mathbb{Z}$ for some divisor d of n.

Exercise 1.0.3 *Show that the number of $(i,j) \in \mathbb{N}^2$, i and j relatively prime, $i + j = n$, is equal to 2 if $n = 1$ and to $\varphi(n)$ if $n \geq 2$.*

Exercise 1.0.4 *Let w be a radical word of length n on the alphabet $\{1, 2, \ldots, k\}$. Let L be its conjugation class, ordered lexicographically, and let $c_i = |w|_i$. Let $K_1, \ldots, K_k$ be the successive intervals of L of length $c_1, \ldots, c_k$. Show that K_j is the set of words in L that begin with j. Show that w, viewed as a sequence, is obtained from the sequence $(w, C(w), \ldots, C^{n-1}(w))$ by replacing in it each x by j if $x \in K_j$.*

Exercise 1.0.5 *Let u and v be two finite factors of equal length of an infinite periodic word s. Suppose that there are infinitely many finite words w such that uw and vw are factors of s. Show that $u = v$.*

2
Words

For the definition of Christoffel words, we follow the geometrical definition of Berstel [B2] (pp. 300–301). This geometrical approach has the advantage of allowing geometrical proofs. We prove the theorem of Borel and Laubie: each Christoffel word is uniquely the product of two Christoffel words. The tree of Christoffel pairs of Berstel and de Luca [BdL] is deduced. Sturmian morphisms are the substitutions which stabilize the set of conjugates of Christoffel words. They play an important role in the theory, allowing us to compute all Christoffel pairs and therefore to describe the previous tree. It will be seen in Section 15.5 that they coincide with the classical Sturmian morphisms.

2.1 Tiling the plane with a parallelogram

The *plane* is the Cartesian plane $\mathbb{R} \times \mathbb{R}$. The *discrete plane* is the set of points with integer coordinates in the plane, that is, the set $\mathbb{Z} \times \mathbb{Z}$. An element of the discrete plane is called an *integral point* or *integral vector*. A *coordinate line* is any line of the form $y = n$, or $x = n$, for some $n \in \mathbb{Z}$.

Recall that a *basis* of $\mathbb{Z} \times \mathbb{Z}$ is a pair of integral vectors such that each vector in $\mathbb{Z} \times \mathbb{Z}$ is uniquely a linear combination over $\mathbb{Z}$ of these two vectors. It is well known that two integral vectors $u = (a, b)$ and $v = (c, d)$ form a basis of $\mathbb{Z} \times \mathbb{Z}$ if and only if the matrix $\begin{pmatrix} a & b \\ c & d \end{pmatrix}$ has determinant ± 1. We denote this determinant by $\det(u, v)$.

Consider a parallelogram in the plane; we assume that it is nontrivial, that is, not included in a line. A *side vector* of the parallelogram is any of the four possible vectors obtained by taking a side and orienting it. Any two side vectors corresponding to adjacent sides form a basis of $\mathbb{R} \times \mathbb{R}$ as vector space over $\mathbb{R}$. The parallelogram defines a tiling of the plane as follows: iteratively translate the parallelogram by its side vectors; see Figure 2.1. Assuming that one of its vertices is the origin, we see that the additive subgroup of $\mathbb{R} \times \mathbb{R}$ spanned by the side vectors of the parallelogram is equal to the set of all vertices of all the parallelograms of the tiling.

Side vectors of a triangle are defined similarly.

A *primitive triangle* (resp. *primitive parallelogram*) is any triangle (resp. any parallelogram) whose vertices are integral points and which does not contain any other integral point. Note that given a parallelogram with integral vertices, the symmetry around its centre preserves the discrete plane; indeed, the centre is half an integral vector, and this symmetry maps any vector u onto twice the centre minus u. Hence, given a diagonal of this parallelogram, the two triangles in the parallelogram, delimited by this diagonal, are simultaneously primitive or not, and this happens if and only if the parallelogram itself is primitive.

From Christoffel Words to Markoff Numbers. Second Edition. Christophe Reutenauer, Oxford University Press.
 DOI: 10.1093/9780197907535.003.0002

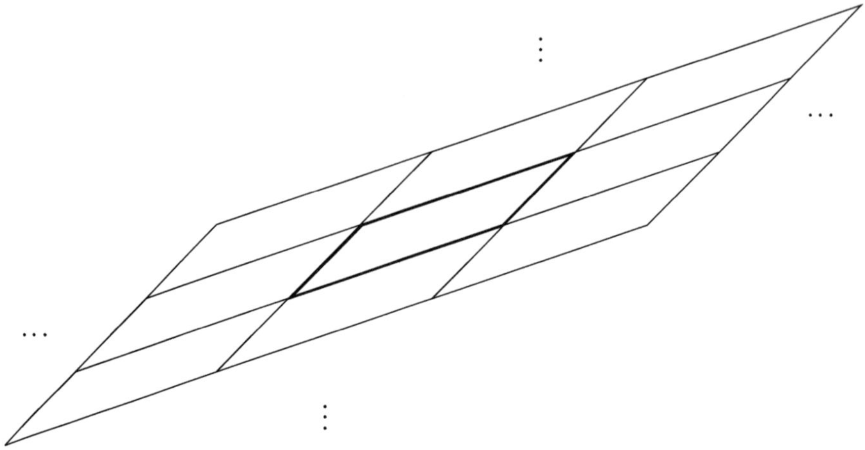

Figure 2.1 Tiling the plane with a parallelogram

The following result holds: *a parallelogram with integral vertices is primitive if and only if any two side vectors, corresponding to adjacent sides, form a basis of* $\mathbb{Z}\times\mathbb{Z}$. This result is proved geometrically as follows: consider the tiling of the plane as above. The parallelograms in the tiling are all primitive if and only if one of them is primitive, in which case there is no integral point outside the set of vertices of all these parallelograms. Hence, the subgroup spanned by the side vectors is all of $\mathbb{Z}\times\mathbb{Z}$ if and only if the parallelogram is primitive. In this case the two vectors span the discrete plane, and they form a basis since they are linearly independent.

From this follows the fact that *a triangle with integral vertices is primitive if and only if any two of its side vectors form a basis of* $\mathbb{Z}\times\mathbb{Z}$.

We also deduce that *given two integral vectors u and v, the parallelogram with vertices* $0, u, v, u+v$ *is primitive if and only if* $\det(u,v) = \pm 1$. We may be more precise, assuming that the two non-zero vectors u and v are in the first quadrant. Then $\det(u,v) > 0$ if and only if the slope α of u is smaller than the slope β of v. Indeed, u (resp. v) is positively proportional to the vector $(1,\alpha)$ (resp. $(1,\beta)$) so that $\det(u,v)$ is positively proportional to the determinant of the matrix $\begin{pmatrix} 1 & \alpha \\ 1 & \beta \end{pmatrix}$, which is $\beta - \alpha$ (we leave to the reader the case where the slope of u or v is infinite). In conclusion, *given two non-zero integral vectors* u, v *in the first quadrant, with* $\text{slope}(u) < \text{slope}(v)$, *the parallelogram constructed on them is primitive if and only if* $\det(u,v) = 1$.

For further use, when we prove that a Christoffel word is uniquely the product of two others, we prove the following lemma.

Corollary 2.1.1 *Let* A and B *be two integral points such that* $B = A + (p,q)$ *with* p and q *relatively prime positive integers. There is a unique integral point M lying in the rectangle with diagonal AB, below this diagonal, such that AMB is a primitive triangle.*

The result and its proof are illustrated in Figure 2.2.

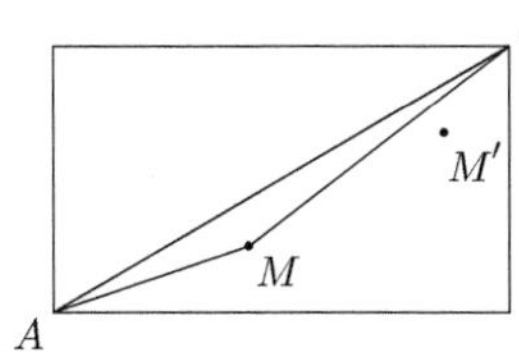

Figure 2.2 The unique point M such that AMB is a primitive triangle

Proof. For M choose the integral point in the rectangle, below the diagonal, and closest to the diagonal. Since p and q are relatively prime, there is no integral point on the open segment (A, B). One deduces that the triangle AMB is primitive.

To prove uniqueness, suppose that there is another point M' such that the triangle $AM'B$ is primitive. Then, since the triangle AMB is primitive, the vectors $\overrightarrow{AM}$ and $\overrightarrow{AB}$ form a basis of $\mathbb{Z}^2$. Their determinant is therefore equal to ± 1, and actually (by the result above), it is 1, since the slope of $\overrightarrow{AM}$ is smaller than that of $\overrightarrow{AB}$. Similarly $\det(\overrightarrow{AM'}, \overrightarrow{AB}) = 1$. Hence, we have $\det(\overrightarrow{AM}, \overrightarrow{AB}) = 1 = \det(\overrightarrow{AM'}, \overrightarrow{AB})$. Thus, taking the difference, $\det(\overrightarrow{MM'}, \overrightarrow{AB}) = 0$ so that MM' and AB are parallel. Since MM' is strictly under the diagonal, it is shorter than AB, and we deduce that there is some integral point on the open segment AB (namely the point $C = A + \overrightarrow{MM'}$), a contradiction. □

2.2 Christoffel words

A *lattice path* is a sequence of consecutive elementary steps in the plane; each *elementary step* is a segment $[(x, y), (x+1, y)]$ or $[(x, y), (x, y+1)]$, with $x, y \in \mathbb{Z}$. In this section, the paths will be finite, but we also consider infinite paths in the sequel to this book.

Let p and q be non-negative relatively prime integers. Consider the segment from some integral point A to $B = A + (p, q)$ and the lattice path from A to B located below this segment, such that the polygon delimited by the segment and the path has no interior integer point.

Given a totally ordered alphabet $\{a < b\}$, the *lower Christoffel word of slope q/p* is the word in the free monoid $\{a, b\}^*$ coding the above path, where a (resp. b) codes a horizontal (resp. vertical) elementary step. See Figure 2.3, which shows the path corresponding to $(p, q) = (7, 4)$, corresponding to the lower Christoffel word *aabaabaabab* of slope 4/7. Note that the slope of a lower Christoffel word is equal to the slope of the segment in the plane delimited by the extreme points of the corresponding discrete path. Note also that for the definition of the lower Christoffel word, the choice of A is immaterial, and we often take $A = (0, 0)$.

We say that the above path and the lower Christoffel word *discretize from below* the segment AB.

The upper Christoffel word of slope q/p is defined similarly by considering the lattice path located above the segment. Since the rectangle with opposite vertices A and B and sides parallel to the coordinate lines has a symmetry around its centre,

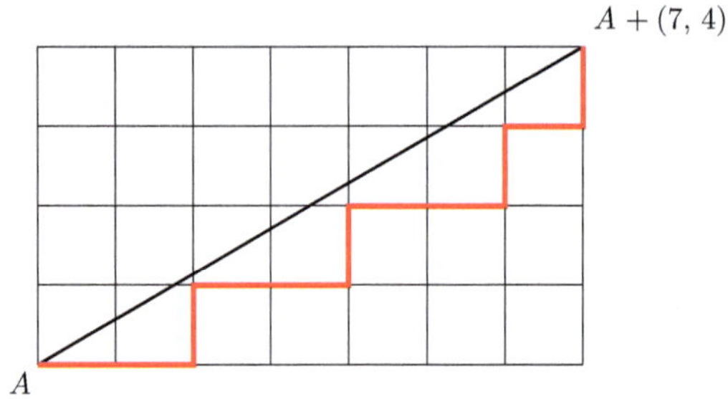

Figure 2.3 The lower Christoffel word *aabaabaabab* of slope 4/7

it follows that *the upper Christoffel word of a given slope is the reversal of the lower Christoffel word of the same slope.*

A *Christoffel word* is a word which is a lower or an upper Christoffel word.

Denote by E the automorphism of the free monoid $\{a, b\}^*$ that exchanges a and b. Then E exchanges lower and upper Christoffel words. It sends the lower Christoffel word of slope q/p onto the upper Christoffel word of slope p/q: to see it, in Figure 2.3 for example, let $A = (0, 0)$, and take the symmetry with respect to the diagonal $y = x$.

Clearly, the number of a's in the lower and upper Christoffel words of slope q/p is p, while the number of b's is q. In particular, $|w|_a$ and $|w|_b$ are relatively prime when w is a Christoffel word and w cannot be a power of another word. Hence, w is radical.

The letters a and b are Christoffel words, simultaneously upper and lower ones, of respective slope 1 and ∞; they correspond to the cases $(p, q) = (1, 0)$ and $(p, q) = (0, 1)$, respectively. The other Christoffel words are called *proper*. The words $a^n b$ and ab^n, for $n \geq 0$, are lower Christoffel words.

2.2.1 Exercises

Exercise 2.2.1 *Show that the number of Christoffel words of length n is* $2\varphi(n)$*, where* φ *is the Euler function, and that the number of lower Christoffel words of length n is 2 if* $n = 1$*, and* $\varphi(n)$ *if* $n \geq 2$*. Hint: use Exercise 1.0.3.*

Exercise 2.2.2 *Show that aa (resp. bb) is a factor of a Christoffel word w if and only if the slope of w is* < 1 *(resp.* > 1*).*

Exercise 2.2.3 *Show that if a lower Christoffel word on the alphabet* $\{a < b\}$ *has ab as prefix (resp. as proper prefix), then its slope is* ≥ 1 *(resp.* > 1*). Conversely, if it has aa as prefix, then its slope is* < 1*. Hint: use the geometrical definition of the word; the segment passes above or under the point* $(1, 1)$*.*

Exercise 2.2.4 *Let w be a Christoffel word. Show that if a word u is a prefix and a suffix of w, then* $u = 1$ *or* $u = w$*.*

Exercise 2.2.5 *Show that an upper Christoffel word on the alphabet $\{a < b\}$ is a lower Christoffel word for the opposite order.*

2.3 Palindromes

Let p and q be relatively prime natural numbers. Call the *cutting word* (or *billiard word*) *associated with* q/p the word m obtained as follows: let A and B be two integral points such that $\overrightarrow{AB} = (p, q)$ and consider the intersections of the open segment AB with the coordinate lines. If the coordinate line is vertical, then code the intersection by a, and if it is horizontal, code it by b. Then m is the word obtained by reading these letters, corresponding to the intersections, successively from left to right. For example, see Figure 2.4, which shows the cutting word associated with 4/7. Note that the cutting word m associated with q/p satisfies $|m|_a = p - 1$, $|m|_b = q - 1$.

It is clear that *any cutting word is a palindrome*: indeed, the defining rectangle (the rectangle of diagonal AB) has a symmetry around its centre, which preserves coordinate lines and the diagonal.

Theorem 2.3.1 *A proper lower Christoffel word of slope r is of the form amb for some palindrome m, which is the cutting word associated with r. The corresponding upper Christoffel word is bma.*

Proof. We prove the first assertion. Note that, clearly, a proper lower Christoffel word begins by a and ends by b. Figure 2.5 shows the bijective correspondence between the steps of the path defining the lower Christoffel words (excluding the first and the last step) and the intersection points defining the cutting word.

For the second assertion, we apply the fact, seen in Section 2.2, that the upper Christoffel word, corresponding to a lower Christoffel word, is the reversal of the latter. □

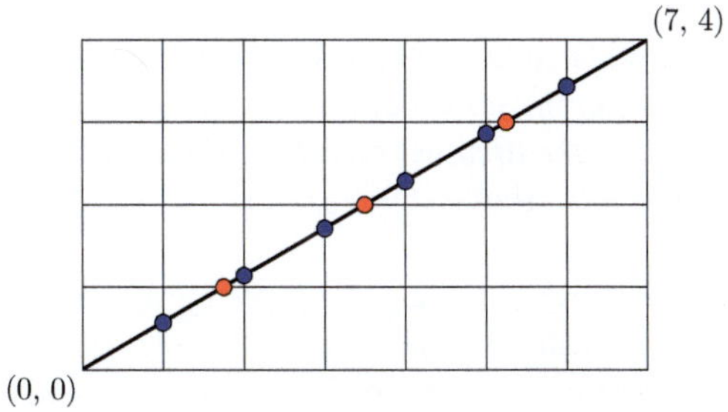

Figure 2.4 The cutting word *abaabaaba* associated with 4/7

Figure 2.5 Bijection between steps and intersections

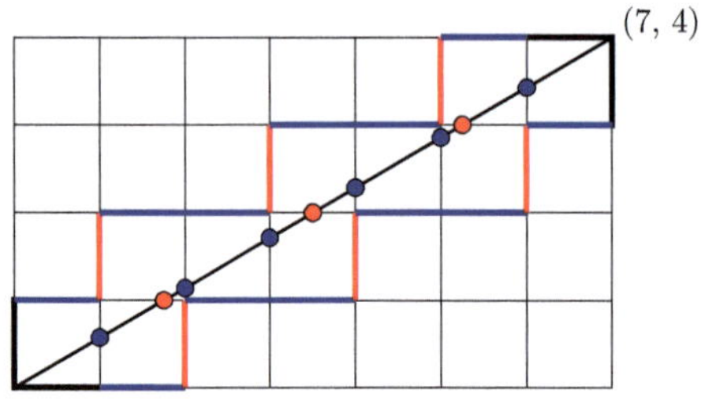

Figure 2.6 Lower and upper Christoffel words and the cutting word

The theorem is illustrated in Figure 2.6, where the lower and upper Christoffel words of slope 4/7 are represented. Their paths together form a snake-like polygon, which illustrates the fact that they are of the form *amb* and *bma* for the same word *m*, which is the cutting word associated with 4/7.

The proof of the next theorem is left to the reader.

Theorem 2.3.2 *Let* p and q *be two positive relatively prime integers. Let* $P = \{ip, 0 < i < q\}$ *and* $Q = \{jq, 0 < j < p\}$. *Then* P *and* Q *are disjoint. Let* $P \cup Q = \{r_1 < r_2 < \ldots < r_{p+q-2}\}$. *Define the word* $m = a_1a_2 \cdots a_{p+q-2}$ *on the alphabet* $\{a < b\}$ *by the following conditions:* $a_i = a$ *if* $r_i \in Q$ *and* $a_i = b$ *if* $r_i \in P$. *Then* m *is the cutting word associated with* $\frac{q}{p}$. *Each cutting word may be obtained in this way.*

We say that *m* is obtained by *superposition of two periodic phenomena.* See [BLRS], Section I.6.5, for concrete examples of this – in particular, one appearing in a novel of Paul Auster.

2.3.1 Exercises

Exercise 2.3.1 *Show that the area of the snake-like polygon defined by the lower and upper Christoffel words of the same slope* q/p *is* $p + q - 1$ *(*p, q *relatively prime; see Figure 2.6). Deduce that the area between the x-axis and the path representing a lower Christoffel word* w *is* $\frac{1}{2}(|w|_a - 1)(|w|_b - 1)$.

Exercise 2.3.2 *For each word* $w \in \{a, b\}^*$, *denote by* $A(w)$ *the area between the x-axis and the path represented by the word w, starting from the origin. Show that* $A(w) = \binom{w}{ba}$, *where the* subword function $\binom{w}{u}$ *denotes the number of occurrences of* u *as subword of* w *(*$u = a_1 \cdots a_k$, *written as a product of letters, is a* subword *of* w *if* $w = w_0a_1w_1 \cdots w_{k-1}a_kw_k$ *for some words* w_i*). Show that* $A(wa) = A(aw) + |wa|_b$ *and* $A(wb) = A(bw) - |bw|_a$. *See [Re5] (Lemma 16.3).*

Exercise 2.3.3 *We consider here paths in the discrete plane* $\mathbb{Z}^2$, *with four elementary steps* $[(x, y), (x + 1, y)]$, $[(x, y), (x, y + 1)]$, $[(x + 1, y), (x, y)]$, $[(x, y + 1), (x, y)]$, $x, y \in \mathbb{Z}$, *respectively coded by the letters* $a, b, \bar{a}$, *and* $\bar{b}$. *Denote by* $w \mapsto \bar{w}$ *the involutive* *automorphism of the free monoid* $\{a, b, \bar{a}, \bar{b}\}^*$ *which exchanges* a *and* $\bar{a}$, b *and* $\bar{b}$. *Let* λ *be the monoid endomorphism of such that* $\lambda(a) = a\bar{b}ab$ *and* $\lambda(b) = ab$, *which commutes with the previous involution. Let* w *be a lower Christoffel word, and define* $W = \lambda(w\bar{w})$.

Show that the path associated with W *is a closed path, which is self-avoiding. Hint:*[1] *let* A *be the origin of the path associated with* W, *and* B *the point attained after the path coded* $\lambda(w)$; *show that the path coded* $\lambda(w)$ *(resp.* $\lambda(\bar{w})$*) is strictly under (resp. above) the segment* AB, *except for* A *and* B; *show that* $\lambda(w)$ *goes from left to right, hence is self-avoiding, and similarly for* $\lambda(\bar{w})$.

Since the path associated with W *is closed and self-avoiding, it defines a subset* T *of the plane, whose frontier is the path. Show that* T *is a* square, *that is,* W *is conjugate to a word of the form* $uv\hat{u}\hat{v}$, *where* $\hat{u}$ *is the reversal of* $\bar{u}$. *Hint: consider the conjugate* $C^{-1}(W)$ *and use the palindromicity in Theorem 2.3.1. Show that one may tile the whole discrete plane by applying appropriate translations to* T. *Hint: take the group of translations generated by the translations corresponding to the paths* u *and* v.

See [BBGL1, BBGL2].

The authors also show that if $w = amb$, m *palindrome, then* $\lambda(w\bar{w})$ *has two factorizations* $uv\hat{u}\hat{v}$ *if and only if* w *is a Christoffel word.*

2.4 Standard factorization

Christoffel words may be constructed recursively. The first step follows from the following result.

Theorem 2.4.1 *(Borel and Laubie [BL], pp. 28–29) Each proper Christoffel word is uniquely the product of two Christoffel words.*

This factorization of a proper Christoffel word is called its *standard factorization.* Note that, as the proof will show, the two elements in the standard factorization of a lower Christoffel word are lower Christoffel words. We say that (u, v) is a *Christoffel pair* if u, v, and uv are all three *lower* Christoffel words.

Proof.

(i) It is enough to show that each proper lower Christoffel word is uniquely the product of two lower Christoffel words: indeed, upper Christoffel words are the reversals of lower Christoffel words. Moreover, if a lower Christoffel word is the product of two Christoffel words, then they must be lower ones. This follows from the fact that, with $a < b$, a proper Christoffel word is a lower one if and only it begins by a and also if and only it ends by b.

(ii) Let q/p be the slope of the proper lower Christoffel word w. Let A and B be integral points such that $B = A + (p, q)$. In the rectangle whose diagonal is the segment $[A, B]$, choose M as in Corollary 2.1.1. Then the triangle AMB is

[1] I am indebted to Sébastien Labbé for these indications.

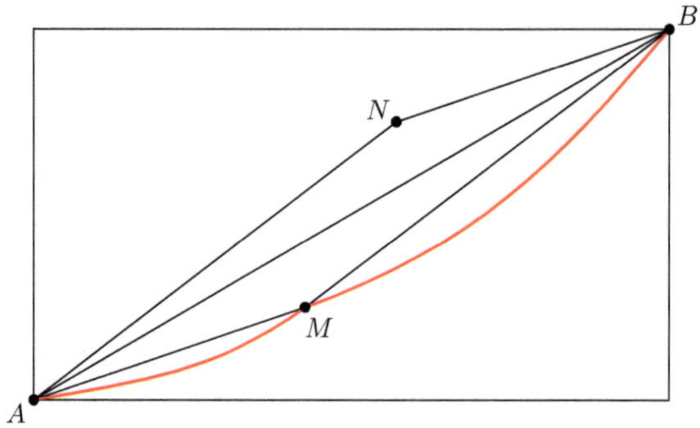

Figure 2.7 Standard factorization

primitive. There is no integral point on the open segment (AM) so that $\overrightarrow{AM} = (i, j)$ with i and j relatively prime. Similarly, $\overrightarrow{MB} = (k, l)$, with k and l relatively prime. Then the lower Christoffel word of slope q/p is the concatenation of the two lower Christoffel words of slope j/i and l/k. See Figure 2.7, where w is represented by the red curve; it is the product of the two words represented by the red curve from A to M and that from M to B.

(iii) Conversely, if a lower Christoffel word w is the product of two such words, $w = uv$ say, then in the defining rectangle of w, with diagonal AB, we define on the path coded by w the point M, corresponding to the factorization $w = uv$; then the commutative image of u (resp. v) is the vector $\overrightarrow{AM}$ (resp. $\overrightarrow{MB}$). We claim that the triangle AMB is primitive. This will imply by Corollary 2.1.1 that M is unique, hence u, v too.

For the claim, since u and v are lower Christoffel words, the paths corresponding to u and v are below the segments AM and MB, respectively; hence, the triangle is contained in the polygon delimited by the path corresponding to w and the segment AB, and we know that this polygon has no interior integer points. Thus, the triangle AMB is primitive. □

In Figure 2.7, the point N is defined by the condition $\overrightarrow{AN} = \overrightarrow{MB}$. Then the parallelogram $AMBN$ is primitive, as follows from Section 2.1.

An example of standard factorization is seen in Figure 2.3: the closest point is $A + (2, 1)$, and therefore the standard factorization of the word is $(aab)(aabaabab)$.

Corollary 2.4.2 *(Borel and Laubie [BL], Proposition 1; see also Cohn [Co2], Section 8; Osborne and Zieschang [OZ], Lemma 2.2) The following conditions are equivalent for two lower Christoffel words u and v of commutative image (i, j) and (k, l), respectively:*

(i) uv is a lower Christoffel word;

(ii) the determinant of the matrix $\begin{pmatrix} i & j \\ k & l \end{pmatrix}$ *is 1;*

(iii) the parallelogram constructed on the vectors (i, j) and (k, l) is primitive, and the slope of (i, j) is smaller than the slope of (k, l).

In this case, $w = uv$ is the standard factorization of w.

Proof. (i) implies (ii): take the notations of the proof of Corollary 2.1.1 and of the third part of the proof of Theorem 2.4.1. We have $(i, j) = \overrightarrow{AM}$, $(k, l) = \overrightarrow{MB}$. Then $1 = \det(\overrightarrow{AM}, \overrightarrow{AB}) = \det(\overrightarrow{AM}, \overrightarrow{AM} + \overrightarrow{MB}) = \det(\overrightarrow{AM}, \overrightarrow{MB})$.

(ii) implies (iii): this follows from Section 2.1.

(iii) implies (i): let A be the origin. Let $\overrightarrow{AM} = (i, j)$, $\overrightarrow{MB} = (k, l)$, $\overrightarrow{AN} = (k, l)$. The parallelogram is $AMBN$. Then the slope condition implies that M is below the segment AB. See Figure 2.7, where the discretization from below of the segment AM (resp. MB) is u (resp. v). Since the parallelogram is primitive, the discretization of AB is uv, which is therefore a lower Christoffel word.

The final assertion follows from Theorem 2.4.1. □

Corollary 2.4.3 *Let $w = uv$ be a lower Christoffel word with its standard factorization.*

(i) Then $u(uv)$ and $(uv)v$ are lower Christoffel words with the indicated standard factorization.
(ii) If w is of length at least 3, then either u is a proper prefix of v, in which case $v = uv'$ is the standard factorization of v; or v is a proper suffix of u, in which case $u = u'v$ is the standard factorization of u.

Proof. Denote by (i, j) and (k, l) the commutative images of u and v respectively. By (ii) in Corollary 2.4.2, we have $il - jk = 1$.

(i) The commutative image of uv is $(i + k, j + l)$. We have $(i + k)l - (j + l)k = 1$; by (i) in the same result, $(uv)v$ is a lower Christoffel word, and this word has the indicated standard factorization. The proof for $u(uv)$ is similar.
(ii) We assume now that the length of w is at least 3, that is, $i + j + k + l \geq 3$. Then $j + k > 0$: otherwise $j = k = 0$, $1 = il - jk = il$ and $i = l = 1$. Then $i + j + k + l = 2$, a contradiction.

It is enough by Theorem 2.4.1 to prove that either $v = uv'$ with v' a lower Christoffel word or that $u = u'v$ with u' a lower Christoffel word.

(a) Suppose first that $i \leq k$. Consider the integral points A, $M = A + (i, j)$, $N = A + (k, l)$, $B = A + (i + k, j + l) = M + (k, l) = N + (i, j)$, as in Figure 2.7 ($M$ is below the diagonal AB since the slope j/i of $\overrightarrow{AM}$ is smaller than the slope l/k of $\overrightarrow{MB}$ because $il - jk = 1$). By the proof of Theorem 2.4.1, the triangle AMB is primitive; hence, the parallelogram $AMBN$ is primitive by Section 2.1. Translate this

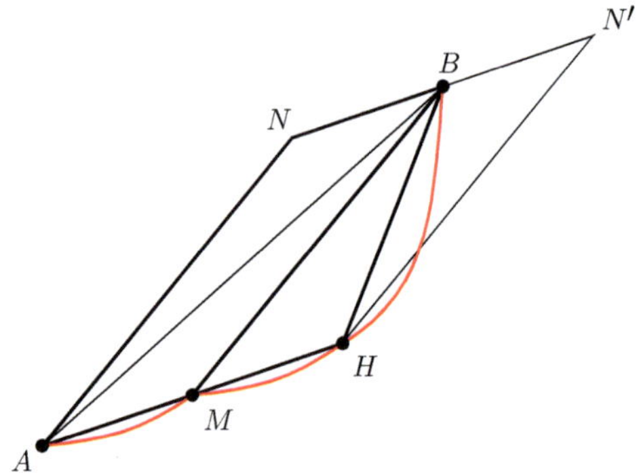

Figure 2.8 u is a prefix of v

parallelogram by the vector $\overrightarrow{AM}$, obtaining the parallelogram $MHN'B$; the condition $i \leq k$ implies that the x-coordinate of $H = A + (2i, 2j)$ is not larger than that of B (see Figure 2.8). The two parallelograms are primitive so that the only integral points in their union are their vertices A, M, H, N', B, and N. In particular, there are no integral points within the interior of the triangle AHB. It follows that the discretization from below of the segment AB is that of AM, followed by that of MH and then by that of HB. But the discretization of AM is u and that of MB is v (proof of Theorem 2.4.1). Hence, the latter is the product of that of MH and that of HB. Since $\overrightarrow{AM} = \overrightarrow{MH}$, the discretizations of AM and MH are equal, which implies that u is a prefix of v; moreover $v = uv'$ where v' is the discretization of HB; hence, v' is a lower Christoffel word.

(b) Suppose now that $i > k$. We have $il - jk = 1$, so we must have $l \leq j$: indeed, otherwise $l > j$, hence $il \geq (k+1)(j+1) = jk + j + k + 1 > jk + 1$ (since $j + k > 0$, as verified at the beginning of (ii)), contradicting $il - jk = 1$. The involutive automorphism E of the free monoid $\{a, b\}^*$ that exchanges a and b also exchanges lower and upper Christoffel words. Moreover, the reversal mapping also exchanges lower and upper Christoffel words, which implies that $E(\tilde{w})$ is a lower Christoffel word. Since $w = uv$, $E(\tilde{w})$ has, by the uniqueness in Theorem 2.4.1, the standard factorization $E(\tilde{v})E(\tilde{u})$. The commutative images of $E(\tilde{v})$ and $E(\tilde{u})$ are (l, k) and (j, i) respectively. Since $l \leq j$, part (a) of the proof shows that $E(\tilde{u}) = E(\tilde{v})m$ for some lower Christoffel word m. Hence, $\tilde{u} = \tilde{v}E(m)$ and $u = u'v$, with $u' = E(\tilde{m})$, and the latter is a lower Christoffel word. □

2.5 The tree of Christoffel pairs

This tree, which is implicit in [BL], appears in [BdL] (Figure 3, p. 200); it appears already in another form in [Co2] (Figure 1, p. 19).

It follows from Corollary 2.4.3 that if (u, v) is a Christoffel pair, then (u, uv) and (uv, v) are both Christoffel pairs. Moreover, (a, b) is clearly a Christoffel pair. Thus,

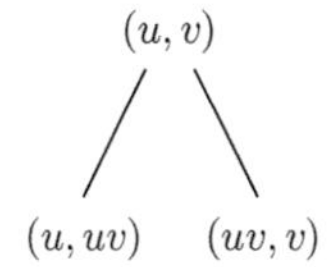

Figure 2.9 The rule for constructing the tree of Christoffel pairs

if we construct an infinite binary tree with root (a, b) and with the rule described in Figure 2.9, then all nodes of this tree are Christoffel pairs. Conversely, each Christoffel pair appears on this tree: this follows from the second part of Corollary 2.4.3. The latter also implies that the Christoffel pairs on the tree are all distinct: indeed, a node (u, v), distinct from the root is either a left child or a right child but not both (indeed, the conditions "u prefix of v" and "v prefix of u" are mutually exclusive), and one concludes recursively.

Note that if (u, v) appears on the tree, then uv is a lower Christoffel word; moreover, each lower Christoffel word appears exactly once in the tree as the product of the two elements in the pair by uniqueness of the standard factorization.

The following result derives easily from the rule for constructing all Christoffel pairs.

Corollary 2.5.1 *The set of pairs $(|u|, |v|)$ for all Christoffel pairs (u, v) is equal to the set of pairs of relatively prime positive natural integers.*

Observe that a proper lower Christoffel word $w = uv$, with its standard factorization, is such that at least one of u or v is not proper if and only if $w = a^n b$ or $w = ab^n$, $n \geq 1$; its standard factorization is $(a, a^{n-1}b)$ or (ab^{n-1}, b) respectively.

For later use, call *ancestor* (resp. *proper ancestor*) of a node N in a rooted tree, each node located on the path from the root to N (resp. and distinct from N).

2.5.1 Exercises

Exercise 2.5.1 *Show that if (u, v) is an ancestor of (f, g) in the tree of Christoffel pairs, then uv is a factor of fg, u is a prefix of f, and v is a suffix of g.*

Exercise 2.5.2 *Let $n \geq 1$. Show that the following conditions are equivalent:*

(i) *(u, v) is a standard pair;*
(ii) *(uv^{n-1}, v) is a standard pair;*
(iii) *$(u, u^{n-1}v)$ is a standard pair.*

Hint: if (u, v) is a standard pair, then either u is prefix of v, $v = uv'$, and then the parent of (u, v) is (u, v'); or v is a suffix of u, $u = u'v$, and then the parent of (u, v) is (u', v).

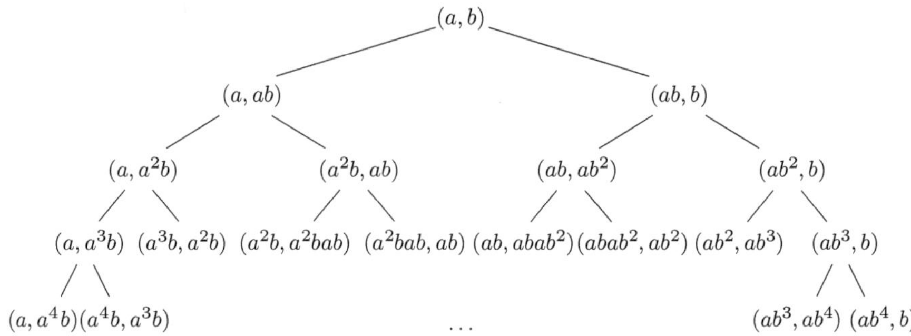

Figure 2.10 The tree of Christoffel pairs

Exercise 2.5.3 *Let $w = uv$ be a Christoffel word with its standard factorization. Show that the prefix of length $|v| - 1$ of v (resp. suffix of length $|u| - 1$ of u) is a prefix (resp. suffix) of w. Hint: use the recursive construction of the standard pairs (Figure 2.9).*

Exercise 2.5.4 *Consider the sequence of pairs $(u_n, v_n)_{n\geq 1}$ obtained by scanning the tree of Christoffel pairs, row by row, starting from the first row and reading each row from left to right. This sequence begins by $(a, b), (a, ab), (ab, b), (a, a^2b), (a^2b, ab), (ab, ab^2), (ab^2, b), (a, a^3b), \ldots$ (see Figure 2.10). Show that for any $n \geq 1$, one has $v_{n-1} = u_n$, except if $n = 2^k$ in which case $v_{n-1} = b$ and $u_n = a$. Show that for any $n \geq 1$, one has*

- $v_{2^k} = a^k b$ *for all* $k \geq 0$*;*
- $v_{2n+1} = v_n$ *for all* $n \geq 1$*;*
- $v_{2n} = v_{n-1}v_n$ *for all* $n \geq 1$*, which is not a power of 2,*

with the convention $v_0 = a$. Call hyperbinary representation *of n a sequence $a_0, \ldots, a_k$ with $a_k \neq 0$, $a_i = 0, 1, 2$ for any i and $n = a_0 2^0 + \cdots + a_k 2^k$; k is the* length *of the representation. Show that for given n, there are at most two lengths of hyperbinary representations of n (the* short *one and the* long *one). Consider the sequence of hyperbinary representations of n, lexicographically sorted; replace in this sequence each long representation by b and each short one by a; show that one obtains the upper Christoffel word $\tilde{v}_n$. See [DLR].*

Define the sequence of natural integers $(b(n))_{n\geq 1}$, with $b(n) = (|v_n|)$. Then the first terms are (1, 2, 1, 3, 2, 3, 1, 4, 3, 5, 2, 5, 3, 4, 1, 5, ...). *It is called the* Stern diatomic sequence *([St], where Stern claims that he follows ideas of his "unforgettable" friend Eisenstein). See also Exercise 19.2.1. For other links with Christoffel words, see [dLDL4].*

2.6 Sturmian morphisms

Consider the totally ordered alphabet $\{a < b\}$. Recall that we denote by (u, v) any endomorphism f of the free monoid $\{a < b\}^*$ defined by $f(a) = u$, $f(b) = v$. We define now four endomorphisms as follows: $G = (a, ab)$, $D = (ba, b)$, $\tilde{G} = (a, ba)$, and $\tilde{D} = (ab, b)$ (we adopt the notations introduced by Berstel and Séébold [BS, Lo]).

We consider in the sequel to this book several infinite complete planar rooted binary trees. For each node in such a tree, there is a unique path from the root to this node; this path is coded by a word $w \in \{a < b\}^*$, with a meaning "left" and b meaning "right". In this way we obtain a bijection $\mathcal{N}$ from $\{a < b\}^*$ onto the set of nodes of the tree, and we say that *w codes the node $\mathcal{N}(w)$*. For example, in the tree of Figure 2.10, the node (a^2bab, ab) corresponds to the word abb: $\mathcal{N}(abb) = (a^2bab, ab)$.

Denote by $\mathrm{End}(\{a,b\}^*)$ the monoid of endomorphisms of the free monoid $\{a,b\}^*$.

Theorem 2.6.1 *Let $\rho : \{a,b\}^* \to \mathrm{End}(\{a,b\}^*)$ be the monoid homomorphism such that $\rho(a) = G$ and $\rho(b) = \tilde{D}$. Let $x \in \{a < b\}^*$ and consider the node $\mathcal{N}(x) = (u,v)$ in the tree of Christoffel pairs, coded by the word x. Then in the monoid of endomorphisms of the free monoid $\{a,b\}^*$, one has the equality $(u,v) = \rho(x)$.*

For example, looking at Figure 2.10, one has $(a^2bab, ab) = \rho(abb) = G \circ \tilde{D} \circ \tilde{D}$: indeed $G \circ \tilde{D} \circ \tilde{D}(a) = G \circ \tilde{D}(ab) = G(abb) = aabab$ and $G \circ \tilde{D} \circ \tilde{D}(b) = G \circ \tilde{D}(b) = G(b) = ab$.

Proof. This is clear for the root, that is, $x = 1$. The general case follows by induction. Suppose indeed that $(u,v) = \rho(x)$. Then $\rho(x)(a) = u$, $\rho(x)(ab) = uv$, and $\rho(x)(b) = v$. Thus, we have $\rho(xa)(a) = \rho(x) \circ G(a) = \rho(x)\,(a) = u$ and $\rho(xa)(b) = \rho(x) \circ G(b) = \rho(x)(ab) = uv$. On the other hand, $\rho(xb)(a) = \rho(x) \circ \tilde{D}(a) = \rho(x)(ab) = uv$ and $\rho(xb)(b) = \rho(x) \circ \tilde{D}(b) = \rho(x)(b) = v$.

In the first (resp. second) case, we have $\rho(xa) = (u, uv)$ (resp. $\rho(xb) = (uv, u)$) in accordance with the construction rule of the tree (see Figure 2.9). □

Corollary 2.6.2 *The image of any lower Christoffel word under the endomorphism G or $\tilde{D}$ is a lower Christoffel word. Conversely, each lower Christoffel word is the image under G or $\tilde{D}$ of a lower Christoffel word.*

Proof. This is clear for the lower Christoffel words of length 1, that is, a and b, since ab is a lower Christoffel word. Consider now a proper lower Christoffel word $w = uv$ with its standard factorization. Then (u,v) appears on the tree of Christoffel pairs and, viewed as an endomorphism of the free monoid $\{a,b\}^*$, is equal to $\rho(x)$ by Theorem 2.6.1 and its notations. In particular, $\rho(x)(a) = u$. Let $\mathcal{N}(ax) = (u', v')$; in particular, $u'v'$ is a lower Christoffel word. Hence, by the theorem $G(u) = G(\rho(x)(a)) = G \circ \rho(x)(a) = \rho(a) \circ \rho(x)(a) = \rho(ax)(a) = u'$ and similarly $G(v) = v'$. Thus $G(w) = G(uv) = G(u)G(v) = u'v'$ is a lower Christoffel word.

The proof that $\tilde{D}(w)$ is a lower Christoffel word is similar.

We prove now the last assertion. It is clear if $w = a, b$ or ab. Otherwise, the standard factorization (u,v) of w is a node of the tree of Christoffel pairs, distinct from the root (a,b). Then (u,v), viewed as an endomorphism of the free monoid, is equal to $\rho(x)$, for some word x, with the notations of Theorem 2.6.1. Then $w = uv = \rho(x)(ab)$. The endomorphism $\rho(x)$ is a product of the endomorphisms G and $\tilde{D}$, and the product is non-empty since $w \neq ab$. If the leftmost endomorphism in this product is G, then $w = G(m)$ for some word m which, by the first part of the proof, is a lower Christoffel word, since ab is. If the leftmost is $\tilde{D}$, we argue similarly, and $w = \tilde{D}(m)$. □

Corollary 2.6.3 *The image of any lower Christoffel word under the endomorphism $\tilde{G}$ or D is a conjugate of some lower Christoffel word.*

Proof. By Corollary 2.6.2, it is enough to prove that for each word w, $D(w)$ and $\tilde{D}(w)$ are conjugate, and similarly for G and $\tilde{G}$. We do it only for D: recall that $D = (ba, b)$ and $\tilde{D} = (ab, b)$. Thus, viewing them as endomorphisms of the free group $\mathrm{F}(a, b)$, we have $D = \iota \circ \tilde{D}$, where ι is the inner automorphism $g \mapsto bgb^{-1}$: indeed this is readily verified on the generators a and b. Note that the image by $\tilde{D}$ of each non-empty word in the free monoid $\{a, b\}^*$ ends with b, so we see that for each non-empty word w in $\{a, b\}^*$, one has $\tilde{D}(w) = ub$ and $D(w) = bu$. Thus, $D(w)$ and $\tilde{D}(w)$ are conjugate. □

A *Sturmian morphism* is an endomorphism of the free monoid $\{a, b\}^*$ which sends each lower Christoffel word onto the conjugate of a lower Christoffel word. Equivalently (since endomorphisms preserve conjugation), it sends each conjugate of a lower Christoffel word onto the conjugate of some lower Christoffel word. Thus, we have shown that the endomorphisms $G, D, \tilde{G}$, and $\tilde{D}$ are Sturmian morphisms; it follows that each product of these morphisms is Sturmian. Sturmian morphisms will be given several characterizations in Section 15.5.

2.6.1 Exercises

Exercise 2.6.1 *Consider the morphisms $G, \tilde{G}, D$, and $\tilde{D}$ defined in this section, and $E = (b, a)$ defined in Section 2.2. Extend all these morphisms to endomorphisms of the free group $\mathrm{F}(a, b)$. Consider the conjugation by a and b in $\mathrm{F}(a, b)$: the first maps each element g onto $a^{-1}ga$, and the second onto $b^{-1}gb$; verify that these are respectively the endomorphims $(a, a^{-1}ba)$ and $(b^{-1}ab, b)$ of $F(a, b)$.*

Show that one has the following identities:

$$D = EGE, \tilde{D} = E\tilde{G}E$$

$$\tilde{G} = (a, a^{-1}ba)G, \tilde{D} = (b^{-1}ab, b)D$$

$$\tilde{G} = \mathcal{R}G\mathcal{R}, \tilde{D} = \mathcal{R}D\mathcal{R}$$

where $\mathcal{R}$ is the involutive anti-automorphism of $\mathrm{F}(a, b)$ which is the identity on A. Show that

$$G^{-1}\tilde{G} = (a, a^{-1}ba), D^{-1}\tilde{D} = (b^{-1}ab, b)$$

Exercise 2.6.2 *Let w be a lower Christoffel word.*

(i) Show that if $w = G(u)$ (resp. $w = \tilde{D}(u)$), then the slope of w is ≤ 1 (resp. ≥ 1).
(ii) Show that if the slope of w is ≤ 1 (resp. ≥ 1), then w is the image by G (resp. by $\tilde{D}$) of a lower Christoffel word u.

Exercise 2.6.3 *Show that for $f = G$ or $\tilde{D}$, if $f(u)$ is a lower Christoffel word, so is u.*

Exercise 2.6.4 *Let w be a word on the alphabet $\{a < b\}$. Show that w is a lower Christoffel word if and only if the following algorithm, with input w, outputs $w = a$ or b:*

(i) Replace w by $G^{-1}(w)$ or $\tilde{D}^{-1}(w)$ if possible; otherwise stop.
(ii) If $w \neq a$ and $w \neq b$, return to 1. See [Se2] (p. 21) and [D'AdLDL].

Exercise 2.6.5 *Let M (resp. $\tilde{M}$) be the submonoid of the monoid of endomorphisms of $\{a < b\}^*$ generated by G and $\tilde{D}$ (resp. by $\tilde{G}$ and D). Show that the set of lower (resp. upper) Christoffel words is the set of all words of the form $f(a)$ or $f(b)$ for some $f \in M$ (resp. $f \in \tilde{M}$). Hint: use Corollary 2.6.2; use the fact, proved in Exercise 2.6.1, that $\mathcal{R}$ conjugates G and $\tilde{G}$ (resp. $\tilde{D}$ and D) and that $\mathcal{R}$ exchanges lower and upper Christoffel words.*

Exercise 2.6.6 *Show that both G* and *$\tilde{D}$, viewed as automorphisms of the free group $F(a, b)$, fix the element $ab^{-1}a^{-1}b$, hence also all automorphisms in M (notation of Exercise 2.6.5). Deduce that for each lower Christoffel word $w = uv$, given its standard factorization, one has $uv^{-1}u^{-1}v = ab^{-1}a^{-1}b$ ([Sc1], Proposition 1.2.11 (i)).*

Exercise 2.6.7 *Let $A = \{a < b\}$. Show that if $u = a^n$ (resp. $u = b^n$), then $\rho(u) = G^n = (a, a^n b)$ (resp. $\rho(u) = \tilde{D}^n = (ab^n, b)$). Show that for all other words $u \in A^*$, one has $\rho(u) = (amb, apb)$ for some words m, p in A^*.*

Exercise 2.6.8 *Show that if u is a lower Christoffel on the alphabet $\{a < b\}$, distinct from b, then $u^n, n \geq 1$ is a prefix of some lower Christoffel word. Hint: show first that there exists a Christoffel pair (u, v); then multiply this pair on the right, viewed as an endomorphism, by the endomorphism $G^n = (a, a^n b)$.*

3
Markoff numbers

The Markoff equation is a special Diophantine equation. A solution is called a Markoff triple. The main result in this chapter, Theorem 3.1.1, is a bijection between lower Christoffel words and Markoff triples. The tree of Markoff triples is obtained by specializing the tree of Christoffel pairs. The Markoff injectivity conjecture, or Frobenius conjecture, is presented.

3.1 Markoff triples and numbers

A *Markoff triple* is a multiset $\{x, y, z\}$ of positive integers satisfying the *Markoff equation*

$$x^2 + y^2 + z^2 = 3xyz$$

Examples are $\{1, 1, 1\}$, $\{1, 1, 2\}$, and $\{1, 2, 5\}$. A Markoff triple is called *proper* if the three numbers are distinct; otherwise, we call it *improper*. A *Markoff number* is an element of a Markoff triple.

Consider the monoid homomorphism μ from the free monoid $\{a, b\}^*$ into $SL_2(\mathbb{Z})$ defined by

$$\mu(a) = \begin{pmatrix} 2 & 1 \\ 1 & 1 \end{pmatrix}, \quad \mu(b) = \begin{pmatrix} 5 & 2 \\ 2 & 1 \end{pmatrix}$$

We assume that the alphabet $\{a, b\}$ is totally ordered: $a < b$.

The matrix construction of the Markoff triples, as shown in the following theorem, was obtained by Cohn [Co1, Co2]. He used the Fricke relation (last equation in Lemma 3.1.4), having noted the striking analogy between Markoff's equation and this relation. Uniqueness was noticed by Bombieri [B], Theorem 26, and independently by the author ([Re4], Theorem 1).

Theorem 3.1.1. *The mapping sending each lower Christoffel word w with standard factorization uv onto the multiset*

$$\left\{\frac{1}{3}\mathrm{tr}(\mu(u)), \frac{1}{3}\mathrm{tr}(\mu(v)), \frac{1}{3}\mathrm{tr}(\mu(w))\right\}$$

is a bijection from the set of proper lower Christoffel words onto the set of proper Markoff triples.

We give the proof after some preparatory lemmas. It is proved in the next lemma that for any lower Christoffel word w,

From Christoffel Words to Markoff Numbers. Second Edition. Christophe Reutenauer, Oxford University Press.
 DOI: 10.1093/9780197907535.003.0003

$$\frac{1}{3}\text{tr}(\mu(w)) = \mu(w)_{12}$$

Thus the triple of the theorem is also equal to $\{\mu(u)_{12}, \mu(v)_{12}, \mu(w)_{12}\}$.

Lemma 3.1.2

(i) ([Re3], Lemma 3.2) If M is a symmetric 2×2 matrix, and $N = \mu(a)M\mu(b)$, then $\frac{1}{3}\text{tr}(N) = N_{12}$.

(ii) If $N = \mu(w)$ for some lower Christoffel word w, then $\frac{1}{3}\text{tr}(N) = N_{12}$.

Proof. One has

$$\begin{pmatrix} 2 & 1 \\ 1 & 1 \end{pmatrix}\begin{pmatrix} p & q \\ q & s \end{pmatrix}\begin{pmatrix} 5 & 2 \\ 2 & 1 \end{pmatrix} = \begin{pmatrix} 2p+q & 2q+s \\ p+q & q+s \end{pmatrix}\begin{pmatrix} 5 & 2 \\ 2 & 1 \end{pmatrix} = \begin{pmatrix} 10p+9q+2s & 4p+4q+s \\ 5p+7q+2s & 2p+3q+s \end{pmatrix}$$

The trace of the latter matrix is $12p + 12q + 3s$ which is three times its 12-entry $4p + 4q + s$. This proves the first assertion.

If w is a proper lower Christoffel word, then $w = amb$ for some palindrome m by Theorem 2.3.1; then $N = \mu(w) = \mu(a)\mu(m)\mu(b)$ and $\mu(m)$ is symmetric, since m is a palindrome and $\mu(a)$ and $\mu(b)$ are symmetric matrices. This proves (ii) by (i) when w is proper. If $w = a$ or b, one sees that (ii) holds by inspection. □

Lemma 3.1.3 *The improper Markoff triples are* $\{1, 1, 1\}$ *and* $\{1, 1, 2\}$.

Proof. Suppose that $\{x, x, y\}$ is a Markoff triple. Then $2x^2 + y^2 = 3x^2y$ so that x^2 divides y^2; thus x divides y: $y = xu$. Then $2x^2 + x^2u^2 = 3x^3u$; thus $2 + u^2 = 3xu$. Thus u divides 2 and $u = 1$ or 2. We have $\frac{2}{u} + u = 3x$; in both cases, $3x = 3$, $x = 1$, and thus $y = u$, and we obtain the triples $\{1, 1, 1\}$ and $\{1, 1, 2\}$. □

Lemma 3.1.4 *(Fricke relations [Fr], (6) p. 91) Let A and B be matrices in $\text{SL}_2(\mathbb{Z})$. Then,* $\text{tr}(A^2B) + \text{tr}(B) = \text{tr}(A)\,\text{tr}(AB)$, $\text{tr}(AB^2) + \text{tr}(A) = \text{tr}(AB)\,\text{tr}(B)$, *and* $\text{tr}(A)^2 + \text{tr}(B)^2 + \text{tr}(AB)^2 = \text{tr}(A)\,\text{tr}(B)\,\text{tr}(AB) + \text{tr}(ABA^{-1}B^{-1}) + 2$.

Proof. We identify as usual any any scalar α with the scalar matrix αI_2 of size 2×2. By the theorem of Cayley-Hamilton, we have $A^2 - \text{tr}(A)A + 1 = 0$ since A has determinant 1. Thus $A^2 + 1 = \text{tr}(A)A$; hence $A^2B + B = \text{tr}(A)AB$ and, taking the trace, $\text{tr}\,(A^2B) + \text{tr}(B) = \text{tr}(A)\text{tr}(AB)$, proving the first relation. The second follows after exchanging A and B.

The Cayley-Hamilton theorem also implies that $A^{-1} = \text{tr}(A) - A$ so that $ABA^{-1}B^{-1} = AB(\text{tr}(A) - A)(\text{tr}(B) - B)$. Thus we have

$$ABA^{-1}B^{-1} = \text{tr}(A)\,\text{tr}(B)AB - \text{tr}(A)AB^2 - \text{tr}(B)ABA + ABAB. \tag{3.1}$$

Now, by the first two relations in the lemma, we have $\text{tr}(AB^2) = \text{tr}(AB)\,\text{tr}(B) - \text{tr}(A)$ and $\text{tr}(ABA) = \text{tr}(A^2B) = \text{tr}(A)\,\text{tr}(AB) - \text{tr}(B)$. Hence, taking the trace in (3.1), we obtain

$$\text{tr}(ABA^{-1}B^{-1}) = \text{tr}(A)\,\text{tr}(B)\,\text{tr}(AB) - \text{tr}(A)\,\text{tr}(AB)\,\text{tr}(B) + \text{tr}(A)^2 - \text{tr}(B)\,\text{tr}(A)\,\text{tr}(AB) + \text{tr}(B)^2 + \text{tr}(ABAB)$$

The last Fricke relation follows by applying once again the theorem of Cayley-Hamilton for the matrix $M = AB$ under the form $\text{tr}(M^2) = \text{tr}(M)^2 - 2$. □

From the first relation, we deduce the following result, which will be used later.

Corollary 3.1.5 *Let A and B be matrices in $\text{SL}_2(\mathbb{Z})$ and $x, y, z \in \mathbb{Z}$. Then the two following conditions are equivalent:*

(i) $x = \frac{1}{3}\text{tr}(A), y = \frac{1}{3}\text{tr}(AB), z = \frac{1}{3}\text{tr}(A^2B)$;

(ii) $x = \frac{1}{3}\text{tr}(A), y = \frac{1}{3}\text{tr}(AB), 3xy - z = \frac{1}{3}\text{tr}(B)$.

Proof. If (i) holds, then $3xy - z = \frac{1}{3}\text{tr}(A)\,\text{tr}(AB) - \frac{1}{3}\text{tr}(A^2B)$. By the first Fricke relation in the lemma, this is equal to $\frac{1}{3}\text{tr}(B)$ and thus (ii) holds. The converse is similar and left to the reader. □

The corollary has a symmetric statement, obtained after exchanging A and B: the condition $x = \frac{1}{3}\text{tr}(B), y = \frac{1}{3}\text{tr}(AB), z = \frac{1}{3}\text{tr}(AB^2)$ is equivalent to the condition $x = \frac{1}{3}\text{tr}(B), y = \frac{1}{3}\text{tr}(AB), 3xy - z = \frac{1}{3}\text{tr}(A)$.

The next lemma will allow us to construct recursively the Markoff triples in the proof of the theorem.

Lemma 3.1.6 *(Markoff [M2], p. 397) Let $\{x, y, z\}$ be a proper Markoff triple with $x < y < z$. Then $\{x, y, 3xy - z\}$ is a Markoff triple and $3xy - z < y$.*

Proof. Define a polynomial in t: $P(t) = t^2 - 3xyt + x^2 + y^2$. Then z is a root of $P(t)$. The other root is $3xy - z = (x^2 + y^2)/z$. Hence this root is a positive integer and $\{x, y, 3xy - z\}$ is a Markoff triple. Now, we have $P(y) = y^2 - 3xy^2 + x^2 + y^2$; since $x \geq 1$, we have $3x \geq 3$, and because $y > x$, we deduce that $P(y) < 0$. Thus y lies strictly between the two roots of P, and since the root z is larger than y, the other one must be smaller, that is, $3xy - z < y$. □

Lemma 3.1.7 *If A, B, and C are $d \times d$ square matrices over the positive integers, with $d \geq 2$, then for any $i, j = 1, \ldots, d$, one has $(AB)_{ij} > A_{ij}, B_{ij}$ and $(ABC)_{ij} > (AC)_{ij}, B_{ij}$.*

Proof. We have $(AB)_{ij} = \sum_k A_{ik}B_{kj}$. By the assumption $d \geq 2$ and by positivity (each entry ≥ 1), this is strictly larger than A_{ij} and B_{ij}. Now, $(ABC)_{ij} = \sum_{k,l} A_{ik}B_{kl}C_{lj} > \sum_k A_{ik}B_{kk}C_{kj} \geq \sum_k A_{ik}C_{kj} = (AC)_{ij}$. Similarly, $(ABC)_{ij} > B_{ij}$. □

Corollary 3.1.8 *If $w = uv$ is a lower Christoffel word with its standard factorization, then $\mu(w)_{12} > \mu(u)_{12}, \mu(v)_{12}$. If, moreover, u is a prefix of v (resp. v is a suffix of u), then $\mu(u)_{12} < \mu(v)_{12}$ and $\mu(uuv)_{12} < \mu(uvv)_{12}$ (resp. $\mu(v)_{12} < \mu(u)_{12}$ and $\mu(uvv)_{12} < \mu(uuv)_{12}$). In particular, $\mu(uuv)_{12} \neq \mu(uvv)_{12}$.*

Proof. Note that if m is a non-empty word, then the 2×2 matrix $\mu(m)$ is square and has positive integer entries. Thus we may apply the previous lemma.

We have $\mu(w) = \mu(u)\mu(v)$ so that the first assertion follows directly from the lemma. Now, if u is a prefix of v, then it is a proper prefix (since w is radical), and we may write $v = uv'$, v' non-empty. Thus the lemma implies that $\mu(u)_{12} < \mu(v)_{12}$; from the lemma also, applied to $A = \mu(u)^2, B = \mu(v'), C = \mu(v)$, we deduce that the 12-entry of $\mu(u)\mu(v)\mu(v) = \mu(u)\mu(u)\mu(v')\mu(v)$ is larger than that of $\mu(u)\mu(u)\mu(v)$.

The other case is treated similarly. The last assertion follows from Corollary 2.4.3 (ii) if $w \neq ab$ and by direct calculation if $w = ab$. □

The next lemma allows us to deduce the Markoff equation from the last Fricke relation.

Lemma 3.1.9 *If* (u, v) *is a Christoffel pair and* $A = \mu(u)$, $B = \mu(v)$, *then* $\mathrm{tr}(ABA^{-1}B^{-1}) = -2$.

Proof. We claim that for any Christoffel pair (u, v), the element $uvu^{-1}v^{-1}$ of the free group $\mathrm{F}(a, b)$ is conjugate to $aba^{-1}b^{-1}$. Using the construction of Christoffel pairs by their tree, it is easy to see that if the claim holds for (u, v), then it also holds for (u, uv) and (uv, v). Indeed, $u(uv)u^{-1}(uv)^{-1} = u(uvu^{-1}v^{-1})u^{-1}$ and $(uv)v(uv)^{-1}v^{-1} = uvu^{-1}v^{-1}$. Finally, the claim is trivial at the root (a, b) on the tree.

The claim implies that $ABA^{-1}B^{-1}$ and $\mu(a)\mu(b)\mu(a)^{-1}\mu(b)^{-1}$ are conjugate matrices; thus $\mathrm{tr}(ABA^{-1}B^{-1}) = \mathrm{tr}(\mu(a)\mu(b)\mu(a)^{-1}\mu(b)^{-1})$. We have $\mu(a)\mu(b)\mu(a)^{-1}\mu(b)^{-1} = \begin{pmatrix} 2 & 1 \\ 1 & 1 \end{pmatrix}\begin{pmatrix} 5 & 2 \\ 2 & 1 \end{pmatrix}\begin{pmatrix} 1 & -1 \\ -1 & 2 \end{pmatrix}\begin{pmatrix} 1 & -2 \\ -2 & 5 \end{pmatrix} = \begin{pmatrix} 12 & 5 \\ 7 & 3 \end{pmatrix}\begin{pmatrix} 3 & -7 \\ -5 & 12 \end{pmatrix} = \begin{pmatrix} 11 & -24 \\ 6 & -13 \end{pmatrix}$, which concludes the proof. □

Proof of Theorem 3.1.1 Let w be a lower Christoffel word with standard factorization uv, and let $(x, y, z) = (\frac{1}{3}\mathrm{tr}(\mu(u)), \frac{1}{3}\mathrm{tr}(\mu(v)), \frac{1}{3}\mathrm{tr}(\mu(w)))$. By the Fricke relation in Lemma 3.1.4 and by Lemma 3.1.9, one has $(3x)^2 + (3y)^2 + (3z)^2 = (3x)(3y)(3z)$; hence $\{x, y, z\}$ is a Markoff triple, since by Lemma 3.1.2 (ii), x, y, and z are positive integers. Suppose that $w \neq ab$; then by Corollary 2.4.3 (ii) and Corollary 3.1.8, x, y, and z are distinct; hence $\{x, y, z\}$ is a proper Markoff triple. If $w = ab$, then, since $\mu(ab) = \begin{pmatrix} 12 & 5 \\ 7 & 3 \end{pmatrix}$, $\{x, y, z\} = \{1, 2, 5\}$ is a proper Markoff triple, too.

Thus, the mapping of the theorem is well defined, and we prove now that it is surjective. Let $\{x, y, z\}$ be any proper Markoff triple. We may assume that $x < y < z$. Then by Lemma 3.1.6, $\{x, y, 3xy - z\}$ is a Markoff triple and $3xy - z < y$. The sum $x + y + (3xy - z)$ is then smaller than $x + y + z$, and we use induction on this sum.

If $\{x, y, 3xy - z\}$ is an improper Markoff triple, then, since x and $3xy - z$ are both $< y$, it must be $\{1, 1, 2\}$ by Lemma 3.1.3; then we have $x = 1, 3xy - z = 1, y = 2$ so that $z = 3 \cdot 2 - 1 = 5$ and the triple $\{x, y, z\}$ corresponds to the lower Christoffel word ab.

If $\{x, y, 3xy - z\}$ is a proper Markoff triple, then by induction it is equal to $\{\frac{1}{3}\mathrm{tr}(\mu(u)), \frac{1}{3}\mathrm{tr}(\mu(v)), \frac{1}{3}\mathrm{tr}(\mu(w))\}$ for some lower Christoffel word with standard factorization $w = uv$. Since the maximum is y, we have by Corollary 3.1.8, $y = \frac{1}{3}\mathrm{tr}$

$(\mu(w)) = \frac{1}{3}\mathrm{tr}(\mu(uv))$. If $x = \frac{1}{3}\mathrm{tr}(\mu(u))$ and $3xy - z = \frac{1}{3}\mathrm{tr}(\mu(v))$, then by Corollary 3.1.5, $\{x, y, z\}$ corresponds to the lower Christoffel word uuv. Otherwise, $x = \frac{1}{3}\mathrm{tr}(\mu(v))$ and $3xy - z = \frac{1}{3}\mathrm{tr}(\mu(u))$ and symmetrically (see the statement after the proof of Corollary 3.1.5), $\{x, y, z\}$ corresponds to the lower Christoffel word uvv.

This proves surjectivity of the mapping, and we now prove injectivity. Suppose that the proper Markoff triple $\{x < y < z\}$ corresponds to the two proper lower Christoffel words $w_1 = u_1v_1$ and $w_2 = u_2v_2$ with their standard factorizations. If $w_1 = ab$, then also $w_2 = ab$: otherwise, we have that $w_1 = ab$ is a proper factor of w_2 and therefore $\mu(w_2)_{12}$ is strictly larger than $\mu(w_1)_{12}$ by Lemma 3.1.7, and the triples corresponding to w_1 and w_2 are not equal since they do not have the same maximum.

Thus, we may assume that $w_1, w_2 \neq ab$. By Corollary 2.4.3, we may assume either that u_1 is a proper prefix of v_1 and u_2 is a proper prefix of v_2; or u_1 is a proper prefix of v_1 and v_2 is a proper suffix of u_2 (the two other cases are similar).

Consider the first case. Let $i = 1$ or $i = 2$. Then by Corollary 2.4.3, we have the standard factorization $v_i = u_iv_i'$. By the prefix condition and Corollary 3.1.8, we have $(\mu u_i)_{12} < (\mu v_i)_{12} < (\mu w_i)_{12}$; since the Markoff triple $\{x < y < z\}$ corresponds to $w_i = u_iv_i$, we have by Lemma 3.1.2 (ii), $x = \frac{1}{3}\mathrm{tr}(\mu(u_i)) = \frac{1}{3}\mathrm{tr}(A)$, $y = \frac{1}{3}\mathrm{tr}(\mu(v_i)) = \frac{1}{3}\mathrm{tr}(\mu(u_iv_i')) = \frac{1}{3}\mathrm{tr}(AB)$ and $z = \frac{1}{3}\mathrm{tr}(\mu(w_i)) = \frac{1}{3}\mathrm{tr}(\mu(u_i^2v_i')) = \frac{1}{3}\mathrm{tr}(A^2B)$ (if we set $A = \mu(u_i)$, $B = \mu(v_i')$). Thus (i) in Corollary 3.1.5 holds, and we deduce (ii): $\{x, y, 3xy - z\} = \{\frac{1}{3}\mathrm{tr}(A), \frac{1}{3}\mathrm{tr}(AB), \frac{1}{3}\mathrm{tr}(B)\}$. The Markoff triple corresponding to $v_i = u_iv_i'$ is $t_i = \{\frac{1}{3}\mathrm{tr}(\mu(u_i)), \frac{1}{3}\mathrm{tr}(\mu(v_i')), \frac{1}{3}\mathrm{tr}(\mu(v_i))\}$. Thus $t_i = \{x, y, 3xy - z\}$. Since this is true for $i = 1$ and $i = 2$, we deduce $t_1 = t_2$. By induction on the length of the words, we deduce that $v_1 = v_2$, that is, $u_1v_1' = u_2v_2'$. By the uniqueness of the standard factorization, $u_1 = u_2$, which implies $w_1 = w_2$.

Consider now the second case. One has $v_1 = u_1v_1'$ and $u_2 = u_2'v_2$. One obtains, similarly using Corollary 3.1.5 and its symmetric statement, that the triple $\{x, y, 3xy-z\}$ corresponds to both lower Christoffel words v_1 and u_2. Thus by induction $u_1v_1' = u_2'v_2$ and, therefore, by uniqueness of the factorization into Christoffel words (Theorem 2.4.1), $u_1 = u_2'$, $v_1' = v_2$. Then we see that the triple $\{x, y, z\}$ corresponds to the words $w_1 = u_1v_1 = u_1^2v_1' = u_1^2v_2$ and $w_2 = u_2v_2 = u_2'v_2^2 = u_1v_2^2$. Hence the maximum of this triple is equal to both $\mu(u_1^2v_2)_{12}$ and $\mu(u_1v_2^2)_{12}$. This contradicts the last assertion in Lemma 3.1.8 applied to the Christoffel word $v_1 = u_1v_2$. □

There is an algebraic analogue of Theorem 3.1.1: in [Ls], Lasnier introduces a family of modules called *Markoff modules*; he shows that these modules are combinatorially similar to Christoffel words, and he constructs a bijective map between the set of Markoff module triples and the set of proper Markoff triples. The Markoff equation also appears in the Fomin-Zelevinsky theory of cluster algebras; see for example the work of Beineke, Brüstle, and Hille [BBH] and that of Lampe [Lm].

3.1.1 Exercises

Exercise 3.1.1 *Show that for any word $w \in \{a, b\}^*$, $\mu(awb)_{12} = \mu(a\widetilde{w}b)_{12}$ [LLS].*

3.2 The tree of Markoff triples

Since all Christoffel pairs appear exactly once on the tree of Christoffel pairs (see Section 2.5), and by Theorem 3.1.1, it follows that if we replace in the previous tree each node (u, v) by the 3-tuple

$$\frac{1}{3}(\mathrm{tr}(\mu(u)), \mathrm{tr}(\mu(uv)), \mathrm{tr}(\mu(v))) \tag{3.2}$$

which is also equal to

$$(\mu(u)_{12}, \mu(uv)_{12}, \mu(v)_{12})$$

By Lemma 3.1.2, we obtain a tree where each proper Markoff triple appears exactly once as a node in the form of a 3-tuple whose underlying set is this triple.

This tree is called the *tree of Markoff triples* (see Figure 3.1).

It also follows from (3.2) and Corollary 3.1.8 that in each 3-tuple (x, z, y) appearing in this tree, the maximum of the three numbers is z; however, the minimum may be x or y.

This tree may be constructed directly, starting from the root $(1, 5, 2)$ and applying the rule given in Figure 3.2. This follows from the Fricke relations. Suppose indeed that $(x, z, y) = \frac{1}{3}(\mathrm{tr}(\mu(u)), \mathrm{tr}(\mu(uv)), \mathrm{tr}(\mu(v)))$. Then $3xz - y = 3\cdot\frac{1}{3}\mathrm{tr}(\mu(u))\cdot\frac{1}{3}\mathrm{tr}(\mu(uv)) - \frac{1}{3}\mathrm{tr}(\mu(v)) = \frac{1}{3}\mathrm{tr}(\mu(u^2v))$ by Lemma 3.1.4; since the standard factorization of u^2v is $u \cdot uv$ by Corollary 2.4.3 (i), we deduce that, by the rule for building the tree of Christoffel pairs (see Figure 2.9), the left child in Figure 3.2 is valid; the proof is similar for the right child.

Corollary 3.2.1 *The three numbers in a Markoff triple are pairwise relatively prime.*

Proof. It is enough to show that the three numbers are relatively prime (since by the Markoff equation, if a prime number divides two of them, it divides the third).

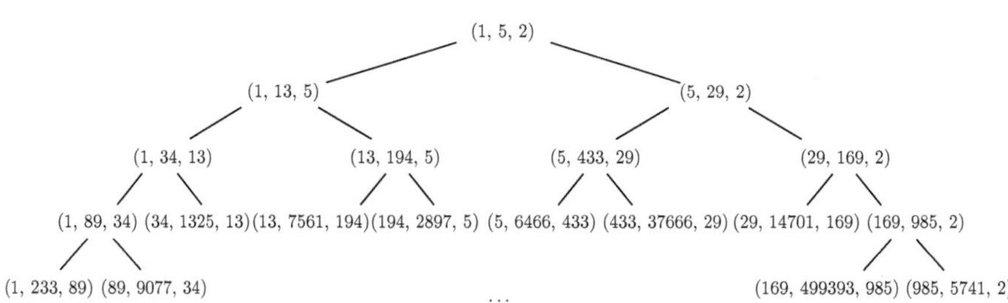

Figure 3.1 The tree of Markoff triples

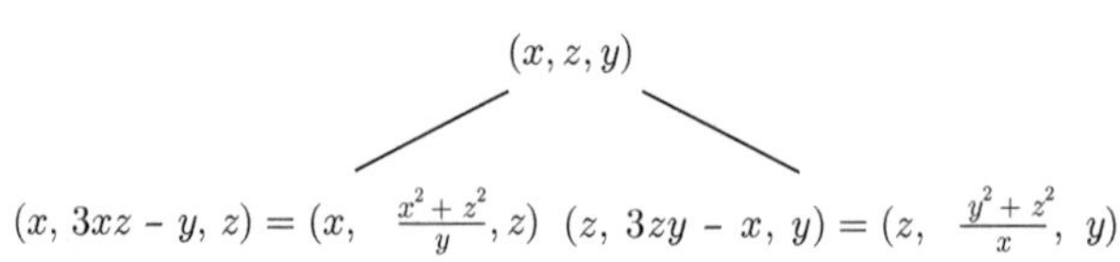

Figure 3.2 The rule for building the tree of Markoff triples

This is clear for the improper triples and for the triple $(1, 5, 2)$, the root of the tree of Markoff triple. If (x, z, y) is some node of this tree, with x, y, z relatively prime, then clearly $x, 3xz - y, z$ are relatively prime, and so are also $z, 3zy - x, y$. By the construction of the tree, the result follows. □

There exists a graph-theoretical construction of Markoff numbers. Indeed, one may construct for each lower Christoffel word a so-called *domino graph*, whose number of *perfect matchings* is equal to the associated Markoff number. This is nicely explained in Aigner's book [A], Theorem 7.12; the result has a long story, including work of Jim Propp, and we refer to [A] for it.

3.3 The Markoff injectivity conjecture

It follows from Theorem 3.1.1 that for each Markoff number m, there exists a lower Christoffel word w such that $m = \frac{1}{3}\mathrm{tr}(\mu(w)) = \mu(w)_{12}$. We say that m is the Markoff number *associated with the Christoffel word* w. In this case, $|w|_a$ and $|w|_b$ are called the *Frobenius coordinates of* m. Denote m by m_w.

It is conjectured that the mapping $w \mapsto m_w$, which is surjective by the theorem, is also injective. This is called the *Markoff injectivity conjecture*. It has many equivalent formulations; see the book by Martin Aigner [A], where one also finds the solution of several particular cases of the conjecture.

The conjecture is attributed to Frobenius. Indeed, it is mentioned twice as open problem in Frobenius's article in 1913 [F], under equivalent formulations. Indeed, on p. 601, Frobenius writes "*Unentschieden ist bis jetzt die Frage, ob einer Markoffschen Zahl p zwei verschiedene Ketten entsprechen können, d.h. ob p in zwei Lösungen p, a, b und p, c, ddie grösste Zahl sein kann*" and on page 614, '*Es ist aber nicht ausgeschlossen, dass zwei verschiedenen Brüchen* $\rho = \frac{\chi}{\chi'}$ *und* $\sigma = \frac{\lambda}{\lambda'}$ *dieselbe Zahl* $p_\rho = p_\sigma$ *entspricht*".

A problem induced by the injectivity conjecture is the following: if the conjecture is true, that is, if there is a bijection between lower Christoffel words and Markoff numbers, determine the order of the set of lower Christoffel words induced by the natural order of Markoff numbers. Equivalently, since proper lower Christoffel words are naturally in bijection with $\mathbb{Q}_+^*$ (via their slope), determine the order of the latter set induced by the bijection between Markoff numbers and this set. Partial answers are provided by Bugeaud, Siksek, and the author [BRS], in Aigner's book ([A], Section 10.2), by Rabideau and Schiffler [RaS], and by Lee, Li, Rabideau, and Schiffler [LLRS]. For generalizations, see Lapointe and the author [LR1] and Labbé, Lapointe, and Steiner [LLS].

4
The Markoff property

The Markoff property, introduced by Markoff, is a certain combinatorial property of bi-infinite words on a binary alphabet $\{a, b\}$. Such a word s has this property if whenever there is a factor xy in s, with $\{x, y\} = \{a, b\}$, then this factor may be extended into a factor of the form $y\tilde{m}xymx$, where the length of m is bounded. Markoff then proves that the word s must be periodic; it turns out that the periodic pattern is equal to some Christoffel word (Theorem 4.2.1).

We introduce here a similar property for (one-sided) infinite words, a variant of Markoff's property, and prove that such a word must be ultimately periodic, with a similar periodic pattern (Theorem 4.1.6); we may bound the length of the non-periodic initial part of the word. The Markoff property for one-sided infinite words is somewhat more technical than the original Markoff property for bi-infinite words; however it is better suited for continued fractions, and it implies easily the latter property.

4.1 Markoff property for infinite words

We say that an infinite word s on the alphabet $\{a, b\}$ satisfies the *Markoff property* if there exist K and N such that for any factorization $s = \tilde{u}xyv$, with $\{x, y\} = \{a, b\}$ and $|u| \geq K$, one has $u = myu', v = mxv'$, and $|m| \leq N$. Note that u is a finite word and v an infinite one.

For example, $s = a(ab)^\infty = aababababab\cdots$ satisfies the Markoff property with $K = 2$ and $N = 0$; it is easily verifed that s does not satisfy the Markoff property with $K = 1$. Another example is $(aab)^\infty = aabaabaab\cdots$ with $K = 2$ and $N = 1$.

The numbers N and K above are natural integers. We extend the definition to $N = -1$: we say that s satisfies the Markoff property for K and $N = -1$ if there is no factorization $s = \tilde{u}xyv$ with $|u| \geq K$; in this case, s is ultimately constant: $s = ux^\infty$, $x = a$ or b and $|u| = K$.

We consider in the whole section *an infinite word s, on the alphabet $\{a, b\}$, that satisfies the Markoff property with the parameters K and N.*

Lemma 4.1.1 *Let $s = ps_1$, for some finite word p, such that $|p| = K - 1$. Then no word of the form $x\tilde{m}xymy$, $\{x, y\} = \{a, b\}$, is a factor of s_1.*

Proof. Suppose that $s_1 = hx\tilde{m}xymy\cdots$. Then $phx\tilde{m}xymy\cdots = ps_1 = s = \tilde{u}xyv$ with $\tilde{u} = phx\tilde{m}$ and $v = my\cdots$. Then $|u| \geq |p| + 1 = K$ so that we obtain from the Markoff property: $u = nyu', v = nxv'$. Thus $mx\tilde{h}\tilde{p} = u = nyu'$ and $my\cdots = v = nxv'$. Then we cannot have $m = n$, since $x \neq y$. If m is shorter than n, then mx and my are both

From Christoffel Words to Markoff Numbers. Second Edition. Christophe Reutenauer, Oxford University Press.
 DOI: 10.1093/9780197907535.003.0004

prefixes of n, a contradiction; similarly if n is shorter than m. This proves the lemma by contradiction. □

Lemma 4.1.2 *If t is an infinite word on the alphabet $\{a, b\}$ such that none of the words $x(xy)^i xy(yx)^i y$, $xy(xy)^j xy(yx)^j y^2$, $\{x, y\} = \{a, b\}$, $i, j \in \mathbb{N}$, is a factor of t, then at most one of the words aa or bb is a factor of t.*

Proof. By contradiction, write $t = hx^2ky^2t_1$, $\{x, y\} = \{a, b\}$. We may assume that k is of minimum length. Then this length is non-zero, otherwise x^2y^2 is a factor of t, contradicting the first condition (with $i = 0$). By minimality of the length of k, we have $k = (yx)^i$, $i \geq 1$. Note that t_1 must begin by x: otherwise, t contains the factor $x^2(yx)^i y^3$, hence also the factor $xyxy^3$, contradicting the second condition (with $j = 0$). Thus we may write $yt_1 = (yx)^j t_2$, for some maximum positive integer j or j infinite. Thus $t = hx^2(yx)^i y(yx)^j t_2 = hx(xy)^i xy(yx)^j t_2$.

We must have $j \leq i$: otherwise, $j > i$ and t contains as a factor the word $x(xy)^i xy(yx)^i y$, a contradiction. In particular, j is not infinite.

Observe that t_2 must begin by x, yx, or y^2. We consider each of these three cases.

Suppose that t_2 begins by x: then t contains $y(yx)^j x = y^2(xy)^{j-1}x^2$, and by minimality of $|k|$, we must have $j - 1 \geq i$, that is, $j > i$, a contradiction.

If t_2 begins by yx, this contradicts the maximality of j.

Suppose that t_2 begins by y^2. Then t contains $x(xy)^i xy(yx)^j y^2$. If $j = i$, then t contains $x(xy)^i xy(yx)^i y$, a contradiction; if on the other hand $j = 1, \ldots, i - 1$, then t contains $xy(xy)^j xy(yx)^j y^2$, a contradiction.

Thus there is a contradiction in all cases, which proves the lemma. □

Recall that G is the endomorphism of the free monoid $\{a, b\}^*$ sending a onto a and b onto ab.

Lemma 4.1.3 *If $s = G(t)$ and $N \geq 0$ then t satisfies the Markoff property for the parameters $K - 1$ and $N - 1$.*

Lemma 4.1.4 *For any $v \in \{a, b\}^*$,*

$$\widetilde{G(va)} = \widetilde{G(v)a} = a\widetilde{G(v)} = G(\tilde{v})a \tag{4.1}$$

In particular, $G(va) = G(v)a$ is a palindrome if and only if v is a palindrome.

Proof. The two first equalities are evident. We prove the third by induction on the length. If $v = 1$, it is evident. Suppose that $v = aw$. Then $a\widetilde{G(v)} = a\widetilde{G(a)G(w)} = a\widetilde{G(w)}a = G(\tilde{w})aa$ (by induction) $= G(\tilde{w})G(a)a = G(\tilde{w}a)a = G(\widetilde{aw})a = G(\tilde{v})a$. Suppose now that $v = bw$. Then $a\widetilde{G(v)} = a\widetilde{G(b)G(w)} = a\widetilde{G(w)}\widetilde{G(b)} = a\widetilde{G(w)}ba = G(\tilde{w})aba$ (by induction) $= G(\tilde{w})G(b)a = G(\tilde{w}b)a = G(\widetilde{bw})a = G(\tilde{v})a$.

For the last assertion: $G(va)$ is a palindrome if and only if $\widetilde{G(va)} = G(va)$; this is equivalent to $G(\tilde{v})a = G(va)$, which in turn is equivalent to $G(\tilde{v}) = G(v)$, and finally, G being injective, to $\tilde{v} = v$; that is, v is a palindrome. □

Proof of Lemma 4.1.3. If t has no factorization $t = \tilde{u}abv$ nor $t = \tilde{u}bav$ with $|u| \geq K-1$, then clearly t satisfies the property for $K-1$ and $N-1$.

Otherwise, t has factorizations of this type. Suppose that $t = \tilde{u}abv$ with $|u| \geq K-1$. Then by (4.1), $s = G(t) = G(\tilde{u})aabG(v) = (\widetilde{G(u)a})abG(v)$ and $|G(u)a| \geq K$. By the Markoff property for s, one has $G(u)a = nbu'$, $G(v) = nav'$, $|n| \leq N$. Then u' cannot be empty so that $u' = u''a$ and $nbu'' = G(u)$ is in the image of G; hence n must end by a, $n = n'a$, and n' is in the image of G, $n' = G(m)$, $n = G(m)a$. Then $G(u) = nbu'' = G(m)abu''$, hence $u = mbu_1$. Moreover, $G(v) = G(m)aav'$ and therefore $v = mav_1$. Finally, $|m| \leq |G(m)| = |n'| = |n| - 1 < |n| \leq N$.

Suppose now that $t = \tilde{u}bav$, with $|u| \geq K-1$ (the Markoff property is a, b-symmetric, but G is not, so we must consider this case, too). Then $s = G(t) = G(\tilde{u})abaG(v) = \widetilde{G(u)}abaG(v)$; by the Markoff property for s, since $|G(u)a| \geq K$, $G(u)a = nau'$, $G(v) = nbv'$, $|n| \leq N$. It follows from the last equation that $n = n'a$ and $n' = G(m)$; hence $G(v) = n'abv' = G(m)abv'$, $v = mbv_1$, and $G(u)a = n'aau' = G(m)aau'$, $u = mau_1$; moreover $|m| \leq |G(m)| = |n'| \leq |n| - 1 < |n| \leq N$.

Thus t satisfies the Markoff property for K and $N-1$. □

Recall that the endomorphism D of the free monoid $\{a, b\}^*$ is defined by $D(a) = ba$ and $D(b) = b$.

Corollary 4.1.5 *If $s = G(t)$ or $s = D(t)$, and if $N \geq 0$, then t satisfies the Markoff property with the parameters $K-1$ and $N-1$.*

Proof. If $s = G(t)$, it is Lemma 4.1.3. If $s = D(t)$, it follows from the lemma by exchanging a and b. □

Theorem 4.1.6 *If an infinite word s on the alphabet $\{a, b\}$ satisfies the Markoff property for the parameters K and N, then $s = uw^\infty$ for some finite word u and some lower Christoffel word w. The length of u is bounded by a function depending only on K and N.*

Proof. The integer function $f(K, N) = 2^{N+3} + K - 2N - 5$, $K \geq 0$ and $N \geq -1$, satisfies $f(K, -1) \geq K$, and if $N \geq 0$, then $K + 2f(N, N-1) + 1 = f(K, N)$; see Exercise 4.1.3.

We prove the theorem by induction on N and show that $|u|$ is bounded by $f(K, N)$. If $N = -1$, then $s = ux^\infty$ with $|u| = K \leq f(K, -1)$, $x = a$ or b, and the result is true.

Suppose now that $N \geq 0$. Let $s = ps_1$ with $|p| = K-1$; then by Lemma 4.1.1, no word of the form $x\tilde{m}xymy$ is a factor of s_1. It follows that s_1 satisfies the hypothesis of Lemma 4.1.2 so that aa and bb are not simultaneously factors of s_1. We claim that s_1 satisfies the Markoff property with parameters $K_1 = N+1$ and N: indeed, if $s_1 = \tilde{u}xyv$, $|u| \geq N+1$, then $s = p\tilde{u}xyv$, $|p\tilde{u}| \geq K+N \geq K$, and therefore $u\tilde{p} = myu'$, $v = mxv'$, $|m| \leq N$; by the length condition on u, we deduce that $u = myu''$, which proves the claim.

Note that if we take a longer word p, with $s = ps_1$, the previous arguments still apply.

Suppose that bb is not a factor of s_1. Then s_1 begins by a or by ba, and we may assume, by taking a word p longer by one letter if necessary, that s_1 begins by a. Then, since s_1 does not contain bb, it implies that s_1 is an infinite product of the words a

and ab; in other words, $s_1 = G(t)$ for some infinite word t. Then by Corollary 4.1.5, t satisfies the Markoff property with parameters N and $N-1$. It follows by induction that $t = uw^\infty$ for some finite word u satisfying $|u| \leq f(N, N-1)$ and some lower Christoffel word w. Then $s = ps_1 = pG(u)G(w)^\infty$. Moreover, by Corollary 2.6.2, $G(w)$ is a lower Christoffel word. The length of $pG(u)$ is bounded by $K + 2|u| \leq K + 2f(N, N-1) \leq f(K, N)$, which ends the proof in this case.

If aa is not a factor of s_1, then, symmetrically, we may assume that $s_1 = D(t)$. Then, similarly, $t = uw^\infty$ for some finite word u satisfying $|u| \leq f(N, N-1)$ and some lower Christoffel word w. Thus $s = ps_1 = pD(u)D(w)^\infty$. We know by Corollary 2.6.2 that $\tilde{D}(w)$ is a lower Christoffel word; moreover, we have seen in the proof of Corollary 2.6.3 that $\tilde{D}(w) = vb$ for some word v and that $D(w) = bv$. Thus $s = pD(u)(bv)^\infty = pD(u)b(vb)^\infty$. We have finally $|pD(u)b| \leq K + 2f(N, N-1) + 1 = f(K, N)$. □

The theorem has a converse: if $s = uw^\infty$ for some lower Christoffel word w, then s satisfies the Markoff property. This may be deduced from the converse of Theorem 4.2.1; it is also a consequence of Theorem 8.3.1 and Theorem 10.1.1.

4.1.1 Exercises

Exercise 4.1.1 *A word p is called a* central factor *of a word q if $q = xpy$ where x* and *y have the same length. Prove that a central factor of a palindrome is a palindrome. Show that if p is a central factor of the palindrome q, then $G(p)a$ is central factor of $G(q)a$; show that the same holds for $a^{-1}G(p)$ if p is non-empty.*

Exercise 4.1.2 *Let $G'(w) = G(w)a$.*

(i) Show that $G'(p)$ is a palindrome if and only if p is a palindrome; show that, for p non-empty, $a^{-1}G(p)$ is a palindrome if and only if p is a palindrome. Hint: use Lemma 4.1.4.

(ii) Let u be a word and let $v = G'(u)$. Show that the mapping $p \mapsto G'(p)$ is a bijection from the set of palindromic factors of u onto the set of palindromic factors of v ending by a. Show that the mapping $p \mapsto a^{-1}G(p)$ is a bijection from the set of palindromic factors of u ending by b onto the set of palindromic factors of v ending by b.

(iii) Deduce that $\mathrm{pal}_a(v) = \mathrm{pal}(u)$ and $\mathrm{pal}_b(v) = \mathrm{pal}_b(u)$, where we denote by $\mathrm{pal}(u)$ the number of palindromic factors of u, and by $\mathrm{pal}_x(u)$ the number of those ending by the letter x. Deduce that

$$\begin{pmatrix}\mathrm{pal}_a(G'(u))\\ \mathrm{pal}_b(G'(u))\end{pmatrix} = \mathrm{Ab}(G)\begin{pmatrix}\mathrm{pal}_a(u)\\ \mathrm{pal}_b(u)\end{pmatrix} + \begin{pmatrix}1\\0\end{pmatrix}$$

where $\mathrm{Ab}(G)$ is the abelianization of G (see Chapter 1).

(iv) By symmetry, show that if $v = b\tilde{D}(u)$, then $\mathrm{pal}_a(v) = \mathrm{pal}_a(u)$ and $\mathrm{pal}_b(v) = \mathrm{pal}(u)$.

Exercise 4.1.3 *Prove the properties of the function f defined at the beginning of the proof of Theorem 4.1.6.*

4.2 Markoff property for bi-infinite words

We say that a bi-infinite word s on the alphabet $\{a, b\}$ satisfies the *Markoff property* if there exists N such that for any factorization $s = \tilde{u}xyv$, with $\{x, y\} = \{a, b\}$, one has $u = myu', v = mxv'$, and $|m| \leq N$. Note that $u, v, u',$ and v' are right infinite words and that $\tilde{u}$ is a left infinite word, the reversal of u, defined in the obvious way.

Theorem 4.2.1 *(Markoff [M1, M2]) If a bi-infinite word on the alphabet $\{a, b\}$ satisfies the Markoff property, then it is equal up to a shift to ${}^{\infty}w^{\infty}$ for some lower Christoffel word w.*

Lemma 4.2.2 *If* u and u' *are words, and* w and w' *are lower Christoffel words on the alphabet $\{a < b\}$, such that $uw^{\infty} = u'w'^{\infty}$, then $w = w'$ and either $u = u'w^i$ or $u' = uw^i$ for some natural number i.*

The *slope* of any word $w \in \{a < b\}^*$ is by definition the quotient $\frac{|w|_b}{|w|_a}$, which is a non-negative rational number or ∞. This is coherent with the definition of the slope of a Christoffel word.

Proof. Let p_n be the prefix of length n of uw^{∞}; the slope of w is equal to the limit when n tends to infinity of the slope of p_n. It follows, therefore, from the hypothesis that w and w' have the same slope. Thus $w = w'$. The hypothesis implies that for some factorization $w = fg$, one has $w = gf$. But this is possible if and only if $f = 1$ or $g = 1$, since the words $C^k(w)$, $k = 0, \ldots, |w| - 1$ are all distinct (see Section 2.2). Thus the lemma follows. □

Proof of Theorem 4.2.1. Let s be a bi-infinite word having the Markoff property. Let s_n denote the right infinite word obtained by truncating s at the left of its term of rank n. Then s_n is a right infinite word satisfying the Markoff property for the parameters $N + 1$ and N (we leave this verification to the reader). It follows from Theorem 4.1.6 that $s_n = u_n w_n^{\infty}$ for some lower Christoffel word w_n and some word u_n whose length is bounded by a constant independent of n. It then follows from the previous lemma that $w_n = w$ is independent of n and that $s = {}^{\infty}w^{\infty}$ up to a shift. □

This theorem has a converse, which may be proved directly; see Exercise 15.2.7. It is interesting to note that the words m that occur in the definition of the Markoff property of bi-infinite words are exactly the central words of Aldo de Luca, studied in the second part of this book; see Exercise 18.5.6.

5
Continued fractions

The theory of continued fractions appears in many textbooks; see, for example, the books by Hardy and Wright [HW], Chapter X; the book of Aigner [A], Chapter 1; or the article of Bombieri [B], Appendix A. We thus do not prove the most basic results of this theory. We define in the first three sections continuant polynomials, finite and infinite continued fractions, and their properties. In Section 5.4, we state the periodicity results on continued fractions. In Section 5.5, we give a result of Legendre (a sufficient condition for being a convergent) and the approximation results for real numbers. In the last section, we introduce the alternating lexicographical order on finite and infinite words; it corresponds to the natural order on real numbers and will be used often in the book.

5.1 Continuant polynomials

We define the Euler *continuant polynomials*, which are useful in the study of continued fractions. They are defined by the initial values $p_{-1} = 0, p_0 = 1$, and for any $k \geq 1$ and any $n_1, \ldots, n_k$ (in some commutative ring) by the *right recursion formula*

$$p_k(n_1, \ldots, n_k) = p_{k-1}(n_1, \ldots, n_{k-1})n_k + p_{k-2}(n_1, \ldots, n_{k-2})$$

It is customary to drop the index k and to write $p(n_1, \ldots, n_k)$ for $p_k(n_1, \ldots, n_k)$

Define the matrix $P(n) = \begin{pmatrix} n & 1 \\ 1 & 0 \end{pmatrix}$. Then

$$P(n_1) \cdots P(n_k) = \begin{pmatrix} p(n_1, \ldots, n_k) & p(n_1, \ldots, n_{k-1}) \\ p(n_2, \ldots, n_k) & p(n_2, \ldots, n_{k-1}) \end{pmatrix} \tag{5.1}$$

The formula is easily proved by induction on n by taking the matrix product from left to right. By taking the product from right to left, one obtains the *left recursion formula*:

$$p(n_1, \ldots, n_k) = n_1 p(n_2, \ldots, n_k) + p(n_3, \ldots, n_k)$$

It follows also, by transposing the product, and using the symmetry of the matrices $P(a)$, that

$$p(n_1, \ldots, n_k) = p(n_k, \ldots, n_1) \tag{5.2}$$

For later use, we show that

$$p(1, n-1, n_1, \ldots, n_k) = p(n, n_1, \ldots, n_k) \tag{5.3}$$

From Christoffel Words to Markoff Numbers. Second Edition. Christophe Reutenauer, Oxford University Press.
 DOI: 10.1093/9780197907535.003.0005

Indeed, if $k = 0$, this follows from $p(n) = n$ and $p(n, m) = nm + 1$ so that $p(1, n-1) = n = p(n)$. If $k \geq 1$, then by the left recursion formula

$$\begin{aligned} p(1, n-1, n_1, \ldots, n_k) &= p(n-1, n_1, \ldots, n_k) + p(n_1, \ldots, n_k) \\ &= (n-1)p(n_1, \ldots, n_k) + p(n_2, \ldots, n_k) + p(n_1, \ldots, n_k) \\ &= np(n_1, \ldots, n_k) + p(n_2, \ldots, n_k) = p(n, n_1, \ldots, n_k) \end{aligned}$$

Similarly, we note the following property of continuant polynomials.

Lemma 5.1.1 *If $i \leq j \leq k \leq l$ and if the u_i are positive integers, then $p(u_i, \ldots, u_l) \geq p(u_j, \ldots, u_k)$.*

With the notations of the previous lemma, say that the sequence $u_j, \ldots, u_k$ is a *factor* of the sequence $u_i, \ldots, u_l$. The lemma thus asserts that if a sequence of positive integers is a factor of another one, then the image under p of the first one is not larger than the image under p of the second one.

Proof. The lemma follows from (5.1); the fact that $p(u_r, \ldots, u_s)$ is the 1,1-entry of $P(u_r) \cdots P(u_s)$, that $P(u_j) \cdots P(u_k)$ is a subproduct of $P(u_i) \cdots P(u_l)$, and that all u_j are ≥ 1. □

5.1.1 Exercises

Exercise 5.1.1 *Show that $p(1, 0, a_1, \ldots, a_n) = p(1, a_1, \ldots, a_n) = p(a_1, \ldots, a_n) + p(a_2, \ldots, a_n)$.*

Exercise 5.1.2 *Show that $p(a_1, \ldots, a_n, a_n, \ldots, a_1) = p(a_1, \ldots, a_n)^2 + p(a_1, \ldots, a_{n-1})^2$. Hint: take the 1, 1-entry in the product $P(a_1) \cdots P(a_n) \times P(a_n) \cdots P(a_1)$.*

5.2 Finite continued fractions

A *finite continued fraction* is an expression of the form

$$a_0 + \cfrac{1}{a_1 + \cfrac{1}{a_2 + \cfrac{1}{\ddots + \cfrac{1}{a_n}}}} \tag{5.4}$$

where $n \in \mathbb{N}$, $a_0 \in \mathbb{Z}$, $a_1, \ldots, a_n \in \mathbb{P}$. This expression is written

$$[a_0, a_1, a_2, \ldots, a_n]$$

We call $n+1$ the *length* of this continued fraction. It represents a well-defined rational number, since by positivity there is no division by 0.

It is well known that each rational number has such an expression. Moreover, a given rational number has exactly two such expressions: if one of them is as previously and $a_n > 1$ or $n = 0$, then the other is $[a_0, a_1, a_2, \ldots, a_{n-1}, a_n - 1, 1]$; if, on the contrary, $a_n = 1$ and $n > 0$, then the other is $[a_0, a_1, a_2, \ldots, a_{n-1} + 1]$. For example, $[-2] = [-3, 1]$, $[1, 1, 2] = [1, 1, 1, 1]$, $[2, 2] = [2, 1, 1]$.

One has

$$[a_0, \ldots, a_n] = \frac{p_n}{q_n} \tag{5.5}$$

where the integers p_n and q_n are defined by the continuant polynomials

$$p_n = p(a_0, \ldots, a_n), \qquad q_n = p(a_1, \ldots, a_n) \tag{5.6}$$

The integers p_n and q_n may be defined recursively by $p_{-1} = 1$, $p_0 = a_0$, $q_{-1} = 0$, $q_0 = 1$, $p_i = a_i p_{i-1} + p_{i-2}$, and $q_i = a_i q_{i-1} + q_{i-2}$ for $i = 1, \ldots, n$.

It follows from (5.1) that for $n \geq 0$

$$\begin{pmatrix} p_n & p_{n-1} \\ q_n & q_{n-1} \end{pmatrix} = P(a_0)P(a_1)\cdots P(a_n) \tag{5.7}$$

Therefore

$$p_n q_{n-1} - q_n p_{n-1} = (-1)^{n+1} \tag{5.8}$$

Moreover, by transposing the matrices, we obtain

$$\begin{pmatrix} p_n & q_n \\ p_{n-1} & q_{n-1} \end{pmatrix} = P(a_n)P(a_{n-1})\cdots P(a_0)$$

We deduce that $p_n/p_{n-1} = [a_n, a_{n-1}, \ldots, a_0]$ and $q_n/q_{n-1} = [a_n, a_{n-1}, \ldots, a_1]$.

5.2.1 Exercises

Exercise 5.2.1 *Consider a Christoffel word w of slope s; call* Slope *of w the number $S = |w|_b/|w|$. Show that $S \in [0, 1]$, $S = s/(1+s)$, $s = S/(1-S)$, $S^{-1} = 1 + s^{-1}$. Show that $S = 0$ if and only if $s = 0$, and $S = 1$ if and only if $s = \infty$. Show that if $S \neq 0, 1$, and if the continued fraction of S is of the form $[0, a_1, \ldots, a_m]$, where the a_i are positive integers, then $s^{-1} = S^{-1} - 1 = [a_1, \ldots, a_m] - 1 = [a_1 - 1, a_2, \ldots, a_m]$, and $s = [0, a_1 - 1, a_2, \ldots, a_m]$ if $a_1 \geq 2$; and if $a_1 = 1$, $s^{-1} = [0, a_2, \ldots, a_m]$ and $s = [a_2, \ldots, a_m]$.*

Exercise 5.2.2 *Call* length *of a positive rational number q/p the sum $p + q$, the fraction being irreducible and $p, q > 0$. Show that the length of a Christoffel word is the length of its slope, and the denominator of its Slope (see Exercise 5.2.1). Show that if the rational number has the continued fraction expansion $[a_0, \ldots, a_n] = [a_0, \ldots, a_n - 1, 1]$, then its length is $p(1, a_0, \ldots, a_n) = p(1, a_0, \ldots, a_n - 1, 1)$; and that if $a_0 = 0$, it is also equal to $p(1, a_1, \ldots, a_n) = p(1, a_1, \ldots, a_n - 1, 1)$.*

5.3 Infinite continued fractions

An *infinite continued fraction* is an expression of the form

$$a_0 + \cfrac{1}{a_1 + \cfrac{1}{a_2 + \cfrac{1}{\ddots + \cfrac{1}{a_n + \cfrac{1}{\ddots}}}}}$$

with $a_0 \in \mathbb{Z}$, $a_n \in \mathbb{P}$ for any $n \in \mathbb{N}$. It is written

$$[a_0, a_1, a_2, \ldots, a_n, \ldots]$$

It is well known that each irrational real number has a unique expression as an infinite continued fraction, in the sense that this number is the limit of the finite continued fractions (5.4) as n goes to infinity. In other words,

$$[a_0, a_1, a_2, \ldots, a_n, \ldots] = \lim_{n\to\infty}[a_0, a_1, a_2, \ldots, a_n]$$

Conversely, such an expression represents an irrational real number.

Starting from $x = [a_0, a_1, a_2, \ldots, a_n, \ldots]$, define $x_n = [a_n, a_{n+1}, a_{n+2}, \ldots]$. Then clearly $x = x_0 = a_0 + \frac{1}{x_1}$ and recursively

$$x = a_0 + \cfrac{1}{a_1 + \cfrac{1}{a_2 + \cfrac{1}{\ddots + \cfrac{1}{a_{n-1} + \cfrac{1}{x_n}}}}}$$

One shows by induction on n that if the numbers p_i and q_i are defined as in Section 5.2, then for $n \geq 1$

$$x = \frac{x_n p_{n-1} + p_{n-2}}{x_n q_{n-1} + q_{n-2}} \tag{5.9}$$

5.4 Periodic expansions yield quadratic numbers

In order to consider *ultimately periodic* continued fractions, it is convenient to use the notation

$$[a_0, \ldots, a_n, \overline{b_1, \ldots, b_p}] := [a_0, \ldots, a_n, b_1, \ldots, b_p, b_1, \ldots, b_p, b_1, \ldots, b_p, \ldots]$$

including the case where the prefix $a_0, \ldots, a_n$ is empty; in this latter case, the continued fraction is called *periodic.*

Recall that a real number x is *quadratic* if it may be written $x = \frac{a+\sqrt{b}}{c}$, where $a, b,$ and c are rational numbers such that b is not a square in $\mathbb{Q}$ and $c \neq 0$. Equivalently, x is a root of some irreducible polynomial P of degree 2 over the rationals;

P is unique up to a non-zero factor. The *conjugate* of the quadratic number x above is $\bar{x} = \frac{a-\sqrt{b}}{c}$. Equivalently, $\bar{x}$ is the other root of the polynomial P. It follows that the conjugate does not depend on the chosen expression.

Proposition 5.4.1 *Let $n, a_0, \ldots, a_{n-1}$ be positive integers, $x = [\overline{a_0, \ldots, a_{n-1}}]$, $y = [\overline{a_{n-1}, \ldots, a_0}]$, and*

$$\begin{pmatrix} p & q \\ r & s \end{pmatrix} = P(a_0) \cdots P(a_{n-1}) = \begin{pmatrix} a_0 & 1 \\ 1 & 0 \end{pmatrix} \cdots \begin{pmatrix} a_{n-1} & 1 \\ 1 & 0 \end{pmatrix}$$

Then x and y are quadratic, $x = \frac{px+q}{rx+s}$, $y = \frac{py+r}{qy+s}$, $y = -1/\bar{x}$, and $y = \frac{p-s+\sqrt{(s-p)^2+4qr}}{2q}$.

Proof. Note that q and r are positive. By (5.1), (5.6), and (5.9), $x = \frac{px+q}{rx+s}$ since $x = x_n$. Thus $rx^2+(s-p)x-q = 0$. The discriminant of this equation is $(s-p)^2+4qr$; hence $x = \frac{p-s+\sqrt{(s-p)^2+4qr}}{2r}$: indeed, x is positive, and the other root is negative, since $q, r > 0$ so that $p - s - \sqrt{(s-p)^2 + 4qr} < 0$. Now, since the matrices in the product are symmetric, the product $\begin{pmatrix} a_{n-1} & 1 \\ 1 & 0 \end{pmatrix} \cdots \begin{pmatrix} a_0 & 1 \\ 1 & 0 \end{pmatrix}$ is equal to $\begin{pmatrix} p & r \\ q & s \end{pmatrix}$. Thus, similarly, we have $y = \frac{py+r}{qy+s}$ and $y = \frac{p-s+\sqrt{(s-p)^2+4qr}}{2q}$.

Now, $-1/\bar{x}$ is equal to $-\frac{2r}{p-s-\sqrt{(s-p)^2+4qr}} = -\frac{2r(p-s+\sqrt{(s-p)^2+4qr})}{(p-s)^2-((s-p)^2+4qr)} = y$. Finally, x and y are irrational since their expansion into continued fractions is infinite. Hence the discriminant is not a square in $\mathbb{Q}$ and x and y are quadratic. □

Concerning the previous result, one has to mention theorems of Lagrange and Galois: Lagrange's theorem asserts that a real number x has an ultimately periodic expansion into continued fractions if and only if x is quadratic; Galois's theorem says that the expansion is periodic (as those in the preceding proposition) if and only if x is quadratic and satisfies $x > 1$ and $-1 < \bar{x} < 0$. A proof of both theorems may be found in Aigner's book [A].

5.5 Approximations of real numbers

Consider an *infinite* continued fraction $x = [a_0, a_1, a_2, \ldots, a_n, \ldots]$. Define p_i and q_i recursively by $p_{-1} = 1, p_0 = a_0, q_{-1} = 0, q_0 = 1, p_i = a_i p_{i-1} + p_{i-2}$, and $q_i = a_i q_{i-1} + q_{i-2}$ for $i \geq 1$.

The number a_n is called the n-th *partial quotient* of x. The rational number p_n/q_n is called the n-th *convergent* of x; n is called the *rank* of the convergent. One has $p_n/q_n = [a_0, \ldots, a_n]$.

The sequence of convergents of even rank is strictly increasing, while the sequence of convergents of odd rank is strictly decreasing. Both converge to x, and one has for any n,

$$|x - p_n/q_n| < 1/q_n^2$$

See [HW] for a proof of these basic results in the theory of continued fractions.

It follows from the definition of the q_n that the sequence $(q_n)_{n\geq 1}$ is strictly increasing. Hence, one deduces Dirichlet's theorem, whose original proof was quite different, using the pigeonhole principle: *for any irrational real number x, there are infinitely many rationals p/q such that* $|x - p/q| < 1/q^2$.

One has a stronger approximation as follows: *for any* $n \geq 1$, *one has either* $|x - p_n/q_n| < 1/2q_n^2$ or $|x - p_{n+1}/q_{n+1}| < 1/2q_{n+1}^2$.

Indeed, by contradiction, suppose that both inequalities are false; then, since x is irrational, $|x - p_n/q_n| > 1/2q_n^2$ and $|x - p_{n+1}/q_{n+1}| > 1/2q_{n+1}^2$.

Note that by (5.8), $q_n(q_{n+1}x - p_{n+1}) - q_{n+1}(q_nx - p_n) = (-1)^{n+1}$. Dividing by q_nq_{n+1}, we obtain $(x - p_{n+1}/q_{n+1}) - (x - p_n/q_n) = (-1)^{n+1}/q_nq_{n+1}$. By a previous statement, $x - p_{n+1}/q_{n+1}$ and $x - p_n/q_n$ have opposite sign so that $|x - p_{n+1}/q_{n+1}| + |x - p_n/q_n| = 1/q_nq_{n+1}$.

Hence, $1/q_nq_{n+1} > 1/2q_n^2 + 1/2q_{n+1}^2$ and multiplying by $2q_n^2q_{n+1}^2$, we deduce that $q_{n+1}^2 + q_n^2 < 2q_nq_{n+1}$ and then $(q_{n+1} - q_n)^2 < 0$, a contradiction.

Let $p/q \in \mathbb{Q}$, $q > 0$. We show that convergents are *optimal* in the following sense: *if* $q < q_{n+1}$, *then* $|qx - p| \geq |q_nx - p_n|$. We may assume that p/q is reduced. The system of two equations in the unknowns u and v

$$p_nu + p_{n+1}v = p, \qquad q_nu + q_{n+1}v = q \tag{5.10}$$

has determinant ± 1. Thus there is a unique integral solution u and v. If we had $u = 0$, then $v = q/q_{n+1} > 0$. Thus $v \geq 1$ and $q \geq q_{n+1}$, against the assumption. If $v = 0$, then $u = p/p_n = q/q_n$ and since the two fractions p/q and p_n/q_n are reduced (the latter because of (5.8)) and $q, q_n > 0$, $p = p_n$ and $q = q_n$; then clearly $|qx - p| \geq |q_nx - p_n|$.

We assume now that u and v are non-zero and show that they have opposite sign. If $v \leq -1$, then $q_nu = q - q_{n+1}v > 0$ and thus $u > 0$. If, on the contrary, $v \geq 1$, then $q_nu = q - q_{n+1}v \leq q - q_{n+1} < 0$ and thus $u < 0$. Since as we saw, $q_nx - p_n$ and $q_{n+1}x - p_{n+1}$ have opposite sign, we see that $u(q_nx - p_n)$ and $v(q_{n+1}x - p_{n+1})$ have the same sign. Now by (5.10), $qx - p = (q_nu + q_{n+1}v)x - (p_nu + p_{n+1}v) = u(q_nx - p_n) + v(q_{n+1}x - p_{n+1})$. The two last summands have the same sign so that $|qx - p| = |u||q_nx - p_n| + |v||q_{n+1}x - p_{n+1}| > |q_nx - p_n|$ ($q_{n+1}x - p_{n+1} \neq 0$ since x is irrational), as required.

We show finally that *if* $|x - p/q| < 1/2q^2$, *then p/q is a convergent of x*. Indeed, since $q_0 = 1$ and $(q_n)_{n\geq 1}$ is strictly increasing, there exists $n \geq 0$ such that $q_n \leq q < q_{n+1}$ (q being positive). Then the previous property shows that $|qx - p| \geq |q_nx - p_n|$. Thus $|x - p_n/q_n| \leq (1/q_n)|qx - p| = (q/q_n)|x - p/q| < 1/2qq_n$. Assuming that $p/q \neq p_n/q_n$, we obtain that $|pq_n - qp_n| \geq 1$ and thus $1/qq_n \leq \frac{|pq_n - qp_n|}{qq_n} = |p/q - p_n/q_n| \leq |x - p/q| + |x - p_n/q_n| < 1/2q^2 + 1/2qq_n$. Thus $1/qq_n < 1/q^2$, hence $q < q_n$, a contradiction. Thus $p/q = p_n/q_n$.

We thus have the following result.

Theorem 5.5.1 *For each convergent p/q of an irrational real number x, one has* $|x - p/q| < 1/q^2$. *Of two successive convergents, one of them, p/q say, satisfies* $|x - p/q| < 1/2q^2$. *If p/q is a rational number satisfying* $|x - p/q| < 1/2q^2$, *then it is a convergent of x.*

The last assertion is due to Legendre [L], (9) pp. 27–29.

For later use, we state and prove the following result. We define the *Fibonacci numbers* (F_n) by the rules:[1] $F_{-1} = 0, F_0 = 1$ and $F_{n+1} = F_n + F_{n-1}$ for any natural number n.

Lemma 5.5.2 *If x and y are two real numbers whose expansions into continued fractions coincide up to the partial quotient of rank $n \geq 0$, then $|x - y| < \frac{1}{F_n^2}$.*

Warning: the numbering starts at 0: a_0 is the partial quotient of rank 0. Hence, the hypothesis of the lemma means that the expansion into continued fractions of x and y coincides for their first $n + 1$ terms.

Proof. If $n = 0$, then x and y have the same integer part so that $|x - y| < 1 = 1/F_0^2$. Let $n \geq 1$. By hypothesis, $x = [a_0, \ldots, a_n, a_{n+1}, \ldots]$ and $y = [a_0, \ldots, a_n, b_{n+1}, \ldots]$. Then, by (5.9), we have $x - y = \frac{x_n p_{n-1} + p_{n-2}}{x_n q_{n-1} + q_{n-2}} - \frac{y_n p_{n-1} + p_{n-2}}{y_n q_{n-1} + q_{n-2}}$ and $x_n = [a_n, a_{n+1}, \ldots]$, $y_n = [a_n, b_{n+1}, \ldots]$ have the same integer part a_n so that $|x_n - y_n| < 1$. Thus

$$x - y = \frac{(x_n p_{n-1} + p_{n-2})(y_n q_{n-1} + q_{n-2}) - (x_n q_{n-1} + q_{n-2})(y_n p_{n-1} + p_{n-2})}{(x_n q_{n-1} + q_{n-2})(y_n q_{n-1} + q_{n-2})}$$
$$= \frac{x_n (p_{n-1} q_{n-2} - q_{n-1} p_{n-2}) - y_n (q_{n-2} p_{n-1} - p_{n-2} q_{n-1}))}{(x_n q_{n-1} + q_{n-2})(y_n q_{n-1} + q_{n-2})}$$

This is by (5.8) equal to $\frac{(-1)^n (x_n - y_n)}{(x_n q_{n-1} + q_{n-2})(y_n q_{n-1} + q_{n-2})}$. Since $n \geq 1$, x_n and y_n are ≥ 1 so that the denominator is greater than or equal to the square of $q_{n-1} + q_{n-2}$. Now it is easy to prove by induction that $q_i \geq F_i$ if $i \geq -1$. Thus $q_{n-1} + q_{n-2} \geq F_{n-1} + F_{n-2} = F_n$, which ends the proof. □

Note that the proof works even if x and y are rational numbers, provided the partial quotients of rank n exist.

5.5.1 Exercises

Exercise 5.5.1 *Show that the generating series $\sum_{n \geq 0} F_n x^n$ of the Fibonacci sequence is $1/(1 - x - x^2)$. Show that F_n is the denominator of the n-th convergent of the* golden ratio $\frac{\sqrt{5}+1}{2}$ *and that $F_n = p(1, \ldots, 1)$, with n 1's. Show that the n-th convergent of the golden ratio is F_{n+1}/F_n.*

5.6 Ordering continued fractions

If real numbers are represented in base k, then one compares them using the lexicographical order $<_{lex}$, as is well known. Indeed, real numbers in the interval $[0, 1]$ are

[1] We take another convention as [HW], Section 10.14, where the Fibonacci sequence (u_n) is defined by $u_1 = u_2 = 1$ and the same recursion.

represented by finite or infinite words over $\{0, 1, \ldots, k-1\}$. For example, with the base $k = 10$, one has $0.1 < 0.1\ldots, 0.91\ldots < 0.92\ldots$.

For real numbers represented by continued fractions, one has to use the *alternating lexicographic order*, denoted $<_{alt}$. Since we want to compare all real numbers, we consider the set of all non-empty finite or infinite words

$$a_0a_1a_2\ldots, a_0 \in \mathbb{Z}, a_1, a_2, \cdots \in \mathbb{P}$$

Then, for two such words, $u <_{alt} v$ if one has one of the two following mutually exclusive conditions:

(i) one of them is a proper prefix of the other: then either u is a proper prefix of odd length of v or if v is a proper prefix of even length of u;
(ii) $u = mau', v = mbv'$ for some words m, u', and v' and integers a and b such that either $a < b$ and m of even length or $a > b$ and m of odd length.

Note that the m in case (ii) and the prefix in case (i) are necessarily finite words. Note also that if $u = au', v = av'$ with $a \in \mathbb{Z}$, and u' and v' are non-empty, then $u <_{alt} v$ if and only if $v' <_{alt} u'$. This is the origin of the word "alternating": one begins by comparing the first digits of both words for the usual order, then, if equality holds, the second ones for the opposite order, then, if equality holds, the third ones for the usual order, and so on. From this, it is also easy to verify that the alternating lexicographical order is a total order. Another consequence is that if $u \le_{alt} v$ and w is a finite word of even (resp. odd) length, then $wu \le_{alt} wv$ (resp. $wu \ge_{alt} wv$).

For $u = a_0a_1a_2 \cdots$, we denote by $[u]$ the continued fraction $[a_0, a_1, a_2, \ldots]$.

Theorem 5.6.1 *The mapping $u \mapsto [u]$ from the set of words as above into $\mathbb{R}$ satisfies: if $u <_{alt} v$, then $[u] \le [v]$. If the latter inequality is an equality, then u and v are finite words.*

The case of equality is illustrated by the example: $11 <_{alt} 2$, but $[11]$ and $[2]$ are equal to 2.

Instances of the theorem are the inequalities $[2] < [2, 1, \ldots], [2, 1, 3, \ldots] < [2, 1], [2, \ldots] < [3, \ldots]$, and $[2, 3, \ldots] < [2, 1, \ldots]$.

Proof of Theorem 5.6.1. Recall that if $u = au', a \in \mathbb{Z}$, then the integer part of $[u]$ is a (warning: the notation $[u]$ does not indicate the integer part of u); moreover, either u' is empty and $[u] = a$, or u' is non-empty and $[u] = a + [u']^{-1}$.

Suppose that $u <_{alt} v$. In what follows, a and b will be in $\mathbb{Z}$ and u' and v' will be non-empty words.

If u is of length 1, $u = a$, then $[u] = a$ and we must have either $v = b$ with $a < b$ or $v = bv'$ and $a \le b$. In the first case, $[u] < [v]$; in the second case, we have the same inequality, since $[v] = b + [v']^{-1}$ and v' has only positive letters so that $[v'] > 0$.

Suppose now that $u = au'$. If v is of length 1, then $v = b$, and we cannot have $b = a$ (otherwise, v is a prefix of odd length of u and $v <_{alt} u$), nor $b < a$ so that $a < b$. Then $a + 1 \le b$ and therefore $[u] = a + [u']^{-1} \le b = [v]$, since $[u'] \ge 1$.

If $v = bv'$, then we cannot have $a > b$. If $a = b$, we have $u' >_{alt} v'$ so that by induction (on the length of the longest common prefix of u and v) $[u'] \geq [v']$ and therefore $[u] \leq [v]$. If $a < b$, we have directly $[u] < [v]$, since their integer parts are respectively a and b.

The last sentence in the statement follows from Sections 5.2 and 5.3. □

Corollary 5.6.2 *Let u and v be words as above such that $[u] \neq [v]$ (this hypothesis is satisfied if one word is finite and the other infinite or if one is a proper prefix of the other). Then $[u] < [v]$ is equivalent to $u <_{alt} v$.*

For later use, we state the following easy lemma, whose proof is left to the reader. Say that in the word u, each letter is *repeated* if $u = a_0 a_0 a_1 a_1 a_2 a_2 \cdots$.

Lemma 5.6.3 *Let u and v be finite or infinite words with repeated letters such that no one is a prefix of the other. Then $u <_{lex} v$ if and only if $u <_{alt} v$.*

Note that the restriction in the lemma is necessary, since, for example, $11 <_{lex} 1122$ and $11 >_{alt} 1122$.

6
Binary quadratic forms

We study here quadratic forms in two variables over the real numbers, which are indefinite; that is, take positive and negative values when the variables are sent to reals (equivalently, the discriminant of the form is positive). These forms appear in the work of Markoff. We define the action of $\mathrm{GL}_2(\mathbb{Z})$ and the corresponding equivalence of forms, roots of a form (especially the plus and minus roots, following Dickson), and the action of the Möbius group on roots. In Section 6.2, we associate a form with each quadratic number, and we evaluate this form at the numerator and denominator of the convergents of the given number.

For books on binary quadratic forms, see [D2, BV, Bu].

6.1 Binary quadratic forms: actions, roots

A *real binary quadratic form* is a non-zero polynomial $f(x, y) = \alpha x^2 + \beta xy + \gamma y^2$ in the variables x and y and real coefficients $\alpha, \beta,$ and γ. Its *discriminant* is $d(f) = \beta^2 - 4\alpha\gamma$. It is called *indefinite* if the discriminant is positive.

Two forms are *proportional* if they differ multiplicatively by a non-zero real factor; we also say that one is a *multiple* of the other.

Given a column vector $v = {}^t(x, y) \in \mathbb{R}^{2\times 1}$ and a quadratic form $f(x, y)$, we write (by a slight abuse of language) $f(v) = f(x, y)$. There is a left action of the group $\mathrm{GL}_2(\mathbb{Z})$ on $\mathbb{R}^{2\times 1}$ by left multiplication $(M, v) \mapsto Mv$. Therefore, the group $\mathrm{GL}_2(\mathbb{Z})$ acts on the right on forms by $(f \cdot M)(v) = f(Mv)$ for any vector $v = {}^t(x, y)$. Concretely, if

$$M = \begin{pmatrix} p & q \\ r & s \end{pmatrix} \in \mathrm{GL}_2(\mathbb{Z}) \tag{6.1}$$

then

$$(f \cdot M)(x, y) = f(px + qy, rx + sy)$$

Two forms are *equivalent* if they are in the same orbit under the action of $\mathrm{GL}_2(\mathbb{Z})$.

Proposition 6.1.1 *Equivalent forms have the same discriminant.*

Proof. Recall that matrices in $\mathrm{GL}_2(\mathbb{Z})$ have determinant ± 1. Let $g = f \cdot M$, with f, M as above; then $g := \alpha' x^2 + \beta' xy + \gamma' y^2 = f(px + qy, rx + sy) = \alpha(px + qy)^2 + \beta(px + qy)(rx + sy) + \gamma(rx + sy)^2$. Therefore

$$\alpha' = \alpha p^2 + \beta pr + \gamma r^2, \quad \beta' = 2\alpha pq + \beta ps + \beta qr + 2\gamma rs, \quad \gamma' = \alpha q^2 + \beta qs + \gamma s^2 \tag{6.2}$$

From Christoffel Words to Markoff Numbers. Second Edition. Christophe Reutenauer, Oxford University Press.
 DOI: 10.1093/9780197907535.003.0006

We have

$$\begin{pmatrix} s & q \\ r & p \end{pmatrix}\begin{pmatrix} \beta & 2\gamma \\ 2\alpha & \beta \end{pmatrix}\begin{pmatrix} p & q \\ r & s \end{pmatrix} = \begin{pmatrix} s\beta + 2q\alpha & 2s\gamma + q\beta \\ r\beta + 2p\alpha & 2r\gamma + p\beta \end{pmatrix}\begin{pmatrix} p & q \\ r & s \end{pmatrix}$$

$$= \begin{pmatrix} s\beta p + 2q\alpha p + 2s\gamma r + q\beta r & s\beta q + 2q\alpha q + 2s\gamma s + q\beta s \\ r\beta p + 2p\alpha p + 2r\gamma r + p\beta r & r\beta q + 2p\alpha q + 2r\gamma s + p\beta s \end{pmatrix}$$

and the latter matrix is equal to $\begin{pmatrix} \beta' & 2\gamma' \\ 2\alpha' & \beta' \end{pmatrix}$. Hence, by taking the determinants of these matrices, we see that the discriminant of g is $\beta'^2 - 4\alpha'\gamma' = (ps - qr)^2(\beta^2 - 4\alpha\gamma)$. □

In particular, the action of $\mathrm{GL}_2(\mathbb{Z})$ on quadratic forms defines an action on indefinite forms.

Given a two-by-two matrix A over $\mathbb{R}$, we associate with it the quadratic form

$$f_A(x, y) = cx^2 + (d - a)xy - by^2, \quad A = \begin{pmatrix} a & b \\ c & d \end{pmatrix} \tag{6.3}$$

It is easy to prove that

$$f_A(x, y) = (-y, x)A\begin{pmatrix} x \\ y \end{pmatrix} \tag{6.4}$$

Proposition 6.1.2 *Let $B = M^{-1}AM$ with $A, B,$ and M being 2 by 2 matrices over $\mathbb{R}$ and M in $GL_2(\mathbb{R})$. Then $f_B = \frac{1}{\det(M)} f_A \cdot M$.*

Proof. Let ${}^t(x', y') = M\,{}^t(x, y)$. Then by definition of $f \cdot M$, we have $(f \cdot M)(x, y) = f(x', y')$. With M as in (6.1), we have $(x', y') = (px + qy, rx + sy)$.

Let $M' = \det(M)M^{-1}$; then $M' = \begin{pmatrix} s & -q \\ -r & p \end{pmatrix}$. It follows that $(-y, x)M' = (-sy - rx, qy + px) = (-y', x')$.

Thus, using (6.4) twice,

$$f_B(x, y) = (-y, x)B\,{}^t(x, y) = \frac{1}{\det(M)}(-y, x)M'AM\,{}^t(x, y)$$

$$= \frac{1}{\det(M)}(-y', x')A\,{}^t(x', y') = \frac{1}{\det(M)} f_A(x', y') = \frac{1}{\det(M)}(f_A \cdot M)(x, y)$$

□

Given a quadratic polynomial $F(x) = ax^2 + bx + c$ over $\mathbb{R}$, with positive discriminant Δ, we call (following Dickson [D2], chapter VII.30) *plus root* (resp. *minus root*) of $F(x)$ the root $\frac{-b+\sqrt{\Delta}}{2a}$ (resp. $\frac{-b-\sqrt{\Delta}}{2a}$). Of course, "plus" and "minus" refer to the sign + or – in the formula for the root. Observe that if x is a root of F, it is the plus root if and only if $2ax + b \geq 0$, that is, if and only if

$$F'(x) \geq 0 \tag{6.5}$$

(see Exercise 6.1.6). We say that the *sign* of the plus root of F is +, and similarly for the minus root. Note that if

$$a > 0 \quad \text{and} \quad c < 0 \tag{6.6}$$

then the sign of a root of F is its sign as real number; this follows since $\Delta = b^2 - 4ac > b^2$; hence $\frac{-b+\sqrt{\Delta}}{2a} > 0$ and $\frac{-b-\sqrt{\Delta}}{2a} < 0$. Clearly, F and dF have the same plus and minus roots if $d > 0$, while they are interchanged if $d < 0$.

Given a quadratic form $f(x, y) = ax^2 + bxy + cy^2$ over $\mathbb{R}$, with positive discriminant Δ, we call *plus root* (resp. *minus root*) of f the corresponding root of the polynomial $f(x, 1)$. A *root* of f is one of these two roots.

With each invertible two-by-two matrix A over $\mathbb{R}$, we associate the *Möbius transformation*

$$\mu_A : x \mapsto A \cdot x = \frac{ax + b}{cx + d} \tag{6.7}$$

where A is given in (6.3). It corresponds to a left action $A \cdot x$ of matrices on $\overline{\mathbb{R}} = \mathbb{R} \cup \infty$ (Exercise 6.1.4). If A has integer coefficients, it sends irrational numbers onto themselves.

Proposition 6.1.3 *([D2], Theorem 72) Let f and g be real binary quadratic forms with positive discriminant, with distinct irrational roots, and $M \in \mathrm{GL}(\mathbb{Z})$. Suppose that $g = f \cdot M$.*

(i) The Möbius transformation μ_M sends the two roots of g onto the two roots of f.

(ii) Moreover, μ_M sends the plus root of g onto the root of f whose sign is that of $\det(M)$.

Proof.

(i) Let θ be a root of g. We have $0 = g(\theta, 1) = (f \cdot M)(\theta, 1) = f(p\theta + q, r\theta + s)$, with M given in (6.1). Since θ is irrational, $r\theta + s \neq 0$; hence $f((p\theta + q)/(r\theta + s), 1) = 1/(r\theta + s)^2 f(p\theta + q, r\theta + s) = 0$. Therefore, $\mu_M(\theta) = (p\theta + q)/(r\theta + s)$ is a root of f.

(ii) The two forms have the same discriminant R^2, $R > 0$. Let ω be the plus root of f. Let $f = \alpha x^2 + \beta xy + \gamma y^2$, $g = \alpha' x^2 + \beta' xy + \gamma' y^2$. Note that $\alpha, \alpha' \neq 0$, since both f and g have two distinct irrational roots. We have $\omega = \frac{-\beta+R}{2\alpha}$.

Let $\epsilon = \det(M) = \pm 1$. Note that $M^{-1} = \epsilon \begin{pmatrix} s & -q \\ -r & p \end{pmatrix}$ and that a matrix and its opposite have the same action as Möbius transformation. Thus

$$\mu_M^{-1}(\omega) = \frac{s\omega - q}{-r\omega + p} = \frac{s\frac{-\beta+R}{2\alpha} - q}{-r\frac{-\beta+R}{2\alpha} + p} = \frac{-s\beta - 2q\alpha + sR}{r\beta + 2p\alpha - rR}$$

$$= \frac{(-s\beta - 2q\alpha + sR)(r\beta + 2p\alpha + rR)}{(r\beta + 2p\alpha - rR)(r\beta + 2p\alpha + rR)}.$$

Note that $R^2 = \beta^2 - 4\alpha\beta$ and $ps - qr = \epsilon$. The denominator is equal to $\beta^2 r^2 + 4\alpha^2 p^2 + 4\alpha\beta pr - (\beta^2 - 4\alpha\gamma)r^2 = 4\alpha(\alpha p^2 + \beta pr + \gamma r^2) = 4\alpha\alpha'$, using (6.2). The numerator is $-\beta^2 rs - 2\alpha\beta ps - \beta rsR - 2\alpha\beta qr - 4\alpha^2 pq - 2\alpha qrR + \beta rsR + 2\alpha psR + rs(\beta^2 - 4\alpha\gamma) = 2\alpha((-qr + ps)R - \beta ps - \beta qr - 2\alpha pq - 2\gamma rs) = 2\alpha(\epsilon R - \beta')$, (using (6.2)). It follows that

$$\mu_M^{-1}(\omega) = \frac{-\beta' + \epsilon R}{2\alpha'}$$

Hence, if $\det(M) = 1$ (resp. $\det(M) = -1$), μ_M sends the plus root (resp. the minus root) of g onto ω. One concludes using (i). □

The next lemma will be useful later.

Lemma 6.1.4 *Let β and β' be the respective plus roots of the quadratic polynomials $px^2 + qx + r$ and $px^2 + qx + r'$ of respective discriminant $\Delta > 0$ and $\Delta' \geq 0$. Then*

$$\beta - \beta' = \frac{r' - r}{\frac{\sqrt{\Delta}}{2}\left(1 + \sqrt{1 + \frac{4p(r-r')}{\Delta}}\right)}$$

Proof. Observe that $\Delta - \Delta' = 4pr' - 4pr$. We have $\beta - \beta' = \frac{-q+\sqrt{\Delta}}{2p} - \frac{-q+\sqrt{\Delta'}}{2p} = \frac{\sqrt{\Delta}-\sqrt{\Delta'}}{2p} = \frac{\Delta-\Delta'}{2p(\sqrt{\Delta}+\sqrt{\Delta'})} = \frac{4p(r'-r)}{2p\sqrt{\Delta}\left(1+\sqrt{\frac{\Delta'}{\Delta}}\right)} = \frac{r'-r}{\frac{\sqrt{\Delta}}{2}\left(1+\sqrt{\frac{\Delta'}{\Delta}}\right)}$. Moreover, $\frac{\Delta'}{\Delta} = 1 + \frac{\Delta'-\Delta}{\Delta} = 1 + \frac{4p(r-r')}{\Delta}$. The formula follows. □

6.1.1 Exercises

Exercise 6.1.1 *Let $f(x, y) = ax^2 + bxy + cy^2$ be a real binary quadratic form.*

(i) *Show that if f is indefinite, then there are infinitely many real points $(x, y) \neq (0, 0)$ such that $f(x, y) > 0$ (resp. $f(x, y) < 0$). Show that these real points may be chosen integral.*

(ii) *Show that if f is* definite *(that is, its discriminant is negative), then for any real point $(x, y) \neq (0, 0)$, $f(x, y)$ is non-zero and has the sign of a.*

Exercise 6.1.2 *Show that the equation $\mu_A(x) = x$ is equivalent to $f_A(x, 1) = 0$. Assuming that f_A is indefinite, deduce that the roots of f_A are the fix points of μ_A.*

Exercise 6.1.3 *Show that*

$$d(f_A) = \text{tr}(A)^2 - 4\det(A) \tag{6.8}$$

Exercise 6.1.4 *The left action of A (given in (6.3) and assumed to be invertible) on $\overline{\mathbb{R}}$ is defined as follows. First suppose that $c \neq 0$; then $A \cdot x = \frac{ax+b}{cx+d}$ if $x \in \mathbb{R}$ and $cx + d \neq 0$, $= \infty$ if $x = -d/c$, and $= a/c$ if $x = \infty$. If $c = 0$, $A \cdot x = \frac{ax+b}{d}$ for x real, and $A \cdot \infty = \infty$.*

Let A and B be two-by-two invertible matrices over $\mathbb{R}$. Show that $\mu_A \circ \mu_B = \mu_{AB}$ or equivalently that for any $x \in \overline{\mathbb{R}}$, $(AB) \cdot x = A \cdot (B \cdot x)$. Show that $\mu_A = \mu_{-A}$.

Exercise 6.1.5 *Denote by $s(w) = |w|_b/|w|_a$ the slope of a word $w \in \{a, b\}^*$. Let f be an endomorphism of this free monoid. Show that $s(f(w))^{-1} = \text{Ab}(f) \cdot s(w)^{-1}$, where* Ab *is defined in Section 5.1.*

Note: Parvaix gives similar results for the Slope (see Exercise 5.2.1) of Sturmian sequences ([Pa], Proposition 1) (see also [Lo], Lemmas 2.2.17 and 2.2.18); using the Slope instead of the slope gives somewhat more complicated formulas.

Exercise 6.1.6 *Consider a quadratic polynomial $F(x) = ax^2 + bx + c$ over $\mathbb{R}$, with positive discriminant Δ. Let α be a root. Show that $F'(\alpha) = \pm\sqrt{\Delta}$. Show that α is a plus root if and only if $F'(\alpha) > 0$.*

6.2 Quadratic numbers and quadratic forms

Consider the quadratic number α whose expansion into continued fractions is $\alpha = [\overline{a_0, \ldots, a_{n-1}}]$ $(n \geq 1)$. Let $A = P(a_0) \cdots P(a_{n-1})$. By (5.1) and (6.3), the quadratic form f_A is given by

$$\begin{aligned} f_A(x, y) = p(a_1, \ldots, a_{n-1})x^2 + (p(a_1, \ldots, a_{n-2}) \\ - p(a_0, \ldots, a_{n-1}))xy - p(a_0, \ldots, a_{n-2})y^2 \end{aligned} \tag{6.9}$$

Proposition 6.2.1 *Let $i \geq 0$ and $i + 1 = nq + r$, $0 \leq r \leq n - 1$ (Euclidean division). Then*

$$f_A(p(a_0, \ldots, a_i), p(a_1, \ldots, a_i)) = (-1)^{i+1} p(a_{r+1}, \ldots, a_{n-1}, a_0, \ldots, a_{r-1})$$

Note that the last expression $(a_{r+1}, \ldots, a_{n-1}, a_0, \ldots, a_{r-1})$ may be obtained as follows: take the cyclic conjugate of $(a_0, \ldots, a_{n-1})$ beginning by a_r and remove this a_r.

Proof. Note that since the determinant of $P(a)$ is -1, the inverse of the matrix $P(a_0)\cdots P(a_i)$ is by (5.1) equal to

$$(P(a_0)\cdots P(a_i))^{-1} = (-1)^{i+1}\begin{pmatrix} p(a_1,\ldots,a_{i-1}) & -p(a_0,\ldots,a_{i-1}) \\ -p(a_1,\ldots,a_i) & p(a_0,\ldots,a_i) \end{pmatrix} \tag{6.10}$$

Let $x = p(a_0,\ldots,a_i), y = p(a_1,\ldots,a_i)$. Then by (5.1), $\begin{pmatrix} x \\ y \end{pmatrix}$ is the first column of $P(a_0)\cdots P(a_i)$, and by (6.10), $(-y, x)$ is the second row of $(P(a_0)\cdots P(a_i))^{-1}$. In view of (6.4), $f(x, y)$ is equal to the 2,1-entry of the matrix

$$(-1)^{i+1}(P(a_0)\cdots P(a_i))^{-1}P(a_0)\cdots P(a_{n-1})P(a_0)\cdots P(a_i)$$

By periodicity of the sequence, the product $P(a_0)\cdots P(a_i)$ is equal to $(P(a_0)\cdots P(a_{n-1}))^q P(a_0)\cdots P(a_{r-1})$. Hence the previous product of matrices is equal to

$$(-1)^{i+1}(P(a_0)\cdots P(a_{r-1}))^{-1}(P(a_0)\cdots P(a_{n-1}))^{-q}P(a_0)\cdots P(a_{n-1})$$

$$(P(a_0)\cdots P(a_{n-1}))^q P(a_0)\cdots P(a_{r-1})$$

$$= (-1)^{i+1}(P(a_0)\cdots P(a_{r-1}))^{-1}P(a_0)\cdots P(a_{n-1})P(a_0)\cdots P(a_{r-1})$$

$$= (-1)^{i+1}P(a_r)\cdots P(a_{n-1})P(a_0)\cdots P(a_{r-1})$$

Thus the proposition follows from (5.1). □

6.2.1 Exercises

Exercise 6.2.1 *By applying Proposition 6.2.1 to $n = 1$ and $a_0 = 1$, show that for the quadratic form $f(x, y) = x^2 - xy - y^2$, one has $f(F_{i+1}, F_i) = (-1)^{i+1}$, where the F_n are the Fibonacci numbers. Use Exercise 5.5.1.*

7
Lagrange number

The Lagrange number is at the heart of Markoff's theory for approximations. It has an expression as upper limit (Hurwitz). Equivalent real numbers have the same Lagrange number. In Section 7.2, we give the exact error term for the approximations of any quadratic real number by its convergents; the formula uses the quadratic form naturally associated with the number (Section 6.2). This formula is useful to characterize convergents which are good with respect to the Lagrange number; this is done in Section 7.3, where the Lagrange number of a quadratic number is given a precise expression in terms of finitely many continuant polynomials.

7.1 Lagrange number of a real number

Let x be an irrational real number. Consider the set of real numbers L such that the inequality $|x - p/q| < 1/Lq^2$ holds for infinitely many rational numbers p/q.

Define $\mathbf{L}(x)$ to be the supremum of all these L. It is called the *Lagrange number* of x.

Note that, by Theorem 5.5.1, there are infinitely many p/q such that $|x - p/q| < 1/2q^2$. Thus $\mathbf{L}(x) \geq 2$. Moreover, if the inequality $|x - p/q| < 1/2q^2$ holds, then p/q is a convergent of x. This shows that $\mathbf{L}(x)$ is the supremum of all numbers $L > 0$ such that $|x - p_n/q_n| < 1/Lq_n^2$ for infinitely many n, where p_n/q_n, $n \geq 0$, are the convergents of x.

Let x be an irrational real number, with its infinite continued fraction $[a_0, a_1, a_2, \ldots]$, and define $x_n = [a_n, a_{n+1}, a_{n+2}, \ldots]$ as in Section 5.3; moreover, for $n \geq 1$, define $y_n = [a_n, \ldots, a_1]$. Finally, for $n \geq 2$, let $\lambda_n(x) = x_n + y_{n-1}^{-1}$. We have

$$\begin{aligned}\lambda_n(x) &= [a_n, a_{n+1}, \ldots] + [a_{n-1}, \ldots, a_1]^{-1} \\ &= a_n + [a_{n+1}, a_{n+2}, \ldots]^{-1} + [a_{n-1}, \ldots, a_1]^{-1} \\ &= [a_{n+1}, a_{n+2}, \ldots]^{-1} + [a_n, a_{n-1}, \ldots, a_1] \qquad (7.1)\end{aligned}$$

For $n = 1$, we define $\lambda_1(x) = [a_2, a_3, \ldots]^{-1} + a_1 = [a_1, a_2, \ldots] = x_1$. The next result is essential for the determination of the Lagrange number of a real number; it is stated by Hurwitz ([H1], p. 283).

Theorem 7.1.1

$$\mathbf{L}(x) = \limsup_{n \to \infty} \lambda_n(x).$$

Note that if (u_n) is a sequence of real numbers, then $\limsup(u_n) = \limsup(u_{n+1})$.

From Christoffel Words to Markoff Numbers. Second Edition. Christophe Reutenauer, Oxford University Press.
 DOI: 10.1093/9780197907535.003.0007

Proof. For $n \geq 0$, we have by (5.9)

$$x - \frac{p_n}{q_n} = \frac{x_{n+1}p_n + p_{n-1}}{x_{n+1}q_n + q_{n-1}} - \frac{p_n}{q_n}$$
$$= \frac{x_{n+1}(p_n q_n - q_n p_n) + p_{n-1}q_n - q_{n-1}p_n}{(x_{n+1}q_n + q_{n-1})q_n} = \frac{(-1)^n}{(x_{n+1} + y_n^{-1})q_n^2}$$

by (5.8), and since by Section 5.2, $q_n/q_{n-1} = [a_n, a_{n-1}, \ldots, a_1] = y_n$. Thus

$$\left|x - \frac{p_n}{q_n}\right| = \frac{1}{\lambda_{n+1}(x)q_n^2} \tag{7.2}$$

Let $L > 0$ be such that there are infinitely many n satisfying $|x - p_n/q_n| < 1/Lq_n^2$. We deduce that for these n one has $1/\lambda_{n+1}(x)q_n^2 < 1/Lq_n^2$ and therefore $L < \lambda_{n+1}(x)$. Thus $L \leq \limsup_{n\to\infty} \lambda_n(x) := M$. It follows that $\mathbf{L}(x) \leq M$.

Furthermore, given $\epsilon > 0$, there are infinitely many n such that $\lambda_{n+1}(x) > M - \epsilon$. Thus, for these n, we have $|x - \frac{p_n}{q_n}| < \frac{1}{(M-\epsilon)q_n^2}$. Therefore, $\mathbf{L}(x) \geq M - \epsilon$. This being true for any ϵ, we deduce that $\mathbf{L}(x) = M$, which was to be shown. □

Two irrational real numbers are called *equivalent* if their expansions into continued fractions are ultimately equal. Formally, let $x = [a_0, a_1, a_2, \ldots], y = [b_0, b_1, b_2, \ldots]$. Then x and y are equivalent if for some $i, j \in \mathbb{N}$, one has $a_{i+n} = b_{j+n}$ for any natural number n.

Corollary 7.1.1 *Equivalent irrational real numbers have the same Lagrange number.*

Proof. It is enough to show that if $x = [a_0, a_1, \ldots]$ and $y = [a_1, a_2, \ldots]$, then $\mathbf{L}(x) = \mathbf{L}(y)$. Now $\lambda_n(y) = [a_{n+1}, a_{n+2}, \ldots] + [a_n, \ldots, a_2]^{-1}$ and $\lambda_{n+1}(x) = [a_{n+1}, a_{n+2}, \ldots] + [a_n, \ldots, a_1]^{-1}$. By Lemma 5.5.2, the difference of the two sequences $[a_n, \ldots, a_2]^{-1} = [0, a_n, \ldots, a_2]$ and $[a_n, \ldots, a_1]^{-1} = [0, a_n, \ldots, a_1]$ has the limit 0. Note that $\limsup(u_n) = \limsup(v_n)$ if $\lim(u_n - v_n) = 0$. Thus the corollary follows. □

Concerning equivalent numbers, one has to recall the following important result: two irrational real numbers x and y are equivalent if and only if for some matrix $\begin{pmatrix} a & b \\ c & d \end{pmatrix} \in GL_2(\mathbb{Z})$, one has $y = \frac{ax+b}{cx+d}$. In other words, two real numbers are equivalent if and only if they lie in the same orbit under the action of $GL_2(\mathbb{Z})$ by Möbius transformations (see (6.7)). See for example [HW], Theorem 175 or [B], Theorem A.1.

Using the concept of *abelian equivalence* of words, the authors of [FLLLMPP-G] characterize the Lagrange number $L(\alpha)$ of any real number α by considering any Sturmian sequence of slope α: $L(\alpha)$ is equal to the upper limit of the sequence k_n/n, where k_n is the maximum of the k such that $w_1 \cdots w_k$ is a factor of s, for some words w_i which are of length n and which have all the same commutative image.

7.2 Exact error term for quadratic numbers

The following result appears in [Re8] Theorem 9.1.

Theorem 7.2.1 *Let β be the quadratic number whose expansion into continued fraction is $\beta = [b_0, \ldots, b_{k-1}, \overline{a_0, \ldots, a_{n-1}}]$. The convergents p_t/q_t of β satisfy for $t \geq k$*

$$\beta - \frac{p_t}{q_t} = \frac{(-1)^t \epsilon_{t-k+1 \mod n}}{q_t^2 \frac{\sqrt{d(f)}}{2}\left(1 + \sqrt{1 + 4(-1)^{t-k+1} \frac{f(e,i)}{d(f) q_t^2} \epsilon_{t-k+1 \mod n}}\right)}$$

with the following notations:

- *for $r \in \{0, \ldots, n-1\}$, ϵ_r is defined by the continuant polynomial*

$$\epsilon_r = p(a_{r+1}, \ldots, a_{n-1}, a_0, \ldots, a_{r-1}) \tag{7.3}$$

- *f is the quadratic form f_A, where $A = P(a_0) \cdots P(a_{n-1})$;*
- *$d(f)$ is the discriminant of f;*
- *${}^t(e, i)$ is the first column of the matrix*

$$M = P(b_0) \cdots P(b_{k-1})^{-1} = \begin{pmatrix} e & h \\ i & j \end{pmatrix}$$

We define $\alpha = [\overline{a_0, \ldots, a_{n-1}}]$ and denote by u_t/v_t its convergents; extend the notation a_t to all natural integers t by $a_t = a_{t \mod n}$. By (5.7), if $t \geq k$, ${}^t(u_{t-k}, v_{t-k})$ is the first column of $P(a_0) \cdots P(a_{t-k})$, and ${}^t(p_t, q_t)$ is the first column of $P(b_0) \cdots P(b_{k-1})P(a_0) \cdots P(a_{t-k}) = M^{-1}P(a_0) \cdots P(a_{t-k})$. Thus we have for $t \geq k$

$$\begin{pmatrix} p_t \\ q_t \end{pmatrix} = M^{-1} \begin{pmatrix} u_{t-k} \\ v_{t-k} \end{pmatrix} \tag{7.4}$$

The following technical lemma will imply that the convergent p_t/q_t appearing in Theorem 7.2.1 is the plus root of a certain quadratic polynomial. Take the notations of Theorem 7.2.1, and let $B = M^{-1}AM$ and $g(x, y) = f_B(x, y)$.

Lemma 7.2.2 *For any $t \geq k$, one has $g'_x(p_t, q_t) \geq 0$.*

Proof. In this proof, we write $P(n_i, \ldots, n_j)$ for $P(n_i) \cdots P(n_j)$, when $i \leq j$. Note that when $j = i - 1$, the latter product is empty. and therefore, it is the identity matrix. Likewise, in view of (5.1), we let $p(b_i, \ldots, b_j) = 1$ if $j = i - 1$, and $= 0$ if $j = i - 2$.

(i) We have by (6.4), $g(x,y) = (-y,x)B\begin{pmatrix} x \\ y \end{pmatrix}$. Thus $g'_x(x,y) = (0,1)B\begin{pmatrix} x \\ y \end{pmatrix} + (-y,x)B\begin{pmatrix} 1 \\ 0 \end{pmatrix}$; thus

$$g'_x(p_t, q_t) = F + S$$

$$F = (0,1)M^{-1}AM\begin{pmatrix} p_t \\ q_t \end{pmatrix}, \quad S = (-q_t, p_t)M^{-1}AM\begin{pmatrix} 1 \\ 0 \end{pmatrix} \tag{7.5}$$

By (7.4), $M\begin{pmatrix} p_t \\ q_t \end{pmatrix} = \begin{pmatrix} u_{t-k} \\ v_{t-k} \end{pmatrix}$; the latter is by (5.1) and (5.6) the first column of $P(a_0, \ldots, a_{t-k})$. Thus, F is the 2, 1-entry of the product

$$P(b_0, \ldots, b_{k-1})P(a_0, \ldots, a_{n-1})P(a_0, \ldots, a_{t-k})$$
$$= P(b_0, \ldots, b_{k-1}, a_0, \ldots, a_{t-k+n})$$

Hence

$$F = \begin{cases} p(b_1, \ldots, b_{k-1}, a_0, \ldots, a_{t-k+n}) & \text{if } k \geq 1 \\ p(a_1, \ldots, a_{t+n}) & \text{if } k = 0 \end{cases}$$

By (5.6) and (6.10), $(-q_t, p_t)$ is the second row of $(-1)^{t+1}P(b_0, \ldots, b_{k-1}, a_0, \ldots, a_{t-k})^{-1}$. Hence S is the 2, 1-entry of

$$(-1)^{t+1}P(b_0, \ldots, b_{k-1}, a_0, \ldots, a_{t-k})^{-1}P(b_0, \ldots, b_{k-1})$$
$$P(a_0, \ldots, a_{n-1})P(b_0, \ldots, b_{k-1})^{-1}$$
$$= (-1)^{t+1}P(a_0, \ldots, a_{t-k})^{-1}P(a_0, \ldots, a_{n-1})P(b_0, \ldots, b_{k-1})^{-1}$$

In the case where $t - k \geq n - 1$, this matrix product is equal to

$$P(a_n, \ldots, a_{t-k})^{-1}P(b_0, \ldots, b_{k-1})^{-1} = P(b_0, \ldots, b_{k-1}, a_n, \ldots, a_{t-k})^{-1}$$
$$= P(b_0, \ldots, b_{k-1}, a_0, \ldots, a_{t-k-n})^{-1} \text{ (by periodicity of the sequence } (a_t)\text{)}.$$

Hence in this case, S is by (6.10) equal to

$$-(-1)^{t+1}(-1)^{k+t-k-n+1}p(b_1, \ldots, b_{k-1}, a_0, \ldots, a_{t-k-n})$$
$$= (-1)^{n+1}p(b_1, \ldots, b_{k-1}, a_0, \ldots, a_{t-k-n})$$

in the case where $k \geq 1$, and in the case $k = 0$, it is $(-1)^{n+1}p(a_1, \ldots, a_{t-n})$.

In the case where $t-k<n-1$, the matrix product is $P(a_{t-k+1},\dots,a_{n-1})P(b_0,\dots,b_{k-1})^{-1}$. Hence S is by (5.1) and (6.10) equal to

$$(-1)^{t+1}(-1)^k(p(a_{t-k+2},\dots,a_{n-1})p(b_1,\dots,b_{k-2})$$
$$-p(a_{t-k+2},\dots,a_{n-2})p(b_1,\dots,b_{k-1}))$$

if $k\geq 1$, and it is $(-1)^{t+1}p(a_{t+2},\dots,a_{n-1})$ if $k=0$.

(ii) We now may show that for $t\geq k$, $g'_x(p_t,q_t)\geq 0$. Suppose first that $t-k\geq n-1$; then $g'_x(p_t,q_t) = F+S$ is equal to $p(b_1,\dots,b_{k-1},a_0,\dots,a_{t-k+n}) + (-1)^{n+1}p(b_1,\dots,b_{k-1},a_0,\dots,a_{t-k-n})$ if $k\geq 1$, and to $p(a_1,\dots,a_{t+n})+(-1)^{n+1}p(a_1,\dots,a_{t-n})$ if $k=0$. In the two cases, it follows from Lemma 5.1.1 and the remark following it, that $g'_x(p_t,q_t)\geq 0$.

Suppose now that $t-k<n-1$. Then $g'_x(p_t,q_t)=F+S$ is equal to

$$p(b_1,\dots,b_{k-1},a_0,\dots,a_{t-k+n})+(-1)^{t+1+k}(p(a_{t-k+2},\dots,a_{n-1})p(b_1,\dots,b_{k-2})$$
$$-p(a_{t-k+2},\dots,a_{n-2})p(b_1,\dots,b_{k-1}))$$

if $k\geq 1$ and to $p(a_1,\dots,a_{t+n})+(-1)^{t+1}p(a_{t+2},\dots,a_{n-1})$ if $k=0$. In this latter case, we use Lemma 5.1.1 and deduce that $g'_x(p_t,q_t)\geq 0$. We have $P(b_1,\dots,b_{k-1},a_0,\dots,a_{t-k+n}) = P(b_1,\dots,b_{k-1})P(a_0,\dots,a_{t-k+n})$ so that by (5.1), $p(b_1,\dots,b_{k-1},a_0,\dots,a_{t-k+n}) = p(b_1,\dots,b_{k-1})p(a_0,\dots,a_{t-k+n}) + p(b_1,\dots,b_{k-2})p(a_1,\dots,a_{t-k+n})$. Hence, in the case $k\geq 1$, $g'_x(p_t,q_t)$ is equal to

$$p(b_1,\dots,b_{k-1})p(a_0,\dots,a_{t-k+n})+p(b_1,\dots,b_{k-2})p(a_1,\dots,a_{t-k+n})$$
$$+(-1)^{t+1+k}(p(a_{t-k+2},\dots,a_{n-1})p(b_1,\dots,b_{k-2})$$
$$-p(a_{t-k+2},\dots,a_{n-2})p(b_1,\dots,b_{k-1}))$$
$$=p(b_1,\dots,b_{k-1})(p(a_0,\dots,a_{t-k+n})-(-1)^{t+1+k}p(a_{t-k+2},\dots,a_{n-2}))$$
$$+p(b_1,\dots,b_{k-2})(p(a_1,\dots,a_{t-k+n})+(-1)^{t+1+k}p(a_{t-k+2},\dots,a_{n-1}))$$

It follows from Lemma 5.1.1 that $g'_x(p_t,q_t)\geq 0$. □

Proof of Theorem 7.2.1. By (7.4) ${}^t(u_{t-k},v_{t-k}) = M\,{}^t(p_t,q_t)$. By Proposition 6.2.1, $f(u_t,v_t)=(-1)^{t+1}\epsilon_{t+1 \bmod n}$.

Recall that we have $B=M^{-1}AM$ and $g=f_B$. By Lemma 6.1.2, $g=(-1)^k f\cdot M$. It follows that $g(p_t,q_t)=(-1)^k(f\cdot M)(p_t,q_t)=(-1)^k f(u_{t-k},v_{t-k})=(-1)^k(-1)^{t-k+1}\epsilon_{t-k+1 \bmod n} = (-1)^{t+1}\epsilon$, with $\epsilon=\epsilon_{t-k+1 \bmod n}$.

Writing $g=a'x^2+b'xy+c'y^2$, we obtain $a'(\frac{p_t}{q_t})^2+b'\frac{p_t}{q_t}+c'+(-1)^t\epsilon/q_t^2=0$: p_t/q_t is a root of the polynomial $g(x,1)+(-1)^t\epsilon/q_t^2$.

We have $\beta = [b_0, \ldots, b_{k-1}, \alpha] = P(b_0) \cdots P(b_{k-1}) \cdot \alpha = M^{-1} \cdot \alpha$. Since $A \cdot \alpha = \alpha$, we have $B \cdot \beta = M^{-1}AM \cdot \beta = M^{-1}A \cdot \alpha = M^{-1} \cdot \alpha = \beta$, and this implies that β is a root of $g(x, 1)$.

By Exercise 6.1.2, α is a root of f; note that it is the plus root of f: this follows from (6.6); because the first coefficient of f is positive, its last coefficient is negative and α is positive. Let θ be the plus root of g. Since $g = (-1)^k f \cdot M$, Lemma 6.1.3 implies that $M \cdot \theta$ is the root of $(-1)^k f$ whose sign is that of $det(M)$. The latter being $(-1)^k$, $M \cdot \theta$ is the plus root of f, that is, α. Hence $\theta = M^{-1} \cdot \alpha = \beta$, which is therefore the plus root of g.

We claim that for $t \geq k$, p_t/q_t is the plus root of the polynomial $g(x, 1) + (-1)^t \epsilon/q_t^2$. It amounts to prove that $2a'p_t/q_t + b' \geq 0$ by (6.5). This follows from Lemma 7.2.2, which states that $2a'p_t + b'q_t \geq 0$.

Let $\Delta = b'^2 - 4a'c'$ and $\Delta' = \Delta - 4a'(-1)^t\epsilon/q_t^2$, the discriminants of the polynomials $g(x, 1)$ and $g(x, 1) + (-1)^t\epsilon/q_t^2$.

By Proposition 6.1.1, the discriminants of f and g are equal; hence $\Delta = d(g) = d(f)$. Moreover, a' being the first coefficient of $g = (-1)^k f \cdot M$, it is equal to $(-1)^k f(e, i)$, because $(f \cdot M)(1, 0) = f(e, i)$.

Thus the theorem follows from Lemma 6.1.4. □

7.2.1 Exercises

Exercise 7.2.1 *Using Theorem 7.2.1, show that when $k = 0$, that is, $\alpha = \beta$, and therefore the p_t/q_t are the convergents of α, one has*

$$\alpha - \frac{p_t}{q_t} = \frac{(-1)^t \epsilon_{t+1 \bmod n}}{q_t^2 \frac{\sqrt{d(f)}}{2}\left(1 + \sqrt{1 + 4(-1)^{t+1}\epsilon_0\epsilon_{t+1 \bmod n}/d(f)q_t^2}\right)}$$

Hint: show that $f(1, 0) = \epsilon_0$. See [Re8] (Corollary 9.1).

Exercise 7.2.2 *Let $\alpha = [\bar{a}]$, $a \in \mathbb{P}$, with convergents p_t/q_t. Using Exercise 7.2.1, show that*

$$\alpha - \frac{p_t}{q_t} = \frac{(-1)^t}{q_t^2 \frac{\sqrt{a^2+4}}{2}\left(1 + \sqrt{1 + 4(-1)^{t+1}/(a^2 + 4)q_t^2}\right)} \tag{7.6}$$

Show that

$$p_t = \frac{1}{2}aq_t + \frac{1}{2}\sqrt{(a^2 + 4)q_t^2 + 4(-1)^{t+1}} \tag{7.7}$$

Show that $p_t = q_{t+1}$, $q_t = p(a, \ldots, a)$ (t a's). In particular ($a = 1$), one has the recursion

$$F_{t+1} = \frac{1}{2}F_t + \frac{1}{2}\sqrt{5F_t^2 + 4(-1)^{t+1}}$$

for Fibonacci numbers. Show similarly ($a = 2$) that one has the recursion

$$P_{t+1} = P_t + \frac{1}{2}\sqrt{8P_t^2 + 4(-1)^{t+1}}$$

for the Pell numbers P_n *defined by* $P_{-1} = 0$, $P_0 = 1$ *and* $P_{n+1} = 2P_n + P_{n-1}$ *for any natural number n. See [Re8] (8); for* $a = 1, 2$, *(7.6) appears in [Ha1, Ha2].*

Exercise 7.2.3 *Show, with the notations Exercise 7.2.2, that the number*

$$(a^2 + 4)q_t^2 + 4(-1)^{t+1}$$

is a perfect square in $\mathbb{N}$; *denote its square root in* $\mathbb{N}$ *by* u_t. *Show that* $u_{t-1} = \alpha^t + \overline{\alpha}^t = \mathrm{tr}(P(a)^t)$ *and that* u_t *satisfies the recursion* $u_{t+2} = au_{t+1} + u_t$, *also satisfied by the* q_t. *For* $a = 1$, *the* u_t *are the* Lucas numbers *(see [HW], 10.14).*

Exercise 7.2.4 *Take the notations of Theorem 7.2.1 and its proof, and assume that* $t \geq k$. *One has in particular*

$$a'p_t^2 + b'p_tq_t + c'q_t^2 + (-1)^t\epsilon = 0$$

so that p_t *is a root of the polynomial* $F(x) = a'x^2 + b'q_tx + c'q_t^2 + (-1)^t\epsilon = 0$. *Show that the discriminant of* $F(x)$ *is*

$$\Delta_t = d(f)q_t^2 + 4(-1)^{t-k+1}f(e,i)\epsilon_{t-k+1 \bmod n}$$

(Hint: f and g have the same discriminant, and the value of a' *is given in the proof.) Show that* $\Delta_t = F'(p_t)^2$ *(use Exercise 6.1.6) and that* Δ_t *is a perfect square in* $\mathbb{N}$ *(this extends to Exercise 7.2.3). Show that* $F'(p_t) = 2a'p_t + b'q_t = g'_x(p_t, q_t)$ *and that* p_t *is the plus root of* $F(x)$ *(use Lemma 7.2.2). Show that the sequence* $r_t = F'(p_t)$ *satisfies the same recursion as* p_t *and* q_t: $r_t = c_tr_{t-1} + r_{t-2}$, *for* $t \geq k + 2$, *where* $[c_0, c_1, c_2, \ldots,] = \beta = [b_0, \ldots, b_{k-1}, \overline{a_0, \ldots, a_{n-1}}]$.

7.3 Good convergents of quadratic numbers

For a quadratic number, we may give a precise expression for its Lagrange number. We take the notations of Theorem 7.2.1.

Corollary 7.3.1 *The Lagrange number of* $\beta = [b_0, \ldots, b_{k-1}, \overline{a_0, \ldots, a_{n-1}}]$ *is equal to* $\frac{\sqrt{d(f)}}{m}$, *where m is the minimum of the n numbers* $\epsilon_r = p(a_{r+1}, \ldots, a_{n-1}, a_0, \ldots, a_{r-1})$, $r = 0, \ldots, n-1$.

This result is equivalent to known results; for example, Lemma 17 in [B] and Proposition 1.29 in [A].

Proof. By Theorem 7.2.1 and (7.2),

$$\lambda_{t+1}(\beta) = \frac{\sqrt{d(f)}}{2}\frac{1+\sqrt{1+4(-1)^{t-k+1}f(e,i)\epsilon_{t-k+1 \bmod n}/d(f)q_t^2}}{\epsilon_{t-k+1 \bmod n}} \tag{7.8}$$

Since $q_t \to \infty$ and since the sequence $\epsilon_{t-k+1 \bmod n}$ is periodic, we see that the limit of the second numerator is 2, when $t \to \infty$. Thus the upper limit of $\lambda_{t+1}(\beta)$ is equal to the maximum of the numbers $\sqrt{d(f)}/\epsilon_r$, which implies the result by Theorem 7.1.1 and the remark following it. □

Given a real number ξ, we say that the rational p/q is a *good approximation of* ξ if

$$|\xi - p/q| < 1/\mathbf{L}(\xi)\, q^2 \tag{7.9}$$

The reason why we do not put a large inequality in this definition is explained in Exercise 7.3.1. The question is wether a given irrational ξ has infinitely many good approximations. Not all reals, even quadratic ones, have this property; see the example of Perron ([P1], pp. 7–8) given later.

Note that a good approximation is necessarily a convergent (which we call then a *good convergent*, and we say that its rank is *good*). Indeed, we know that $\mathbf{L}(\xi) \geq 2$. Hence a good approximation p/q satisfies $|\xi - p/q| < 1/2q^2$ and hence by Legendre's theorem (Theorem 5.5.1), it is a convergent.

Theorem 7.3.2 *Take the notations of Theorem 7.2.1 and Corollary 7.3.1. For t large enough, a convergent p_t/q_t of β is good if and only if $\epsilon_{t-k+1 \bmod n} = m$ and $(-1)^{t-k+1}f(e,i) > 0$.*

Corollary 7.3.3 *The quadratic number β has infinitely many good approximations if and only if either n is odd, or n is even and there exists some $r \in \{0,\ldots,n-1\}$ such that $\epsilon_r = m$ and that $(-1)^r f(e,i) > 0$.*

We give the examples of Perron. Let $b > c$. Consider $\xi = [\overline{b,c}]$. We have $n = 2$, $k = 0$, $\epsilon_0 = c$, $\epsilon_1 = b$, and therefore $m = c$ and the unique r is 0. Since $P(b)P(c) = \begin{pmatrix} bc+1 & b \\ c & 1 \end{pmatrix}$, the quadratic form f is $f(x,y) = cx^2 - bcxy - by^2$ and moreover $(e,i) = (1,0)$; hence $(-1)^r f(e,i) = c > 0$ so that there are by the corollary infinitely many good approximations.

Consider now $\xi = [\overline{c,b}]$. Then $r = 1$, $f(x,y) = bx^2 - bcxy - cy^2$ and $(e,i) = (1,0)$; then $(-1)^r f(e,i) = -b < 0$, and there are only finitely many good approximations.

Proof of Theorem 7.3.2. We know by Corollary 7.3.1 that $\mathbf{L}(\beta) = \sqrt{d(f)}/m$. By (7.2), p_t/q_t is a good convergent if and only if $\lambda_{t+1}(\beta) > \mathbf{L}(\beta)$. By (7.8), $\lambda_{t+1}(\beta)$ is equal to

$$\frac{\mathbf{L}(\beta)}{2}\left(1+\sqrt{1+4(-1)^{t-k+1}f(e,i)\,\epsilon_{t-k+1 \bmod n}/d(f)\,q_t^2}\right)\frac{m}{\epsilon_{t-k+1 \bmod n}}$$

Therefore, if $\epsilon_{t-k+1 \bmod n} = m$ and if $(-1)^{t-k+1}f(e,i) > 0$, then $\lambda_{t+1}(\beta) > \mathbf{L}(\beta)$, and t is good. If $\epsilon_{t-k+1 \bmod n} = m$, but $(-1)^{t-k+1}f(e,i) < 0$, then $\lambda_t(\beta) < \mathbf{L}(\beta)$. Finally, let t be such that $\epsilon_{t-k+1 \bmod n} > m$; for these t, there are finitely many values $m/\epsilon_{t-k+1 \bmod n}$ (since there are finitely many ϵ_r), and let u be the largest of them; then $u < 1$. Since $q_t \to \infty$ when $t \to \infty$, there exists T such that $\forall t \geq T$, the above expression $(1/2)(1+\sqrt{*})$ is $< u^{-1}$. Then $\lambda_t(\beta) < \mathbf{L}(\beta)$. □

Proof of Corollary 7.3.3. Suppose that n is odd. There exists $r \in 0, \ldots, n-1$ such that $\epsilon_r = m$. If $t-k+1 \equiv r$ modulo n and if t is large enough, then by the theorem, t is good as soon as $(-1)^{t-k+1}f(e,i) > 0$. Suppose that this number is negative. Then replacing t by $t+n$, we find a good t, since the sequence ϵ_t has period n.

Suppose now that n is even. If there exists r such that $\epsilon_r = m$ and $(-1)^r f(e,i) > 0$, then t is good if t is large enough and if $t-k+1 \equiv r$. Then there are infinitely many good t, since t good implies $t+n$ good.

Conversely, if there are infinitely many good t, then there exists t such that $t-k+1$ mod $n = r$ with $\epsilon_r = m$ and $(-1)^{t-k+1}f(e,i) > 0$. Hence $(-1)^r f(e,i) > 0$, since $t-k+1$ and r have the same parity, n being even. □

We say that a subset of $\mathbb{N}$ is *ultimately periodic* if it is a union of a finite set and of finitely many arithmetic progressions.

Corollary 7.3.4 *The set of good t is ultimately periodic.*

Proof. This is because the sequence $\epsilon_{t-k+1 \bmod n}$ has period n, while the set of t such that $(-1)^{t-k+1}f(e,i) > 0$ has period 2. □

7.3.1 Exercises

Exercise 7.3.1 *Let ξ be real irrational. Show that if there are infinitely many rational numbers p/q such that $|\xi - p/q| \leq 1/\mathbf{L}(\xi)q^2$, then there are infinitely many rational numbers p/q satisfying (7.9).*

8
Words and quadratic numbers

Among the remarkable combinatorial properties, which make Markoff's theory work, are the following facts: if w is a lower Christoffel word, then the corresponding upper Christoffel word is the reversal $\tilde{w}$ of w; these two words are conjugate;[1] their conjugation class is closed under reversal; and in this class, w is the smallest element and $\tilde{w}$ the largest for the alphabetical order. This is Theorem 8.2.3. It is used in order to compute the Lagrange number of Markoff irrationalities and the Markoff supremum of periodic bi-infinite sequences, which have $\chi(w)$ as a periodic pattern. Here, χ is the homomorphism from the free monoid $\{a,b\}^*$ into $\{1,2\}^*$, which sends a onto 11 and b onto 22; this morphism plays an important role, since the real numbers involved in the theory of Markoff have continued fractions where the partial quotients are only 1s and 2s coming in pairs.

8.1 Continued fractions associated with Christoffel words

Lemma 8.1.1 *Let $u = a_0 \cdots a_{n-1}$ be a non-empty word in $\mathbb{P}^*$, $a_i \in \mathbb{P}$. Let x and y be the real numbers whose expansions in continued fractions are respectively $[u^\infty]$ and $[\tilde{u}^\infty]$. Let*

$$M = P(a_0) \cdots P(a_{n-1}) = \begin{pmatrix} p & q \\ r & s \end{pmatrix} \tag{8.1}$$

so that $q = p(a_0, \ldots, a_{n-2})$. Then

$$y + x^{-1} = \frac{\sqrt{\operatorname{tr}(M)^2 - 4(-1)^n}}{q}$$

Proof. The equality for q follows from (5.1). By Proposition 5.4.1, x is quadratic, $y = -1/\bar{x}$, and $y = \frac{py+r}{qy+s}$. We deduce that $x^{-1} = -\bar{y}$ and that $qy^2 + (s-p)y - r = 0$. The discriminant of this quadratic polynomial in y is $\Delta = (s-p)^2 + 4qr = (s+p)^2 - 4(ps - qr) = \operatorname{tr}(M)^2 - 4(-1)^n$, since the determinant of $P(a)$ is -1, thus that of M is $(-1)^n$. Since $y > 0$, y is by (6.6) the plus root of the polynomial; therefore $y = \frac{p-s+\sqrt{\Delta}}{2q}$, and thus $\bar{y}$ is the minus root. Hence $y + x^{-1} = y - \bar{y} = \sqrt{\Delta}/q$. □

Let w be a lower Christoffel word and $\tilde{w}$ be the associated upper Christoffel word (see Section 2.2). Denote by $m = m_w$ the Markoff number associated with w (see

[1] In general, a word and its reversal are not conjugate.

From Christoffel Words to Markoff Numbers. Second Edition. Christophe Reutenauer, Oxford University Press.
 DOI: 10.1093/9780197907535.003.0008

Section 3.3). Denote by χ the monoid homomorphism from the free monoid $\{a, b\}^*$ into the free monoid $\{1, 2\}^*$, sending a onto 11 and b onto 22. Recall from Section 3.1 that the monoid homomorphism $\mu : \{a, b\}^* \to SL_2(\mathbb{Z})$ is defined by $\mu(a) = \begin{pmatrix} 2 & 1 \\ 1 & 1 \end{pmatrix}$ and $\mu(b) = \begin{pmatrix} 5 & 2 \\ 2 & 1 \end{pmatrix}$. Let

$$\mu(w) = \begin{pmatrix} p & q \\ r & s \end{pmatrix}$$

Then $q = m = (p + s)/3$; see Section 3.3.

Theorem 8.1.2 *Let x and y be the real numbers whose expansions in continued fractions are respectively $[\chi(w)^\infty]$ and $[\chi(\tilde{w})^\infty]$. Then $x = \frac{px+q}{rx+s}$, $y = \frac{py+r}{qy+s} = \frac{p-s+\sqrt{(s-p)^2+4qr}}{2q} = \frac{p-s+\sqrt{9m^2-4}}{2m}$, $x = -1/\bar{y}$ and $y + x^{-1} = \sqrt{9 - 4/m^2}$.*

Proof. Let $\chi(w) = a_0 \cdots a_{n-1}$. Note that n is even. One has

$$P(1)^2 = \begin{pmatrix} 1 & 1 \\ 1 & 0 \end{pmatrix}\begin{pmatrix} 1 & 1 \\ 1 & 0 \end{pmatrix} = \begin{pmatrix} 2 & 1 \\ 1 & 1 \end{pmatrix} = \mu(a)$$

$$P(2)^2 = \begin{pmatrix} 2 & 1 \\ 1 & 0 \end{pmatrix}\begin{pmatrix} 2 & 1 \\ 1 & 0 \end{pmatrix} = \begin{pmatrix} 5 & 2 \\ 2 & 1 \end{pmatrix} = \mu(b)$$

Therefore

$$P(a_0) \cdots P(a_{n-1}) = \begin{pmatrix} a_0 & 1 \\ 1 & 0 \end{pmatrix} \cdots \begin{pmatrix} a_{n-1} & 1 \\ 1 & 0 \end{pmatrix} = \mu(w) = \begin{pmatrix} p & q \\ r & s \end{pmatrix}$$

Note that, with the notations of (8.1), $\mathrm{tr}(M)^2 - 4(-1)^n = 9m^2 - 4$. We then apply Proposition 5.4.1 and Lemma 8.1.1. □

8.2 Markoff supremum of a bi-infinite sequence

Let $s = (a_n)_{n \in \mathbb{Z}}$ be a bi-infinite sequence over $\mathbb{P}$. For $i \in \mathbb{Z}$, let

$$\lambda_i(s) = [a_{i-1}, a_{i-2}, a_{i-3}, \ldots]^{-1} + a_i + [a_{i+1}, a_{i+2}, a_{i+3}, \ldots]^{-1}$$

Note that one has two other formulas for $\lambda_i(s)$:

$$\begin{aligned} \lambda_i(s) &= [a_{i-1}, a_{i-2}, a_{i-3}, \ldots]^{-1} + [a_i, a_{i+1}, a_{i+2}, a_{i+3}, \ldots] \\ &= [a_i, a_{i-1}, a_{i-2}, a_{i-3}, \ldots] + [a_{i+1}, a_{i+2}, a_{i+3}, \ldots]^{-1} \end{aligned} \tag{8.2}$$

Define $\mathbf{M}(s) = \sup_{i \in \mathbb{Z}}(\lambda_i(s))$. We call this number the *Markoff supremum of s*.

Lemma 8.2.1 *Let s be as above, with period n, and let the matrix M be as in (8.1). For $i \in \mathbb{Z}$, one has*

$$\lambda_i(s) = \frac{\sqrt{\operatorname{tr}(M)^2 - 4(-1)^n}}{p(a_{i+1}, \ldots, a_{i+n-1})}$$

Proof. By periodicity of the sequence and by matrix conjugation, it is enough to prove the lemma for $i = -1$. We have $\lambda_{-1}(s) = [a_{-1}, a_{-2}, a_{-3}, a_{-4}, \ldots] + [a_0, a_1, a_2, \ldots]^{-1} = y + x^{-1}$, where $y = [a_{-1}, a_{-2}, \ldots]$ and $x = [a_0, a_1, \ldots]$. It follows from Lemma 8.1.1 that $\lambda_{-1}(s) = \frac{\sqrt{\operatorname{tr}(M)^2-4(-1)^n}}{p(a_0,\ldots,a_{n-2})}$. □

The definition of χ has been recalled in Section 8.1.

Theorem 8.2.2 *Let w be some lower Christoffel word with associated Markoff number m. Let s be the bi-infinite sequence $^\infty\chi(\tilde{w})^\infty$. Then $\mathbf{M}(s) = \sqrt{9 - \frac{4}{m^2}}$.*

The theorem will be a consequence of Theorem 8.1.2 and of the following result, which gives a combinatorial property of Christoffel words.

Theorem 8.2.3 *Let w be a lower Christoffel word and $\tilde{w}$ be the corresponding upper Christoffel word. Then w and $\tilde{w}$ are conjugate, and in their conjugation class, w is the smallest and $\tilde{w}$ the largest for the alphabetical order.*

It follows that the conjugation class of a Christoffel word is closed under reversal. See Theorem 14.2.1 for another property of the conjugation between the lower and upper Christoffel words of the same slope.

We need a preliminary result. We extend the definition of discretization to half-lines (rays) and infinite words. We consider pairs (A, L) of an integral point A and a half-line of non-negative slope; we say that this pair is in *vertical position* if A and the point H where the half-line begins have the same x-coordinate and if A is below H (see Figure 8.1). The *vertical distance between A and L* is the length of AH.

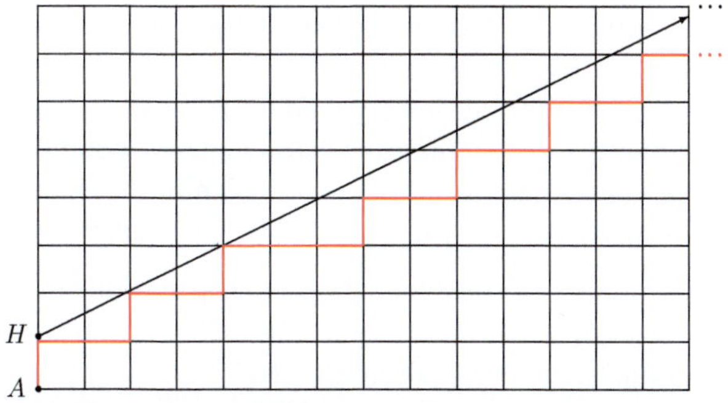

Figure 8.1 Discretization of a half-line

Now the *discretization from below of the pair* (A, L) is the infinite word on the alphabet $\{a < b\}$ that encodes, as in Section 2.2, the infinite lattice path starting from A and such that there is no integral point between the half-line and the path, except those on the half-line and on the path. See Figure 8.1, where the infinite word is $baabaabaaabaabaabaaba\cdots$.

Lemma 8.2.4 *Let (A, L) and (A, L') be two pairs in vertical position, where L and L' are parallel half-lines, and the vertical distance between A and L is smaller than that between A and L'. Then the discretization from below of (A, L) is $\leq$ the discretization from below of (A, L') for the alphabetical order.*

Note that when the slope of the two half-lines is rational, then the two discretizations may be equal; a trivial example (which the reader may verify) is given by the half-lines in the positive quadrant, supported by the two lines of equations $y = 1/3$ and $y = 2/3$, with $A = (0, 0)$. However, when the slope is irrational, it may be shown that there always exist integral points located strictly between the two half-lines so that the discretizations are distinct and one is therefore strictly smaller than the other.

Proof. The half-line L is by assumption below the half-line L'. Consider the path discretizing (A, L) and the path discretizing (A, L'); they have the same origin. Consider also the first step where they differ (if it exists). This step is horizontal for the first and vertical for the second' hence, the first word is smaller than the second one. □

Corollary 8.2.5 *Suppose that (A, L) and (A', L') are pairs as above, in vertical position, with L parallel to L'. Suppose that the vertical distance between A and L is smaller than that between A' and L'. Then the discretization from below of (A, L) is not larger than the discretization from below of (A', L') for the alphabetical order.*

Proof. By translation over the integers, we are reduced to the case $A = A'$, which is the previous lemma. □

Proof of Theorem 8.2.3. Consider Figure 8.2 (disregarding for the moment the points E and F and the segment between them), where A and B are as in Section 2.2, and where C is defined by $\overrightarrow{AB} = \overrightarrow{BC} = (p, q)$, p and q relatively prime. The red curve AB represents a lattice path, which encodes the lower Christoffel word w; the latter is also represented by the red curve BC, since $\overrightarrow{AB} = \overrightarrow{BC}$.

We define the point H as the integral point on the red path AB that is furthest from the line AB. It is unique; indeed if there is a second integral point H' on the path AB at the same distance as H from the line AB, then, assuming that H' is at the right of H (the other case is similar), the vector $\overrightarrow{HH'}$ is integral and parallel to AB, and we deduce that that the point $A + \overrightarrow{HH'}$ is an integral point on the open segment AB, contradicting that p and q are relatively prime. Similarly, we define K on the red path BC to be the furthest integral point from line AB. Then $\overrightarrow{AB} = \overrightarrow{HK}$.

Clearly, the word represented by the curve HBK is a conjugate of w. We show that this word is equal to $\tilde{w}$, the upper Christoffel word associated with w. For this, it is

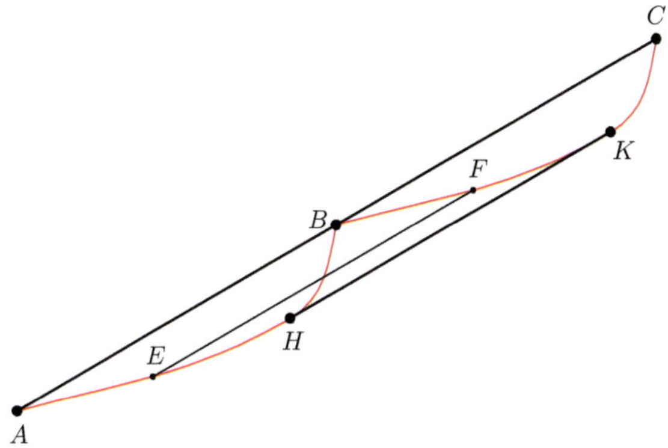

Figure 8.2 The lower and upper Christoffel words are conjugate

enough by Section 2.2 to show that the path HBK is the discretization from above of the segment HK; in other words, there are no integral points in the interior of the polygon $\mathcal{P}$ delimited by HK and the path HBK.

To begin, the *diagonal distance* between two parallel lines of positive finite slope is the length of any segment obtained by intersecting these lines and a line of slope -1. The diagonal distance between the two lines ABC and HK is less than $\sqrt{2}$, since there is no integral point in the interior of the polygon delimited by the path AHB and the segment AB so that the point $H + (-1, 1)$ is not in this polygon.

Now suppose there were an integral point M in the interior of $\mathcal{P}$. This point is strictly below the path HBK; hence, for some integer $i \geq 1$, $M + (-i, i)$ is on the path; in particular, $M + (-1, 1)$ is then in the interior of the polygon or on its frontier HBK; since $\mathcal{P}$ lies between the lines ABC and HK, this implies that the diagonal distance between these lines is at least $\sqrt{2}$, contradicting the previous bound on diagonal distance.

This proves that w and $\tilde{w}$ are conjugate and, moreover, that $\tilde{w}$ is represented by the path HBK.

Now consider a conjugate u of w, different from w and $\tilde{w}$. Referring to Figure 8.2, there exists a point E on the path AHB, with a corresponding point F on the path BKC (that is, $\overrightarrow{AB} = \overrightarrow{EF}$), such that u encodes the red path EBF.

By construction, H is the integral point on the path AHB that is furthest from the line AB; since the usual distance of a point from this line and the vertical distance are proportional, it follows that H is also the unique integral point on the path AHB that is furthest from the line AB for the vertical distance.

The vertical distance of E to the line AB is larger than that of A (which is 0) and smaller than that of H by uniqueness of H. Corollary 8.2.5 then implies that we have $w^\infty <_{lex} u^\infty <_{lex} \tilde{w}^\infty$. Since the three words $w, u,$ and $\tilde{w}$ have the same length, we deduce that $w <_{lex} u <_{lex} \tilde{w}$. □

Corollary 8.2.6 *Let w be a lower Christoffel word, $\tilde{w}$ be the corresponding upper Christoffel word, and u any conjugate of w, $u \neq w$. Then $[\chi(u)^\infty]^{-1} + [\chi(\tilde{u})^\infty] < [\chi(w)^\infty]^{-1} + [\chi(\tilde{w})^\infty] = \sqrt{9 - 4/m^2}$, where m is the Markoff number associated with w.*

Proof. If two words are conjugate, then their reversals are conjugate. It follows then from Theorem 8.2.3 that $\tilde{u}$ is a conjugate of w.

We have by Theorem 8.2.3, $w <_{lex} u$ and $\tilde{u} <_{lex} \tilde{w}$. Therefore, using the fact that χ is increasing for the lexicographical order, we obtain $\chi(w) <_{lex} \chi(u)$ and $\chi(\tilde{u}) <_{lex} \chi(\tilde{w})$. Since these words have the same length, it follows that $\chi(w)^\infty <_{lex} \chi(u)^\infty$ and $\chi(\tilde{u})^\infty <_{lex} \chi(\tilde{w})^\infty$. Note that each word in the image of χ has repeated letters. It follows then from Lemma 5.6.3 that $\chi(w)^\infty <_{alt} \chi(u)^\infty$ and $\chi(\tilde{u})^\infty <_{alt} \chi(\tilde{w})^\infty$. This implies, by Theorem 5.6.1, that $[\chi(u)^\infty]^{-1} + [\chi(\tilde{u})^\infty] < [\chi(w)^\infty]^{-1} + [\chi(\tilde{w})^\infty]$. The last equality follows from Theorem 8.1.2. □

Proof of Theorem 8.2.2. Let $n = 2|w|$. Let $s = (a_k)_{k\in\mathbb{Z}} = {}^\infty\chi(\tilde{w})^\infty$. The latter notation implies that s has period $n = |\chi(\tilde{w})|$, $a_0 \cdots a_{n-1} = \chi(\tilde{w})$ and in particular that a_0 is the first letter of $\chi(\tilde{w})$.

Inspection of the mapping χ and of the bi-infinite word s shows that for each *even* integer i, there is some conjugate u of $\tilde{w}$ such that $\chi(\tilde{u}) = a_i \cdots a_{i+n-1}$ (since by Theorem 8.2.3 the conjugation class of w is closed under reversal); we then say that u *belongs to i*. Clearly w belongs to 0.

If u belongs to the even integer i, then we have $a_i a_{i+1} a_{i+2} \cdots = \chi(\tilde{u})^\infty$ and $\cdots a_{i-3} a_{i-2} a_{i-1} = {}^\infty\chi(\tilde{u})$; then $\chi(u)^\infty = a_{i-1} a_{i-2} a_{i-3} \cdots$. Therefore, by (8.2), $\lambda_i(s) = [a_{i-1}, a_{i-2}, a_{i-3}, \ldots]^{-1} + [a_i, a_{i+1}, a_{i+2}, \ldots] = [\chi(u)^\infty]^{-1} + [\chi(\tilde{u})^\infty]$. In particular, $\lambda_0(s) = [\chi(w)^\infty]^{-1} + [\chi(\tilde{w})^\infty]$. It follows from Corollary 8.2.6 that $\lambda_i(s) \le \lambda_0(s)$.

Suppose now that i is an odd integer. Then by (8.2), $\lambda_i(s) = [a_i, a_{i-1}, a_{i-2}, \ldots] + [a_{i+1}, a_{i+2}, a_{i+3}, \ldots]^{-1}$. Now $i+1$ is even; let v belong to $i+1$, that is, $\chi(\tilde{v}) = a_{i+1} \cdots a_{i+n}$. Then $\chi(\tilde{v})^\infty = a_{i+1} a_{i+2} a_{i+3} \cdots$, and as above, we also have $\chi(v)^\infty = a_i a_{i-1} a_{i-2} \cdots$. Thus $\lambda_i(s)$ is equal to $[\chi(v)^\infty] + [\chi(\tilde{v})^\infty]^{-1}$. Since $\tilde{v}$ is conjugate to w, we deduce by Corollary 8.2.6 that $\lambda_i(s) \le \lambda_0(s)$.

Since by Corollary 8.2.6, $\lambda_0(s) = \sqrt{9 - 4/m^2}$, we conclude that $\mathbf{M}(s)$ is equal to this latter number.

Finally, for further use, observe that, in the two cases i even and i odd, one has the strict inequality $\lambda_i(s) < \lambda_0(s)$, when $u \ne w$ and $\tilde{v} \ne w$; this follows indeed from Corollary 8.2.6. □

Let w be a lower Christoffel word. By Theorem 8.2.3, w and the corresponding upper Christoffel word $\tilde{w}$ are conjugate. Let $h = 0, \ldots, |w| - 1$ be such that $w = C^h(\tilde{w})$. Note that h is unique since w is radical.

Corollary 8.2.7 *Suppose that $w \ne a, b$. Let $\chi(\tilde{w}) = a_0 \cdots a_{n-1}$, so that $n = 2|w|$. The minimum m among the numbers $\epsilon_r = p(a_{r+1}, \ldots, a_{n-1}, a_0, \ldots, a_{r-1})$, $r = 0, \ldots, n-1$, is attained exactly for $r = 0$ and $r = 2h - 1$. This minimum is equal to the Markoff number associated to w, that is, $\mu(w)_{12}$.*

An example: $w = aabab$, $\tilde{w} = babaa$, $h = 3$, $\chi(\tilde{w}) = 2211221111$. For $r = 0$, $m = p(2, 1, 1, 2, 2, 1, 1, 1, 1)$; for $r = 2h - 1 = 5$, $m = p(1, 1, 1, 1, 2, 2, 1, 1, 2)$.

In the corollary, the case $w = a$ or b was excluded. It is left to the reader: in this case $n = 2$ and the minimum of ϵ_r is attained for $r = 0, 1$.

We need the following lemma, whose proof is left to the reader.

Lemma 8.2.8 *Let w be a radical word in $\{a, b\}^* \setminus \{a, b\}$. Then $\chi(w)$ is radical.*

Proof of Corollary 8.2.7. Define a_i for $i \in \mathbb{Z}$ by defining $a_i = a_{i \bmod n}$, and define the bi-infinite sequence $s = (a_i)_{i \in \mathbb{Z}} = {}^\infty\chi(\tilde{w})^\infty$. Since w, hence $\tilde{w}$, is radical, so are $\chi(w)$ and $\chi(\tilde{w})$ by Lemma 8.2.8 (the lemma applies since $w \neq a, b$). From this follows that the i such that $a_i \cdots a_{i+n-1} = \chi(\tilde{w})$ (resp. $= \chi(w)$) are exactly those satisfying $i \equiv 0$ (resp. $i \equiv 2h$) modulo n.

Since the sequence s has the minimal non-zero period n, the corollary is equivalent to the following assertion: the minimum of the numbers $p(a_{i+1}, \ldots, a_{i+n-1})$ is attained exactly when $i \equiv 0$ or $2h - 1 \mod n$. Equivalently, by Lemma 8.2.1, the maximum of the numbers $\lambda_i(s)$ is attained exactly for these i's.

We refer to the proof of Theorem 8.2.2, its notations, and in particular the final observation in this proof. If i is even and if $\tilde{u} \neq \tilde{w}$, one has $\lambda_i(s) < \lambda_0(s)$. But $\chi(\tilde{u}) = a_i \cdots a_{i+n-1}$ so that the condition $\tilde{u} = \tilde{w}$, equivalently $\chi(\tilde{u}) = \chi(\tilde{w})$, is equivalent to the condition $i \equiv 0 \mod n$ by the first paragraph of the proof.

If i is odd, and if $v \neq \tilde{w}$, then $\lambda_i(s) < \lambda_0(s)$. Since $\chi(\tilde{v}) = a_{i+1} \cdots a_{i+n}$, the condition $v = \tilde{w}$ is equivalent to $\tilde{v} = w$, equivalent to $a_{i+1} \cdots a_{i+n} = \chi(w)$, and this is equivalent to $i + 1 \equiv 2h \mod n$.

We prove now the final assertion. We have proved that $m = p(a_1, \ldots, a_{n-1})$. It follows from the equality $\chi(w) = a_{n-1} \cdots a_0$, from the equality $\mu(w) = P(a_{n-1}) \cdots P(a_0)$ (see the proof of Theorem 8.1.2, where the notation is reversed), from (5.1), and from (5.2) that $\mu(w)_{12} = p(a_{n-1}, \ldots, a_1) = p(a_1, \ldots, a_{n-1})$. □

General results on extremal values of continuant polynomials among rearrangements of a given word are given by De Luca, Edson, and Zamboni [DLEZ].

8.2.1 Exercises

Exercise 8.2.1 *Let $s = (a_n)$ be a bi-infinite sequence over $\mathbb{P}$. Show that $\mathbf{M}(s)$ is invariant under shift and that $\mathbf{M}(s) = \mathbf{M}(\tilde{s})$, where $\tilde{s} = (a_{-n})$.*

Exercise 8.2.2 *With the notation of Theorem 2.3.1, show that the conjugation class C of $w = amb$ (which is also that of $\tilde{w} = bma$ by Theorem 8.2.3) is equal to $\{amb\} \cup \{m_2bam_1 \mid m = m_1m_2\}$ and also to $\{bma\} \cup \{m_2abm_1 \mid m = m_1m_2\}$. Deduce that there is a mapping f from $C \setminus \tilde{w}$ into $C \setminus w$ defined by $f(m_2abm_1) = m_2bam_1$ and that $u <_{lex} f(u)$. Deduce that the elements of C in lexicographical order are $w, f(w), f^2(w), \ldots, f^{|w|-1}(w)$ and that the last element of this sequence is $\tilde{w}$. See [LR2].*

Exercise 8.2.3 *Consider two parallel lines of positive slope in the plane. Suppose that their diagonal distance (see the proof of Theorem 8.2.3) is smaller than $\sqrt{2}$. Show that there exists at most one bi-infinite lattice path such that each integral vertex of this path is between these lines or on them.*

8.3 Markoff irrationalities

Let w be a lower Christoffel word. Define x_w to be the quadratic number whose continued fraction expansion is $[\chi(\tilde{w})^\infty]$. For example, $x_{ab} = [\overline{2, 2, 1, 1}]$, $x_{aabab} = [\overline{2, 2, 1, 1, 2, 2, 1, 1, 1, 1}]$.

We call *Markoff irrationality*[2] each real number β which is equivalent to x_w, for some lower Christoffel word w.

Theorem 8.3.1 *Let w be a lower Christoffel word and m be the associated Markoff number. If β is equivalent to x_w, then the Lagrange number of β is $\sqrt{9 - \frac{4}{m^2}}$.*

The proof is very similar to the proof of Theorem 8.2.2, with some changes due to the fact that we deal here with one-sided infinite words.

Proof. By Corollary 7.1.2, it is enough to show the result for $\beta = x_w$. Define $a_0 \cdots a_{n-1} = \chi(\tilde{w})$, and consider the bi-infinite sequence $s = (a_i)_{i \in \mathbb{Z}}$, where $a_i = a_{i \bmod n}$ for any $i \in \mathbb{Z}$. One has $s = {}^\infty\chi(\tilde{w})^\infty$. It is therefore enough by Theorem 8.2.2 to show that $\mathbf{L}(x_w) = \mathbf{M}(s)$. By definition, $\mathbf{M}(s)$ is the supremum of the numbers $\lambda_i(s)$; since they are finitely many, $\mathbf{M}(s)$ is their maximum. By (7.1), for any $i \geq 1$, $\lambda_i(x_w) = \lambda_i(s) + \eta_i$, where by Proposition 5.5.2, η_i tends to 0 when i tends to infinity. Thus the upper limit of the sequence $\lambda_i(x_w)$ is equal to $\mathbf{M}(s)$. That is, $\mathbf{L}(s) = \mathbf{M}(s)$. □

It follows from the definition of equivalence that each number β equivalent to x_w is of the form

$$\beta = [b_0, \ldots, b_{k-1}, \overline{a_0, \ldots, a_{n-1}}] \tag{8.3}$$

with $a_0 \cdots a_{n-1} = \chi(\tilde{w})$. Note that we could equivalently take w in this formula (by changing the b_i's), since w and $\tilde{w}$ are conjugate.

Theorem 8.3.2 *Let β as in (8.3), and take the notations of Theorem 7.2.1 and Corollary 8.2.7. Let $\sigma = \pm 1$ be such that $\sigma(-1)^{k-1} f(e, i) > 0$. There exists a unique $t_0 \in \{0, \ldots, n-1\}$ such that $t_0 \equiv k - 1$ or $2h + k - 2 \bmod n$ and $(-1)^{t_0} = \sigma$. Then, for t large enough, p_t/q_t is a good convergent if and only if $t \equiv t_0 \mod n$.*

This result is a variant of a theorem of Bombieri ([B], Theorem 29; see also [BFPRS], Theorem 1.5 and [Re8], Theorem 18.1).

An example: let $w = aabab, \tilde{w} = babaa, \beta = x_{aabab} = [(2211221111)^\infty]$ so that $k = 0, n = 10$; next, $(e, i) = (1, 0)$ so that $f(e, i) > 0$ and $\sigma = (-1)^{k-1} = -1$, and $C^3(babaa) = aabab$; hence $h = 3$. Thus t_0 is odd; since $k - 1 = -1, 2h + k - 2 = 4$, we have $t_0 = 9$. Thus the good convergents are those of large enough rank $t \equiv 9 \mod 10$, that is (since the ranks begin at 0), the numbers $[(2211221111)^m]$ for some $m \geq 1$, large enough.

[2] It will be shown that the Markoff irrationalities are exactly those real numbers β with $L(\beta) < 3$ (Theorem 10.1.1).

Proof of Theorem 8.3.2. The equality $(-1)^{t_0} = \sigma$ determines uniquely the parity of t_0. The numbers $k-1$ and $2h+k-2$ have opposite parity. Since $n = 2|w|$ is even, this proves uniqueness of t_0. For existence, take the parity of t_0 determined by the above equality and choose $t_0 \equiv k-1 \bmod n$ or $2h+k-2 \bmod n$ depending on this parity.

Suppose that $t \equiv t_0 \bmod n$. Since n is even, t and t_0 have the same parity; hence $(-1)^t = \sigma$. We have $t \equiv k-1$ or $2h+k-2 \bmod n$, thus $t-k+1 \equiv 0$ or $t-k+1 \equiv 2h-1 \bmod n$. Then by Corollary 8.2.7, $\epsilon_{t-k+1} = m$, the Markoff number associated with w, and the minimum of the numbers ϵ_r. Since $(-1)^t = \sigma$ and $\sigma(-1)^{k-1}f(e,i) > 0$, we have $(-1)^{t-k+1}f(e,i) > 0$. Then t is good for t large enough by Theorem 7.3.2.

Conversely, if t is good, by Theorem 7.3.2 we must have $t-k+1 \equiv r \mod n$ with $\epsilon_r = m$ and $(-1)^{t-k+1}f(e,i) > 0$. Thus $r \equiv 0$ or $r \equiv 2h-1 \mod n$ by Corollary 8.2.7, and therefore $t \equiv k-1$ or $2h+k-2 \mod n$, and $\sigma = (-1)^t$. Thus $t \equiv t_0 \bmod n$. □

Corollary 8.3.3 *Each Markoff irrationality β has infinitely many convergents p_t/q_t such that $|\beta - p_t/q_t| < 1/L(\beta)q_t^2$.*

8.3.1 Exercises

Exercise 8.3.1 *Show directly, by using Exercise 7.2.2 and its notations, that $L(\alpha) = a^2+4$. Show that if n is odd, then $|\alpha - p_n/q_n| < 1/L(\alpha)q_n^2$ and $\alpha < p_n/q_n$, and if n is even, $|\alpha - p_n/q_n| > 1/L(\alpha)q_n^2$ and $\alpha > p_n/q_n$.*

Apply this to x_a and x_b and to their convergents, which are the F_{n+1}/F_n and P_{n+1}/P_n, where the F_n and the P_n are the Fibonacci and the Pell numbers; see Exercises 5.5.1 and 7.2.2.

9
Lagrange numbers less than three

This is the heart of Markoff's theory. We prove that if the Lagrange number of a sequence is less than 3, then many combinatorial consequences happen: except for some initial part of the sequence, the letters of the sequence must be 1 and 2; it has no occurrences of the factors 121 or 212; the factors 1122 and 2211 are preceded and followed by words which have some lexicographical property (see Lemma 9.1.3); a lot of special factors are forbidden; and the 1s and 2s appear in pairs. This series of lemmas culminates in Theorem 9.1.7: the sequence is the image under the substitution $\chi : a \mapsto 11, b \mapsto 22$ of a sequence having the Markoff property. A similar result is established for bi-infinite sequences (Theorem 9.2.1).

9.1 From $L(s) < 3$ to the Markoff property

In accordance with Theorem 7.1.1, the *Lagrange number* of an infinite word $s = a_0a_1a_2\cdots$ on the alphabet $\mathbb{Z}$, with $a_i > 0$ for $i > 0$, is equal to $\limsup_{i\to\infty} \lambda_i(s)$, where $\lambda_i(s)$ is defined for $i \geq 2$ by

$$\begin{aligned}\lambda_i(s) &= [a_{i-1}, a_{i-2}, ..., a_1]^{-1} + [a_i, a_{i+1}, a_{i+2}, ...]\\ &= [a_{i-1}, a_{i-2}, ..., a_1]^{-1} + a_i + [a_{i+1}, a_{i+2}, ...]^{-1}\\ &= [a_i, a_{i-1}, \ldots, a_1] + [a_{i+1}, a_{i+2}, \ldots]^{-1}\end{aligned}$$

This latter formula also defines it for $i = 1$: $\lambda_1(s) = [a_1] + [a_2, a_3, ...]^{-1} = [a_1, a_2, ...]$.

In the sequel, if we want to indicate the position i in an infinite word $s = a_0a_1a_2\cdots$, then we underline the letter in position i, that is, a_i. Recall also that a finite word w is a factor of s if $s = uwt$ for some finite word u and some infinite word t.

Throughout the whole section, we consider an infinite word $s = a_0a_1a_2\cdots$ whose Lagrange number is < 3. This implies the existence of an integer $I \geq 1$ and of a real number $\theta < 3$ such that for any $i \geq I$, one has $\lambda_i(s) \leq \theta$.

Lemma 9.1.1 *Let u be a finite or infinite word over $\mathbb{P}$ and n an integer ≥ 2. Then* $[0nu] + [01(n-1)u] = 1$.

Note that we use here the bracket notation introduced in Section 5.6.

Proof. Suppose that u is non-empty and let $x = [u]$. Then the lemma follows from the equalities:

From Christoffel Words to Markoff Numbers. Second Edition. Christophe Reutenauer, Oxford University Press.
 DOI: 10.1093/9780197907535.003.0009

$$\frac{1}{n+\frac{1}{x}}+\frac{1}{1+\frac{1}{n-1+\frac{1}{x}}}=\frac{1}{n+\frac{1}{x}}+\frac{1}{1+\frac{x}{(n-1)x+1}}=\frac{x}{nx+1}+\frac{(n-1)x+1}{(n-1)x+1+x}=1.$$

If u is empty, then one obtains the result by a straightforward computation or by letting x tend to ∞ in the previous equalities. □

Lemma 9.1.2

(i) *If $i \geq I$, then $a_i \in \{1,2\}$;*
(ii) *if $s = a_0 \cdots 1\underline{2}1 \cdots$, with i being the underlined position, then $i < I$;*
(iii) *if $s = a_0 \cdots 21\underline{2} \cdots$, with i being the underlined position, then $i < I$.*

Proof.

(i) Since $\lambda_i(s) = [a_{i+1}, a_{i+2}, \ldots]^{-1} + [a_i, a_{i-1}, \ldots, a_1] \geq a_i$ and $\lambda_i(s) < 3$ for $i \geq I$, we see that $a_i \in \{1,2\}$ for $i \geq I$.
(ii) We have $s = a_0\tilde{u}1\underline{2}1v$; then $\lambda_i(s) = 2 + [1u]^{-1} + [1v]^{-1}$. Clearly $[1u] < 2$ and $[1v] < 2$; thus their inverses are $> 1/2$ and therefore $\lambda_i(s) > 3$. Thus we must have $i < I$.
(iii) We have $s = a_0\tilde{u}21\underline{2}v$. Suppose by contradiction that $i \geq I$. Then $\lambda_i(s) < 3$. By what we have shown in (ii), this occurrence of the factor 212 is not followed by 1 so that $s = a_0\tilde{u}21\underline{2}2v'$ and $\lambda_i(s) = 2+[12u]^{-1}+[2v']^{-1} = 2+[012u]+[02v']$. By the definition of the alternating lexicographical order (Section 5.6), we have $012u \geq_{alt} 012$, since 012 is a prefix of odd length of $012u$; for the same reason, $02v' >_{alt} 021$ if the first letter of v' is 1, and if this letter is ≥ 2, the inequality also holds. By Corollary 5.6.2, we deduce $[012u] \geq [012]$ and $[02v'] > [021]$. Thus $\lambda_i(s) \geq 2 + [012] + [021] = 2 + [012] + [03] = 2 + 1$ (by Lemma 9.1.1) $= 3$, a contradiction. Thus we must have $i < I$. □

Lemma 9.1.3 *Let s have the factorization $s = a_0\tilde{u}11\underline{2}2v$ with u non-empty and $[u] \leq [v]$ (resp. $s = a_0\tilde{u}22\underline{1}1v$ with u non-empty and $[v] \leq [u]$) and let i be the underlined position. Then $i < I$.*

Proof. Consider the case $s = a_0\tilde{u}11\underline{2}2v$ with u non-empty and $[u] \leq [v]$. Then $\lambda_i(s) = 2 + [011u] + [02v]$. Since u is a finite word and v an infinite one, we have by Corollary 5.6.2 $u \leq_{alt} v$ and therefore $02u \leq_{alt} 02v$. By Corollary 5.6.2 again, $[02u] \leq [02v]$, hence $\lambda_i(s) \geq 2 + [011u] + [02u] = 3$, by Lemma 9.1.1. Thus $i < I$.

The case $s = a_0\tilde{u}22\underline{1}1v$ is quite similar by using $[02v] \leq [02u]$. □

This yields the following useful result, whose proof is left to the reader.

Corollary 9.1.4 *If s has a factor $111\underline{2}22$ or $22\underline{2}111$, with i being the underlined position, then $i < I$.*

A *block* in a finite or infinite word s is any finite factor w of s with the following properties: w is a power of a single letter a, $s = uwv$, u does not end with a, and v does not begin with a.

Lemma 9.1.5 *There is a factorization $s = a_0 h t$ with $|h| = I - 2$ or $I - 1$, such that t has no block of odd length.*

Proof. First, we prove the property $(*)$: s has no factorization $s = a_0\tilde{u}21^m2v$, with m odd and $|u| \geq I - 3$. By contradiction, suppose that s has such a factorization. Choose m minimum. If $m = 1$, then $s = a_0\tilde{u}21\underline{2}v$, and this contradicts Lemma 9.1.2 (iii) since the underlined position is $\geq I$. Thus $m \geq 3$. We cannot have $s = a_0\tilde{u}21^m\underline{2}1\cdots$ by Lemma 9.1.2 (ii), since there would be a factor $1\underline{2}1$ with an underlined position $\geq I$. We cannot have $s = a_0\tilde{u}21^m\underline{2}22$, since this contradicts Corollary 9.1.4 because the underlined position is $\geq I$. Thus we have $s = a_0\tilde{u}21^m221^p\cdots$, where $p \geq 1$ is chosen to be maximal and may be infinite (that is, either $s = a_0\tilde{u}21^m221^p2\cdots$ or $s = a_0\tilde{u}21^m221^\infty$). Then $s = a_0\tilde{u}21^{m-2}11\underline{2}21^p\cdots$. Since the underlined position is $\geq I$, we have by Lemma 9.1.3,

$$[1^p\cdots] < [1^{m-2}2u]$$

Since $m-2$ is odd, we must have $p \leq m-2$ (otherwise, $1^p\cdots = 1^{m-2}1\cdots >_{alt} 1^{m-2}2u$, and therefore $[1^p\cdots] > [1^{m-2}2u]$ by Corollary 5.6.2). In particular, p is finite and $s = a_0\tilde{u}21^m221^p2\cdots$. If p is even, then $p < m-2$ (since $p \leq m-2$ and $m-2$ are odd), and $1^p\cdots = 1^p2\cdots >_{alt} 1^p11^{m-3-p}2u = 1^{m-2}2u$, a contradiction by Corollary 5.6.2. Thus p is odd, which contradicts the minimality of m. This proves $(*)$.

We now prove property $(**)$: s has no factorization $s = a_0\tilde{u}12^m1v$, with m odd and $|u| \geq I - 2$. The arguments are similar to the previous ones, and we go somewhat faster. By contradiction, suppose s has such a factorization. Choose m minimum. We cannot have $m = 1$ by Lemma 9.1.2 (ii), as otherwise $s = a_0\tilde{u}1\underline{2}1\cdots$. Thus $m \geq 3$. We cannot have $s = a_0\tilde{u}12^m1\underline{2}\cdots$ by Lemma 9.1.2 (iii). We cannot have $s = a_0\tilde{u}12^{m-1}\underline{2}111\cdots$ by Corollary 9.1.4. Thus we have $s = a_0\tilde{u}12^m112^p\cdots$, with $p \geq 1$, maximal and possibly infinite. Then $s = a_0\tilde{u}12^{m-2}2\underline{2}112^p\cdots$; then we must have

$$[2^{m-2}1u] < [2^p\cdots]$$

by Lemma 9.1.3, since the underlined position is $\geq I$. If $p > m-2$, we have a contradiction by Corollary 5.6.2 (using the definition of $<_{alt}$ and the fact that $m-2$ is odd). Thus $p \leq m-2$, p is finite, and we have $s = a_0\tilde{u}12^m112^p1\cdots$, by maximality of p. If p is even, then $p < m-2$, a contradiction again, since 2^p is followed in s by a 1. Thus p is odd, which contradicts the minimality of m.

We conclude from $(*)$ and $(**)$ that if $s = a_0ht$ with $|h| = I - 2$, then t has no blocks of odd length, except perhaps the first block in t. In this case we take h longer by one letter, which proves the lemma. □

Lemma 9.1.6 *There exist $L, M \in \mathbb{N}$, $L \geq I$, depending only on θ and I, such that for each factorization $s = a_0\tilde{u}1122v$ (resp. $s = a_0\tilde{u}2211v$) with $|u| \geq L$, one has $u = w2u'$, $v = w1v'$ (resp. $u = w1u'$, $v = w2v'$) for some word w of length at most M.*

Proof. According to Lemma 9.1.5, we may write $s = gs_1$, where the length of g is I or $I - 1$, in such a way that s_1 has only blocks of even length.

Recall that if $i \geq I$, then $\lambda_i(s) \leq \theta$; moreover, I is an integer ≥ 1, and $\theta < 3$. Choose $M \in \mathbb{N}$ having the following property: for any $n \in \mathbb{N}$ such that the Fibonacci number F_n satisfies $F_n^2 < 1/(3-\theta)$, one has $n < M$. Define $L = I + M$.

We claim that if $s = a_0\tilde{u}1122v$ and $|u| \geq L$, with w the longest common prefix of u and v, then $|w| \leq M$. The claim is proved below.

We have $[u] > [v]$ by Lemma 9.1.3. Since $I \geq 1$, we have $|u| > |w|$, and consequently $u = wyu'$, $v = wxv'$ with $\{x, y\} = \{1, 2\}$. We want to show that $x = 1, y = 2$. By contradiction, suppose that $u = w1u'$, $v = w2v'$; since $[w1u'] > [w2v']$, w must be of odd length by Corollary 5.6.2; then $s = a_0\tilde{u}'1\tilde{w}1122w2v'$. We have $|u| \geq I + M$, $u = w1u'$, and $|wy| \leq M + 1$ so that $|u'| \geq I - 1$; hence $|a_0\tilde{u}'| \geq I \geq |g|$, since $|g| = I$ or $I - 1$. Thus $1\tilde{w}1122w2$ is a factor of s_1. Hence s_1 has a block of odd length (Exercise 9.1.2), a contradiction. This proves that $x = 1, y = 2$.

We now prove the claim. If w is empty, then $|w| \leq M$. If w is non-empty, let $n + 1 = |w|$. Then $s = a_0(\widetilde{wu'})11\underline{2}2wv'$. Let $a = [wv']$, $b = [wu']$. Then $\lambda_i(s) = 2 + [02a] + [011b]$. Thus

$$3 - \lambda_i(s) = 1 - \frac{1}{2 + \frac{1}{a}} - \frac{1}{1 + \frac{1}{1+\frac{1}{b}}} = \frac{b - a}{(2a+1)(2b+1)}$$

as follows from Exercise 9.1.1. By Lemma 5.5.2 (and the warning after it), $|b - a| < 1/F_n^2$. Since $a, b > 0$, we deduce that $3 - \lambda_i(s) < 1/F_n^2$ and $3 - 1/F_n^2 < \lambda_i(s)$. Since $\lambda_i(s) \leq \theta < 3$, we deduce $3 - 1/F_n^2 < \theta$ and $F_n^2 < 1/(3-\theta)$, and therefore $n < M$ and $|w| = n + 1 \leq M$, which proves the claim.

The case $s = a_0\tilde{u}2211v$ is treated similarly. □

Theorem 9.1.7 *Let $s = a_0a_1a_2\cdots$ be an infinite word over $\mathbb{Z}$. Assume that $a_i > 0$ for $i > 0$ and that the Lagrange number of s is < 3. Then there exists a factorization $s = a_0hs'$ and an infinite word t over $\{a, b\}$ with the following properties: t satisfies the Markoff property, with parameters K and N, and $s' = \chi(t)$. Assuming that $\lambda_i(s) \leq \theta < 3$ for any $i \geq I$, then K and N depend only on θ and I, and the length of h may be bounded by a function depending only on θ and I.*

Recall that χ is the substitution sending a onto 11 and b onto 22.

Proof. If an infinite word over $\{1, 2\}$ has only blocks of even length, then it is in the image of χ. Let L and M be as in Lemma 9.1.6; in particular, $L \geq I$. Write $s = a_0\tilde{h}s'$ with $|a_0\tilde{h}| = L$ or $L+1$, where the length is chosen so that, according to Lemma 9.1.2 (i) and Lemma 9.1.5, s' has only 1s and 2s, and only blocks of even length. Then $s' = \chi(t)$ for some t.

We show that t satisfies the Markoff property. This will prove the theorem (with h replaced by $\tilde{h}$). Let $t = \tilde{u}xyv$ with $\{x, y\} = \{a, b\}$ and with $|u| \geq K = \lceil (M+1)/2 \rceil$. Note that $K \geq (M+1)/2$; hence $2|u| \geq M + 1$. Consider the case $x = a, y = b$, the other case being similar. Then $s = a_0\tilde{h}s' = a_0\tilde{h}\chi(t) = a_0\tilde{h}\chi(\tilde{u})1122\chi(v)$. Since $\chi(\tilde{u}) = \widetilde{\chi(u)}$, we have $s = a_0(\widetilde{\chi(u)h})1122\chi(v)$. Since the length of $\chi(u)h$ is at least L (because $|h| \geq L - 1$ and $|u| \geq 1$), Lemma 9.1.6 implies that $\chi(u)h = p2u'$ and $\chi(v) = p1v'$ for some finite word p of length at most M.

We show that the length of p is even. Indeed, otherwise, since the length $2|u|$ of $\chi(u)$ is larger than M, $p2$ is in the image of χ; hence the last letter of p is 2, but $p1$ is also in the image of χ, and thus the last letter of p is 1, a contradiction.

Thus p is in the image of χ, $p = \chi(m)$ and m is of length at most $M/2$. Since $|\chi(u)| \geq M + 1 \geq |p2|$ and $\chi(u)h = p2u'$, we obtain $\chi(u) = p2u_1' = \chi(m)22\chi(u'')$. Similarly, $\chi(v) = \chi(m)11\chi(v'')$. This implies that $u = mbu''$, $v = mav''$, with m of length at most $M/2$. We conclude by taking $N = \lceil M/2 \rceil$. □

We conclude by a fundamental result, which is the consequence of the previous chapters and which will be used to prove Markoff's theorem for approximations.

Corollary 9.1.8 *Let s be an infinite word over $\mathbb{Z}$, with $a_i > 0$ for $i > 0$, such that for some I and some $\theta < 3$ one has $\lambda_i(s) < \theta$ for any $i \geq I$. Then $s = u\chi(w)^\infty$ for some lower Christoffel word w, and some word u whose length is bounded by a function depending only on I and θ.*

Proof. By Theorem 9.1.7, we have $s = a_0hs'$, $s' = \chi(t)$, where t satisfies the Markoff property for the parameters K and N and where the length of h and the numbers K and N are each bounded by some function depending only on I and θ. By Theorem 4.1.6, we have $t = vw^\infty$, where w is a lower Christoffel word and the length of v is bounded by some function depending only on K and N. Consequently, $s = a_0h\chi(v)\chi(w)^\infty$, which proves the result. □

9.1.1 Exercises

Exercise 9.1.1 *Show that*

$$1 - \frac{1}{2 + \frac{1}{a}} - \frac{1}{1 + \frac{1}{1+\frac{1}{b}}} = \frac{b - a}{(2a + 1)(2b + 1)}$$

Exercise 9.1.2 *Let w be a word in $\{1, 2\}^*$ of odd length, and s_1 an infinite word having $1\tilde{w}1122w2$ as factor. Show that s_1 has a block of odd length.*

9.2 Bi-infinite sequences

Recall that the Markoff supremum of a bi-infinite sequence $s = (a_n)_{n\in\mathbb{Z}}$ has been defined in Section 8.2.

Theorem 9.2.1 *(Markoff [M1, M2]) If the Markoff supremum of a bi-infinite sequence $s = (a_n)_{n\in\mathbb{Z}}$ is < 3, then s is equal up to a shift to $\chi(t)$ for some bi-infinite sequence t over $\{a, b\}$ satisfying the Markoff property.*

Proof. We have $\mathbf{M}(s) = \sup_{i\in\mathbb{Z}}(\lambda_i(s)) < 3$. Choose some real number θ such that $\mathbf{M}(s) < \theta < 3$ and some $I \geq 1$ such that the Fibonacci number F_{I-1} satisfies $\mathbf{M}(s) + \frac{1}{F_{I-1}^2} \leq \theta$.

Fix some integer k, and consider the infinite word $t_k = (b_n)_{n\in\mathbb{N}}$ defined by $b_n = a_{k+n}$. We show that for $i \geq I$, one has $\lambda_i(t_k) \leq \theta$. Indeed, $\lambda_i(t_k) = [b_i, \ldots, b_1] + [b_{i+1}, b_{i+2}, \ldots]^{-1} = [a_{k+i}, \ldots, a_{k+1}] + [a_{k+i+1}, a_{k+i+2}, \ldots]^{-1}$ and $\lambda_{k+i}(s) = [a_{k+i}, \ldots, a_{k+1}, a_k, a_{k-1}, \ldots] + [a_{k+i+1}, a_{k+i+2}, \ldots]^{-1}$. By hypothesis, we have $\lambda_{k+i}(s) \leq \mathbf{M}(s)$. The expansions into continued fractions of the real numbers $[a_{k+i}, \ldots, a_{k+1}]$ and $[a_{k+i}, \ldots, a_{k+1}, a_k, a_{k-1}, \ldots]$ coincide for their $i \geq I$ first terms so that by Lemma 5.5.2, these two numbers differ by at most $\frac{1}{F_{I-1}^2}$. Thus $\lambda_i(t_k) \leq \mathbf{M}(s) + \frac{1}{F_{I-1}^2} \leq \theta$.

It follows from Theorem 9.1.7 that $t_k = w_k\chi(u_k)$ for some word w_k and some infinite word u_k satisfying the Markoff property with parameters K and N; moreover, the lengths of w_k, K, and N are bounded by constants which are independent of k. Since this is true for any k, it follows that $s = \chi(t)$ for some bi-infinite sequence t; furthermore, t satisfies the Markoff property with parameter N. The details are left to the reader. □

Corollary 9.2.2 *(Markoff [M1, M2]) Let s be a bi-infinite sequence over $\mathbb{P}$. The following conditions are equivalent:*

(i) the Markoff supremum $\mathbf{M}(s)$ of s is < 3;
(ii) s is equal up to a shift to $\chi(t)$ for some bi-infinite sequence t over $\{a, b\}$ satisfying the Markoff property;
(iii) s is equal up to a shift to ${}^{\infty}\chi(w)^{\infty}$ for some lower Christoffel word w.

Proof. (i) implies (ii) by Theorem 9.2.1. If (ii) holds, then by Theorem 4.2.1, t is equal up to a shift to ${}^{\infty}w^{\infty}$, for some lower Christoffel word w; then $\chi(t)$ is equal up to a shift to $\chi({}^{\infty}w^{\infty}) = {}^{\infty}\chi(w)^{\infty}$, and (iii) holds. The implication (iii)⇒(i) follows from Theorem 8.2.2, since the shift does not change $\mathbf{M}(s)$. □

10
Markoff's theorem for approximations

In this chapter we prove Markoff's theorem for approximations: say that an irrational number x is *badly approximated* if its Lagrange number $L(x)$ is < 3; the theorem then asserts that the continued fraction of x is ultimately periodic and has as a periodic pattern a Christoffel word on the alphabet $\{11, 22\}$. Moreover, x has then infinitely many good approximations. See Theorem 10.1.1 and Corollary 10.1.2.

10.1 Markoff's theorem for approximations

Theorem 10.1.1 *Let x be an irrational real number. Then its Lagrange number $\mathbf{L}(x)$ is less than 3 if and only if x is equivalent to some number $x_w = [\chi(\tilde{w})^\infty]$ (or equivalently to $[\chi(w)^\infty]$) for some lower Christoffel word w. In this case, let m be the Markoff number $\mu(w)_{12} = \frac{1}{3}\mathrm{tr}(\mu(w))$ associated with w. Then $\mathbf{L}(x) = \mathbf{L}(x_w) = \sqrt{9 - \frac{4}{m^2}}$ and $x_w = \frac{p-s}{2m} + \frac{1}{2}\sqrt{9 - \frac{4}{m^2}}$ with $\mu(w) = \begin{pmatrix} p & q \\ r & s \end{pmatrix}$ and therefore $m = q = \frac{1}{3}(p + s)$.*

Proof. Note first that since w and $\tilde{w}$ are conjugate (Theorem 8.2.3), so are their images under χ, and therefore $[\chi(\tilde{w})^\infty]$ and $[\chi(w)^\infty]$ are equivalent real numbers, since their expansions into continued fractions are ultimately equal. In particular, they have the same Lagrange number by Corollary 7.1.2.

The same result shows that if x is equivalent to x_w, then $\mathbf{L}(x) = \mathbf{L}(x_w)$, and by Theorem 8.3.1, $\mathbf{L}(x_w) = \sqrt{9 - \frac{4}{m^2}} < 3$, where m is the Markoff number as in the statement. Then $\mu(w)_{12} = \frac{1}{3}\mathrm{tr}(\mu(w))$ by Lemma 3.1.2. Moreover, by Theorem 8.1.2 (where y corresponds to x_w), $x_w = \frac{p-s}{2m} + \frac{1}{2}\sqrt{9 - \frac{4}{m^2}}$.

Conversely, let x be such that $\mathbf{L}(x) < 3$. Let $s = a_0 a_1 a_2 \cdots$ be the infinite word such that x has the continued fraction expansion $[a_0, a_1, a_2, \ldots]$. Then $\mathbf{L}(s) = \mathbf{L}(x) < 3$. Thus there exist I and θ as in Corollary 9.1.8. We deduce that $s = u\chi(w)^\infty$ for some lower Christoffel word w; since w and $\tilde{w}$ are conjugate, we also have $s = v\chi(\tilde{w})^\infty$ and therefore x is equivalent to x_w. □

Markoff's theorem is often stated as a series of progressively better approximations, with exceptions, the first exception being the golden ratio and the numbers equivalent to it. Formally, this is the following result.

Corollary 10.1.2 *Let M be a finite set of Markoff numbers, such that for any Markoff numbers m_1 and m_2, if $m_1 < m_2$ and $m_2 \in M$, then $m_1 \in M$. Let W be the finite set of all lower Christoffel words corresponding to the Markoff numbers in M (in other*

From Christoffel Words to Markoff Numbers. Second Edition. Christophe Reutenauer, Oxford University Press.
© Christophe Reutenauer (2026). DOI: 10.1093/9780197907535.003.0010

words, W is the set of all lower Christoffel words w such that $\mu(w)_{12} \in M$). Let m be the smallest Markoff number not in M. Then for each irrational real number not equivalent to any x_w, $w \in W$, there are infinitely many rational approximations p/q of x, such that $|x - p/q| < 1/\sqrt{9 - \frac{4}{m^2}}q^2$.

We give three examples. We use the following fact: 1, 2, and 5 are the three smallest Markoff numbers and correspond respectively to the lower Christoffel words a, b, and ab.

First, let $M = \emptyset$. Then $m = 1$, $W = \emptyset$, and $\sqrt{9 - \frac{4}{m^2}} = \sqrt{5}$. We obtain that each real irrational number has infinitely many rational approximations satisfying $|x - p/q| < 1/\sqrt{5}q^2$. This is a theorem of Hurwitz [H1] (Satz 1, p. 279), which we state as a corollary.

Corollary 10.1.3 *For each irrational real number x there are infinitely many rational fractions $\frac{p}{q}$, such that $|x - p/q| < 1/\sqrt{5}q^2$.*

Note that there is a straightening of the previous result: of any three consecutive convergents of x, one of them, p/q say, satisfies the inequality in the corollary (Emile Borel [Bl], p. 344; see also [A], proof of Theorem 1.21, pp. 21–22). Compare to the similar result for two consecutive convergents and the constant 2 at the denominator, the second statement in Theorem 5.5.1.

Now let $M = \{1\}$. Then $m = 2$, $W = \{a\}$, $\sqrt{9 - \frac{4}{m^2}} = \sqrt{8}$. Moreover, $x_a = \frac{\sqrt{5}+1}{2}$, the *golden ratio*. We obtain that each real irrational number x not equivalent to x_a has infinitely many rational approximations satisfying $|x - p/q| < 1/\sqrt{8}q^2$. Note that '$x$ equivalent to x_a' means that in the expansion into continued fractions of x, only 1s occur after some rank.

Finally, let $M = \{1, 2\}$. Then $W = \{a, b\}$, $m = 5$, $\sqrt{9 - \frac{4}{m^2}} = \frac{\sqrt{221}}{5}$, $x_b = 1 + \sqrt{2}$. We obtain that each irrational real number that is equivalent neither to x_a nor to x_b has infinitely many rational approximations satisfying $|x - p/q| < 1/\frac{\sqrt{221}}{5}q^2$. Note that '$x$ equivalent to x_b' means that in the expansion into continued fractions of x, only 2s occur after some rank.

We proceed now to the proof of the corollary. This requires several lemmas.

Lemma 10.1.4 *If $\lambda_{n+1}(x) > \alpha$, then $|x - p_n/q_n| < 1/\alpha q_n^2$. If $\lambda_{n+1}(x) < \alpha$, then $|x - p_n/q_n| > 1/\alpha q_n^2$.*

Proof. It follows directly from (7.2). □

Corollary 10.1.5 *If $\alpha > 0$ and $\mathbf{L}(x) > \alpha$, then x has infinitely many rational approximations p/q of x, such that $|x - p/q| < 1/\alpha q^2$.*

Proof. Since by Theorem 7.1.1, $\mathbf{L}(x) = \limsup_n(\lambda_n(x))$, there are infinitely many n such that $\lambda_{n+1}(x) > \alpha$, and we apply the previous lemma. □

Lemma 10.1.6 *Let x be a real irrational number such that $\mathbf{L}(x) < 3$. Then there are infinitely many rational approximations p/q of x such that $|x - p/q| < 1/\mathbf{L}(x)\, q^2$.*

Proof. By Theorem 10.1.1, x is equivalent to x_v for some lower Christoffel word; hence x is a Markoff irrationality. Thus the lemma follows from Corollary 8.3.3. □

Proof of Corollary 10.1.2. If w is a word of length n, then each entry in $\mu(w)$ is at least equal to 2^{n-1}: indeed, the entries of the generating matrices $\mu(a)$ and $\mu(b)$ are all ≥ 1, and the n-th power of $\begin{pmatrix} 1 & 1 \\ 1 & 1 \end{pmatrix}$ is equal to $\begin{pmatrix} 2^{n-1} & 2^{n-1} \\ 2^{n-1} & 2^{n-1} \end{pmatrix}$; thus each entry of $\mu(w)$ is at least equal to 2^{n-1}. Therefore, if $\mu(w)_{12} \in M$, the length of w is bounded. Thus W is finite.

Let x be a real irrational number not equivalent to any x_w, $w \in W$. Let $\alpha = \sqrt{9 - \frac{4}{m^2}}$. Then $\alpha < 3$. If $\mathbf{L}(x) \geq 3$, then $\mathbf{L}(x) > \alpha$, and by Corollary 10.1.5, we find infinitely many rational approximations p/q of x, such that $|x - p/q| < 1/\alpha q^2$. If $\mathbf{L}(x) < 3$, then by Theorem 10.1.1, x is equivalent to x_v for some lower Christoffel word v, with associated Markoff number n and $\mathbf{L}(x) = \sqrt{9 - \frac{4}{n^2}}$. By hypothesis, $v \notin W$ so that n is not in M and therefore $n \geq m$. If $n > m$, then $\mathbf{L}(x) > \alpha$, and by Corollary 10.1.5, we also find infinitely many approximations as above. If $n = m$, then we use Lemma 10.1.6. □

11
Markoff's theorem for quadratic forms

In this chapter, we give Markoff's theorem for quadratic forms: consider a form f which is real, binary, and indefinite and satisfies the inequality $L(f) > \frac{1}{3}\sqrt{d(f)}$ ($d(f)$ is the discriminant of f, and the infimum $L(f)$ is defined later); the conclusion is that f is, up to a constant, equivalent to a so-called Markoff form.

The first two sections are concerned with results which are reminiscent of the work of Gauss; in the last section, Markoff's theorem is deduced using the results of previous chapters.

11.1 Indefinite real binary quadratic forms

Consider a real binary quadratic form $f(x, y) = \alpha x^2 + \beta xy + \gamma y^2$. Recall that if its discriminant $d(f) = \beta^2 - 4\alpha\gamma$ is positive, the form is called indefinite.

We are interested here in the *infimum* over the integers of such a form, defined by

$$L(f) = \inf\{|f(x, y)|, x, y \in \mathbb{Z}, (x, y) \neq (0, 0)\}$$

We say that the infimum is *attained* if there exists $(x, y) \in \mathbb{Z}^2 \setminus (0, 0)$ such that $L(f) = f(x, y)$. This means that the infimum is the minimum of the previous set. This happens when the coefficients of f are integers.

For two equivalent forms f and g, the sets $\{f(x, y), (x, y) \in \mathbb{Z}^2 \setminus (0, 0)\}$ and $\{g(x, y), (x, y) \in \mathbb{Z}^2 \setminus (0, 0)\}$ are equal. Thus, two equivalent forms have the same infimum, and if it is attained for one of them, it is attained for the other, too.

We consider in this section, and the next one, only indefinite quadratic forms $f(x, y) = \alpha x^2 + \beta xy + \gamma y^2$, such that for any rational numbers (equivalently, integers) x and y *with $(x, y) \neq (0, 0)$, one has $f(x, y) \neq (0, 0)$.* Note that this condition is invariant under equivalence of forms. Moreover f has then two real irrational roots, which are distinct since the discriminant of f is positive.

Theorem 11.1.1 *Let $f(x, y)$ be an indefinite binary quadratic form, such that $f(x, y) \neq 0$ for any integers* x and y *that are not both zero. Then f is equivalent to a form having two distinct irrational real roots, one* > 1 *and the other between* -1 *and* 0.

Proof. Let u and v be the two roots of f. Then they are distinct. Consider their expansion into continued fractions $u = [a_0, a_1, \ldots]$, $v = [b_0, b_1, \ldots]$. Let n be the smallest integer such that $a_n \neq b_n$. We show by induction on n that, taking an equivalent form, we may assume that $u > 0 > v$. If $n = 0$, then we may assume that $a_0 > b_0$ and therefore $u > a_0 > v$; thus $u - a_0 > 0 > v - a_0$. By Proposition 6.1.3 (i), these two numbers are

From Christoffel Words to Markoff Numbers. Second Edition. Christophe Reutenauer, Oxford University Press.
 DOI: 10.1093/9780197907535.003.0011

the roots of the quadratic form g such that $f = g \cdot \begin{pmatrix} 1 & -a_0 \\ 0 & 1 \end{pmatrix}$. Suppose now that $n > 0$, that is, $a_0 = b_0$; then with $h(x, y) = f(a_0 x + y, x)$, the roots of h are the numbers $\frac{1}{u-a_0}$ and $\frac{1}{v-a_0}$. Their expansions into continued fractions are $[a_1, a_2 ...]$ and $[b_1, b_2 ...]$, and we conclude by induction on n.

Thus we are reduced to a quadratic form f having a positive root u and a negative root v. If one of them at least has absolute value > 1, then the interval $[v, u]$ contains at least two integers. If their absolute values are both < 1, we replace $f(x, y)$ by $f(y, x)$, obtaining an equivalent quadratic form whose roots are the inverses of the previous ones, hence having different signs and both having absolute values > 1.

Thus we may assume that, the two roots being $u > v$, the interval $[v, u]$ contains at least two integers. Let h be the upper integer part of v. Then $-1 < v - h < 0$ and $u - h > 1$. These two numbers are, by Proposition 6.1.3 (i), the roots of the quadratic form g such that $f = g \cdot \begin{pmatrix} 1 & -h \\ 0 & 1 \end{pmatrix}$, which concludes the proof. □

11.2 Infimum

Let $f_0 = f = \alpha x^2 + \beta xy + \gamma y^2$ be an indefinite binary quadratic form, such that $f(x, y) \neq 0$ for any integers x and y that are not both zero; let ξ and $-\frac{1}{\eta}$ be the roots of the polynomial $f(x, 1)$, and assume that $\xi, \eta > 1$, as allowed by the conclusion of Theorem 11.1.1. By hypothesis, ξ and η are irrational real numbers, and we consider their expansions into continued fractions: $\xi = [a_0, a_1, a_2, ...]$, $\eta = [a_{-1}, a_{-2}, a_{-3}, ...]$, which define the bi-infinite word $(a_n)_{n \in \mathbb{Z}}$. Note that all the a_n are positive.

Recall that $P(a) = \begin{pmatrix} a & 1 \\ 1 & 0 \end{pmatrix}$. Consider the bi-infinite sequence of quadratic forms f_i, such that for any $n \in \mathbb{Z}$ one has $f_{n+1} = f_n \cdot P(a_n)$. It exists and is unique, since we must have $f_1 = f_0 \cdot P(a_0)$, $f_2 = f_1 \cdot P(a_1)$, ..., and $f_{-1} = f_0 \cdot P(a_{-1})^{-1}$, $f_{-2} = f_{-1} \cdot P(a_{-2})^{-1}$, ... Denote $f_n = f_n(x, y) = \alpha_n x^2 + \beta_n xy + \gamma_n y^2$.

Theorem 11.2.1 *(Markoff [M1], p. 388) The infimum of the form f is equal to the infimum of the real numbers $|\alpha_n|$, $n \in \mathbb{Z}$.*

Lemma 11.2.2 *For $n \in \mathbb{Z}$, let $\xi_n = [a_n, a_{n+1}, ...]$ and $\eta_n = [a_{n-1}, a_{n-2}, ...]$. Then ξ_n and $\frac{-1}{\eta_n}$ are the roots of the quadratic form f_n.*

Proof. One has $\xi_{n+1} = \frac{1}{\xi_n - a_n}$ and $f_{n+1} = f_n \cdot P(a_n)$, so that $f_n = f_{n+1} \cdot P(a_n)^{-1} = f_{n+1} \cdot \begin{pmatrix} 0 & 1 \\ 1 & -a_n \end{pmatrix}$. This shows by Proposition 6.1.3 (i) that if ξ_n is a root of f_n, then ξ_{n+1} is a root of f_{n+1}. We conclude by induction that ξ_n is a root of f_n for any $n \in \mathbb{N}$. For $n < 0$ and $\frac{-1}{\eta_n}$ the proof is similar. □

Proof of Theorem 11.2.1, following Markoff [M1], pp. 386–388. Since the f_n are all equivalent, they have the same infimum. We have $f_n(1, 0) = \alpha_n$ so that the infimum of f is $\leq |\alpha_n|$.

Suppose first that all a_n are equal to 1. We have $\xi = \frac{\sqrt{5}+1}{2} = \eta$. The two numbers ξ and $\frac{-1}{\eta} = \frac{1-\sqrt{5}}{2}$ are conjugate and are the roots of $x^2 - x - 1$ so that, multiplying if necessary by a non-zero factor, we may assume that $f(x, y) = x^2 - xy - y^2$. The infimum of this form is 1, since the values of f for integers x and y are integers. Hence the theorem follows in this case, since all α_n are non-zero integers and $\alpha_0 = 1$.

Suppose now that at least one of the a_n is ≥ 2. By shifting the bi-infinite sequences (a_n) and (f_n), using Lemma 11.2.2, we may assume that $a_0 \geq 2$, so that $\xi > 2$. We claim that if $|f(x, y)| \leq |\alpha_0|$ for some integers x and y, which are not both zero, then $|f(x, y)| \geq |\alpha_n|$ for some $n \in \mathbb{Z}$. This will imply the theorem: indeed, each element of $\{f(x, y), (x, y) \in \mathbb{Z}^2 \setminus (0, 0)\}$ is bounded from below by some α_n, itself an element of this set.

We prove the claim. If $y = 0$, we have $f(x, y) = \alpha_0 x^2$ and therefore $|f(x, y)| \geq |\alpha_0|$, since x is a non-zero integer.

Suppose that $x = 0$. Note that $f(x, y) = f_{-1} \cdot P(a_{-1}) = f_{-1} \cdot \begin{pmatrix} a_{-1} & 1 \\ 1 & 0 \end{pmatrix} = f_{-1}(a_{-1}x + y, x)$. Thus $f(x, y) = f(0, y) = f_{-1}(y, 0) = \alpha_{-1}y^2$. Thus $|f(x, y)| \geq |\alpha_{-1}|$, since y is a non-zero integer.

We may therefore assume that x and y are both non-zero, and we use the equation

$$f(x, y) = \alpha_0 y^2 \left(\frac{x}{y} - \xi\right)\left(\frac{x}{y} + \frac{1}{\eta}\right) \tag{11.1}$$

Suppose first that $\frac{x}{y} > 0$. We may assume that $x, y > 0$. By contradiction, assume that $\frac{x}{y} < 2$; then $\frac{x}{y} < \xi$. Thus $|f(x, y)| = |\alpha_0 y^2(\frac{x}{y} - \xi)(\frac{x}{y} + \frac{1}{\eta})| > |\alpha_0| y^2 (2 - \frac{x}{y})\frac{x}{y}$, since $|\frac{x}{y} - \xi| = \xi - \frac{x}{y} > 2 - \frac{x}{y}$ and $\frac{1}{\eta} > 0$. We conclude that $|f(x, y)| > |\alpha_0|(2y - x)x \geq |\alpha_0|$ (because x and y are integers and since $2y - x > 0$), contradicting $|f(x, y)| \leq |\alpha_0|$.

We therefore have $\frac{x}{y} \geq 2$. It follows that $\frac{x}{y} + \frac{1}{\eta} > 2$. The hypothesis $|f(x, y)| \leq |\alpha_0|$ with (11.1) then implies that $y^2|\xi - \frac{x}{y}|(\frac{x}{y} + \frac{1}{\eta}) \leq 1$; hence $|\xi - \frac{x}{y}| < \frac{1}{2y^2}$. Hence, by Theorem 5.5.1, $\frac{x}{y}$ is a convergent of ξ. By Section 5.2, using the notation $\frac{p_n}{q_n}$ for the convergents of ξ, we have $\frac{x}{y} = \frac{p_n}{q_n}$, and by (5.7) ${}^t(p_n, q_n) = M\,{}^t(1, 0)$, where $M = P(a_0)P(a_1) \cdots P(a_n)$. Hence $f(p_n, q_n) = f(M\,{}^t(1, 0)) = (f \cdot M)(1, 0) = f_n(1, 0) = \alpha_n$. Now, since the convergent is an irreducible fraction, we have $(x, y) = d(p_n, q_n)$ for some non-zero integer d so that $f(x, y) = d^2 f(p_n, q_n)$ and therefore $|f(x, y)| \geq |\alpha_n|$.

Suppose now that $\frac{x}{y} < 0$. Then, since $\xi > 2$, (11.1) and the claim's hypothesis $|f(x, y)| \leq |\alpha_0|$ imply that $|\frac{x}{y} + \frac{1}{\eta}| < \frac{1}{2y^2}$ so that $-\frac{x}{y}$ is a convergent of $\frac{1}{\eta} = [0, a_{-1}, a_{-2}, a_{-3}, \ldots]$. Denote by p_n/q_n the convergents of this number. Then $(x, y) = d(-p_n, q_n)$ for some non-zero integer d and some $n \geq 1$ ($n \neq 0$, since $x/y \neq 0$ and $p_0/q_0 = 0$). Then $f(x, y) = d^2 f(-p_n, q_n)$, and we show next that $f(-p_n, q_n) = \alpha_{-n-1}$, which will imply that $|f(x, y)| \geq |\alpha_{-n-1}|$.

By applying (5.7) to the continued fraction of $\frac{1}{\eta}$, we have $\begin{pmatrix} p_{n+1} & p_n \\ q_{n+1} & q_n \end{pmatrix} = P(0)P(a_{-1}) \cdots P(a_{-n-1})$. Transposing and taking inverses, we obtain, since the matrices $P(a)$ are symmetric and of determinant -1,

$$(-1)^n \begin{pmatrix} q_n & -q_{n+1} \\ -p_n & p_{n+1} \end{pmatrix} = P(0)^{-1}P(a_{-1})^{-1}\cdots P(a_{-n-1})^{-1} \tag{11.2}$$

Since

$$\begin{pmatrix} 0 & 1 \\ 1 & 0 \end{pmatrix}\begin{pmatrix} q_n & -q_{n+1} \\ -p_n & p_{n+1} \end{pmatrix}\begin{pmatrix} 1 \\ 0 \end{pmatrix} = \begin{pmatrix} 0 & 1 \\ 1 & 0 \end{pmatrix}\begin{pmatrix} q_n \\ -p_n \end{pmatrix} = \begin{pmatrix} -p_n \\ q_n \end{pmatrix}$$

multiplying (11.2) on the left by $P(0)$ and on the right by $\begin{pmatrix} 1 \\ 0 \end{pmatrix}$, we see that

$$(-1)^n \begin{pmatrix} -p_n \\ q_n \end{pmatrix} = P(a_{-1})^{-1}\cdots P(a_{-n-1})^{-1}\begin{pmatrix} 1 \\ 0 \end{pmatrix}$$

Now, we have $f_{-n-1} = f \cdot P(a_{-1})^{-1}\cdots P(a_{-n-1})^{-1}$, so that

$$f_{-n-1}(1,0) = f\left(P(a_{-1})^{-1}\cdots P(a_{-n-1})^{-1}\begin{pmatrix} 1 \\ 0 \end{pmatrix}\right)$$

This implies that $\alpha_{-n-1} = f_{-n-1}(1,0) = f(-p_n, q_n)$. □

Corollary 11.2.3 *Let f be as in Theorem 11.2.1. Then $\frac{\sqrt{d(f)}}{L(f)} = \mathbf{M}(s)$, the Markoff supremum of the bi-infinite sequence $s = (a_n)_{n\in\mathbb{Z}}$.*

For the definition of the Markoff supremum, see Section 8.2.

Proof. By the theorem, $L(f) = \inf(\{|\alpha_n|, n \in \mathbb{Z}\}$. Since by Lemma 11.2.2, ξ_n and $\frac{-1}{\eta_n}$ are the roots of the polynomial $\alpha_n x^2 + \beta_n x + \gamma_n$, their difference is equal, up to the sign, to $\frac{\sqrt{d(f_n)}}{|\alpha_n|}$; thus one has $\xi_n + \frac{1}{\eta_n} = \frac{\sqrt{d(f_n)}}{|\alpha_n|} = \frac{\sqrt{d(f)}}{|\alpha_n|}$, since the first sum is positive. Thus, if $L(f) > 0$, we obtain that $\frac{\sqrt{d(f)}}{L(f)} = \sup_{n\in\mathbb{Z}} \frac{\sqrt{d(f)}}{|\alpha_n|} = \sup_{n\in\mathbb{Z}}(\xi_n + \frac{1}{\eta_n}) = \mathbf{M}(s)$. If $L(f) = 0$, then $\inf(|\alpha_n|) = 0$; hence $\mathbf{M}(s) = \sup(\xi_n + \frac{1}{\eta_n}) = \sup \frac{\sqrt{d(f_n)}}{|\alpha_n|} = \infty$, and the result is still true. □

11.3 Markoff's theorem for quadratic forms

Let w be a lower Christoffel word and let $\mu(w) = \begin{pmatrix} p & q \\ r & s \end{pmatrix}$, where $q = m$ is the Markoff number associated with w and therefore $m = \frac{1}{3}(p + s)$; see Section 3.3. Define the associated *Markoff quadratic form* by $f_w(x, y) = mx^2 + (s - p)xy - ry^2$.

Theorem 11.3.1 *Let $f(x, y)$ be an indefinite binary quadratic form. Assume that its infimum $L(f)$ and its discriminant $d(f)$ satisfy the inequality $L(f) > \frac{1}{3}\sqrt{d(f)}$. Then f is equivalent to a multiple of some Markoff form f_w. Let m be the Markoff number* *associated with the lower Christoffel word w. One has* $d(f_w) = 9m^2 - 4$, $L(f_w) = f_w(1,0) = m$ *so that the infima of* f_w *and* f *are attained, and* $\frac{\sqrt{d(f)}}{L(f)} = \frac{\sqrt{d(f_w)}}{L(f_w)} = \sqrt{9 - \frac{4}{m^2}}$.

Proof. The hypothesis implies that $L(f) \neq 0$; in particular, $f(x,y) \neq 0$ for any integers x and y, not both zero. By Theorem 11.1.1, we may assume that the roots of f are distinct, and that one is in $(-1,0)$ and the other is > 1. Taking the notations of Section 11.2, we may therefore assume that $f = f_0$. Let $s = (a_n)_{n\in\mathbb{Z}}$ and define its shift s_k by $s_k = (a_{n+k})_{n\in\mathbb{Z}}$. By Corollary 11.2.3 and the hypothesis, $\mathbf{M}(s) < 3$. From Corollary 9.2.2 (iii) we deduce that for some k, $s_k = {}^\infty\chi(w)^\infty$, where w is a lower Christoffel word. Since w and $\tilde{w}$ are conjugate by Theorem 8.2.3, so are $\chi(w)$ and $\chi(\tilde{w})$, and we also have $s_l = {}^\infty\chi(\tilde{w})^\infty$ for some l. Now f is equivalent to f_l. Moreover, by Lemma 11.2.2, the expansion into continued fractions of the root ξ_l of f_l is $[\chi(\tilde{w})^\infty]$ so that by Theorem 8.1.2 it satisfies the equation $\xi_l = \frac{p\xi_l + r}{q\xi_l + s}$, where $\mu(w) = \begin{pmatrix} p & q \\ r & s \end{pmatrix}$. The other root of f_l is $-\frac{1}{[\chi(w)^\infty]}$, which is the conjugate of ξ_l by the same theorem. Thus f_l is proportional to the form $qx^2 + (s-p)xy - ry^2 = f_w$.

We have $d(f_w) = (s-p)^2 + 4qr = (s+p)^2 - 4ps + 4qr = 9m^2 - 4$, since $p + s = 3m$ and since the matrix $\mu(w)$ has determinant 1. It follows from Theorem 8.2.2 that $\mathbf{M}(s_l) = \sqrt{9 - \frac{4}{m^2}}$. Thus, by Corollary 11.2.3, $\frac{\sqrt{d(f_w)}}{L(f_w)} = \sqrt{9 - \frac{4}{m^2}}$. Thus $L(f_w) = \frac{\sqrt{9m^2-4}}{\sqrt{9-\frac{4}{m^2}}} = m = f_w(1,0)$, which ends the proof. □

Markoff's theorem is often stated as a series of progressively better inequalities relating $d(f)$ and $L(f)$, with exceptions where they do not hold, the first exception being the quadratic form $x^2 - xy - y^2$ and the forms equivalent to a multiple of this form. Formally, this is the following result:

Corollary 11.3.2 *Let M be a finite set of Markoff numbers, such that for any Markoff numbers* m_1 and m_2, *if* $m_1 < m_2$ *and* $m_2 \in M$, *then* $m_1 \in M$. *Let W be the finite set of all lower Christoffel words corresponding to the Markoff numbers in M (in other words, W is the set of all lower Christoffel words w such that* $\mu(w)_{12} \in M$*). Let m be the smallest Markoff number not in M. Then for each indefinite binary quadratic form f(x, y) not equivalent to a multiple of any form* f_w, $w \in W$, *one has* $\sqrt{d(f)} \geq \sqrt{9 - \frac{4}{m^2}} L(f)$.

We give several examples. First, let $M = \emptyset$. Then $W = \emptyset$, $m = 1$, and $\sqrt{9 - \frac{4}{m^2}} = \sqrt{5}$. We obtain a result due to Korkine and Zolotareff [KZ].

Corollary 11.3.3 *Let f(x, y) be an indefinite binary quadratic form. Then* $\sqrt{d(f)} \geq \sqrt{5} L(f)$.

Now let $M = \{1\}$. Then $W = \{a\}$, $m = 2$, $\sqrt{9 - \frac{4}{m^2}} = \sqrt{8}$. We have $\mu(a) = \begin{pmatrix} 2 & 1 \\ 1 & 1 \end{pmatrix}$ so that $f_a = x^2 - xy - y^2$. Thus we obtain the following result, which is also due to Korkine and Zolotareff [KZ].

Corollary 11.3.4 *Let $f(x, y)$ be an indefinite binary quadratic form. If f is not equivalent to a multiple of the form $f_a = x^2 - xy - y^2$, then $\sqrt{d(f)} \geq \sqrt{8}L(f)$.*

These two results are mentioned by Markoff [M1], who gives them as the motivation for his own work. The next example is $M = \{1, 2\}$. Then $W = \{a, b\}$, $m = 5$, $\sqrt{9 - \frac{4}{m^2}} = \frac{\sqrt{221}}{5}$. Since $\mu(b) = \begin{pmatrix} 5 & 2 \\ 2 & 1 \end{pmatrix}$, we have $f_b = 2x^2 - 4xy - 2y^2$. Thus, if a form is neither equivalent to a multiple of $f_a = x^2 - xy - y^2$, nor equivalent to a multiple of $f_b = 2x^2 - 4xy - 2y^2$, one has $\sqrt{d(f)} \geq \frac{\sqrt{221}}{5}L(f)$.

Proof of Corollary 11.3.2. Suppose first that $\sqrt{d(f)} \geq 3L(f)$. Then the conclusion holds, since $\sqrt{9 - \frac{4}{m^2}} < 3$. Suppose now that $\sqrt{d(f)} < 3L(f)$. Then by Theorem 11.3.1, f is equivalent to a multiple of some f_v, v a lower Christoffel word, with associated Markoff number n, and $\frac{\sqrt{d(f)}}{L(f)} = \sqrt{9 - \frac{4}{n^2}}$. By hypothesis, we must have $v \notin W$. Hence $n \notin M$. Thus $n \geq m$, and consequently $\sqrt{9 - \frac{4}{n^2}} \geq \sqrt{9 - \frac{4}{m^2}}$, which ends the proof. □

12
Numerology

This chapter gives several examples, which may help the reader work in concrete terms with Markoff numbers, Christoffel words, Markoff constants, and quadratic forms. Some results of Frobenius, Aigner, and Clemens are also given.

12.1 Thirteen Markoff numbers

The 13 smallest Markoff numbers are

$$1, 2, 5, 13, 29, 34, 89, 169, 194, 233, 433, 610, 985$$

which are also those smaller than 1000. The corresponding *Markoff constants* $\sqrt{9 - \frac{4}{m^2}} = \frac{\sqrt{9m^2-4}}{m}$ are respectively

$$\sqrt{5}, \sqrt{8}, \frac{\sqrt{221}}{5}, \frac{\sqrt{1517}}{13}, \frac{\sqrt{7565}}{29}, \frac{\sqrt{10400}}{34} = \frac{\sqrt{2600}}{17}$$
$$\frac{\sqrt{71285}}{89}, \frac{\sqrt{257045}}{169}, \frac{\sqrt{338720}}{194} = \frac{\sqrt{84680}}{97}, \frac{\sqrt{488597}}{233}$$
$$\frac{\sqrt{1687397}}{433}, \frac{\sqrt{3348896}}{610} = \frac{\sqrt{837224}}{305}, \frac{\sqrt{8732021}}{985}$$

The 13 lower Christoffel words corresponding to the 13 previous Markoff numbers are respectively

$$a, b, ab, a^2b, ab^2, a^3b, a^4b, ab^3, a^2bab, a^5b, abab^2, a^6b, ab^4$$

See Figures 2.10 and 3.1.

The *Markoff spectrum* is the set of real numbers of the form $\frac{\sqrt{d(f)}}{L(f)}$ where f is a real binary indefinite quadratic form. The *Lagrange spectrum* is the set of real numbers of the form $\mathbf{L}(x)$ where x is a real irrational number. It follows from Theorems 10.1.1 and 11.3.1 that the intersection of both spectra with the interval $[0, 3)$ is equal to the set of Markoff constants $\sqrt{9 - \frac{4}{m^2}}$ where m is a Markoff number.

There are many results on these spectra, which may be found in the extraordinary book of Cusick and Flahive [CF]. In particular, the Markoff spectrum contains strictly the Lagrange spectrum (*loc. cit.* Theorem 1, p. 35). As shown above, the intersections of the two spectra with the set of real numbers < 3 are both equal to a discrete subset

From Christoffel Words to Markoff Numbers. Second Edition. Christophe Reutenauer, Oxford University Press.
 DOI: 10.1093/9780197907535.003.0012

of the reals. The intersections of both spectra with $[3, \infty)$ have gaps, and they both contain $[\sqrt{21}, \infty)$ (*loc. cit.* Theorems 8 and 9, p. 55).

12.2 The golden ratio and other numbers

The images of the 13 lower Christoffel words of Section 12.1 under the homomorphism μ are the matrices

$$\mu(a) = \begin{pmatrix} 2 & 1 \\ 1 & 1 \end{pmatrix}, \mu(b) = \begin{pmatrix} 5 & 2 \\ 2 & 1 \end{pmatrix}, \mu(ab) = \begin{pmatrix} 12 & 5 \\ 7 & 3 \end{pmatrix}$$

$$\mu(a^2b) = \begin{pmatrix} 31 & 13 \\ 19 & 8 \end{pmatrix}, \mu(ab^2) = \begin{pmatrix} 70 & 29 \\ 41 & 17 \end{pmatrix}, \mu(a^3b) = \begin{pmatrix} 81 & 34 \\ 50 & 21 \end{pmatrix}$$

$$\mu(a^4b) = \begin{pmatrix} 212 & 89 \\ 131 & 55 \end{pmatrix}, \mu(ab^3) = \begin{pmatrix} 408 & 169 \\ 239 & 99 \end{pmatrix}, \mu(a^2bab) = \begin{pmatrix} 463 & 194 \\ 284 & 119 \end{pmatrix}$$

$$\mu(a^5b) = \begin{pmatrix} 555 & 233 \\ 343 & 144 \end{pmatrix}, \mu(abab^2) = \begin{pmatrix} 1045 & 433 \\ 613 & 254 \end{pmatrix}$$

$$\mu(a^6b) = \begin{pmatrix} 1453 & 610 \\ 898 & 377 \end{pmatrix}, \mu(ab^4) = \begin{pmatrix} 2378 & 985 \\ 1393 & 577 \end{pmatrix}$$

For w, a lower Christoffel word, recall that the real quadratic number x_w, whose expansion into continued fractions is $\chi(\tilde{w})^\infty$, is equal to $x_w = \frac{p-s}{2m} + \frac{1}{2}\sqrt{9 - \frac{4}{m^2}} = \frac{p-s+\sqrt{9m^2-4}}{2m}$, with $\mu(w) = \begin{pmatrix} p & m \\ r & s \end{pmatrix}$ and $3m = p + s$; see Theorem 8.1.2. For w equal to one of the 13 words in Section 12.1, the sequence of x_w, in which the first x_a is the *golden ratio*, is

$$x_a = \frac{1+\sqrt{5}}{2}, x_b = 1 + \sqrt{2}, x_{ab} = \frac{9+\sqrt{221}}{10}, x_{a^2b} = \frac{23+\sqrt{1517}}{26}$$

$$x_{ab^2} = \frac{53+\sqrt{7565}}{58}, x_{a^3b} = \frac{15+\sqrt{650}}{17}, x_{a^4b} = \frac{157+\sqrt{71285}}{178}$$

$$x_{ab^3} = \frac{309+\sqrt{257045}}{338}, x_{a^2bab} = \frac{172+\sqrt{84680}}{194}, x_{a^5b} = \frac{411+\sqrt{488597}}{466}$$

$$x_{abab^2} = \frac{791+\sqrt{1687397}}{866}, x_{a^6b} = \frac{538+\sqrt{837224}}{610}, x_{ab^4} = \frac{1801+\sqrt{8732021}}{1970}$$

Observe that Theorem 10.1.1 and Corollary 10.1.2 imply that the golden ratio and the numbers which are equivalent to it (these are sometimes called *noble numbers*) are the irrational real numbers which have, in some sense, the worst approximations by rational numbers; they are the numbers whose Lagrange number is the minimum possible. This property is relevant in phyllotaxis, as shown in [BeRe].

12.3 The matrices $\mu(w)$ and Frobenius congruences

For any lower Christoffel word w, the matrix $\mu(w)$ has a special form which is given by the following result. It is a particular case of Theorem 4.13 in Aigner's book [A] (the case $a = 2$; see also [A], Example 4.9), due to Clemens [Cle]. The existence of the following numbers u and v was shown directly from the Markoff equation by Frobenius ([F], p. 602).

Theorem 12.3.1 *Let w be a proper lower Christoffel word with standard factorization $w = w_1 w_2$. Let m, m_1,* and *m_2 be the Markoff numbers respectively associated with w, w_1,* and *w_2. There exist unique natural integers u* and *v, such that $m_1 u \equiv m_2 \mod m$, $0 < u < \frac{m}{2}$, and $u^2 + 1 = mv$. One has*

$$\mu(w) = \begin{pmatrix} 2m + u & m \\ 2m - u - v & m - u \end{pmatrix}$$

Corollary 12.3.2 *(variant of Markoff [M2])* $f_w(x, y) = mx^2 - (m + 2u)xy - (2m - u - v)y^2$.

Recall that we denote by m_w the Markoff number associated with the lower Christoffel word w. By Theorem 3.1.1 and Lemma 3.1.2, we may write

$$\mu(w) = \begin{pmatrix} a_w & m_w \\ c_w & 3m_w - a_w \end{pmatrix} \tag{12.1}$$

Lemma 12.3.3 *(following Aigner [A], (4.9)) If $w = w_1 w_2$ is a lower Christoffel word with its standard factorization, then $m_{w_1} = -a_w m_{w_2} + m_w a_{w_2}$ and $m_{w_2} = -a_{w_1} m_w + m_{w_1} a_w$.*

Proof. We have $\mu(w) = \mu(w_1)\mu(w_2)$; thus, since these matrices have determinant 1,

$$\mu(w_1) = \mu(w)\mu(w_2)^{-1} = \begin{pmatrix} a_w & m_w \\ c_w & 3m_w - a_w \end{pmatrix} \begin{pmatrix} 3m_{w_2} - a_{w_2} & -m_{w_2} \\ -c_{w_2} & a_{w_2} \end{pmatrix}$$

and

$$\mu(w_2) = \mu(w_1)^{-1}\mu(w) = \begin{pmatrix} 3m_{w_1} - a_{w_1} & -m_{w_1} \\ -c_{w_1} & a_{w_1} \end{pmatrix} \begin{pmatrix} a_w & m_w \\ c & 3m_w - a_w \end{pmatrix}$$

It follows that $m_{w_1} = -a_w m_{w_2} + m_w a_{w_2}$ and $m_{w_2} = (3m_{w_1} - a_{w_1})m_w - m_{w_1}(3m_w - a_w) = -a_{w_1} m_w + m_{w_1} a_w$. □

Following a particular case of Definition 4.10 in [A], we define the *index* of w as the number $I_w = \frac{a_w}{m_w}$. The following result is a particular case of Theorem 4.11 in Aigner's book [A].

Lemma 12.3.4 *For each proper lower Christoffel word w, one has* $I_a = 2 < I_w < \frac{5}{2} = I_b$.

Proof. By inspection, $I_a = 2$, $I_b = \frac{5}{2}$. Say $w = w_1w_2$, the standard factorization of w. Then by the previous lemma, $m_{w_1} = -a_w m_{w_2} + m_w a_{w_2}$ and $m_{w_2} = -a_{w_1} m_w + m_{w_1} a_w$. Dividing the first equation by $m_w m_{w_2}$ and the second by $m_w m_{w_1}$, we obtain $0 < \frac{a_{w_2}}{m_{w_2}} - \frac{a_w}{m_w}$, $0 < \frac{a_w}{m_w} - \frac{a_{w_1}}{m_{w_1}}$. Thus $I_{w_1} < I_w < I_{w_2}$. We may conclude, since by induction, $I_a \leq I_{w_1}$ and $I_{w_2} \leq I_b$. □

Proof of Theorem 12.3.1. We have $m = m_w$, $m_1 = m_{w_1}$, $m_2 = m_{w_2}$. Let $a = a_w$, $c = c_w$.

We have $m_2 = -a_{w_1} m + m_1 a$. Hence $m_1 a \equiv m_2 \mod m$. Now, by Lemma 12.3.4, we have $2 < \frac{a}{m} < \frac{5}{2} = 2 + \frac{1}{2}$. Thus $2m < a < 2m + \frac{1}{2}m$. We deduce that $a = 2m + u$ for some u with $0 < u < \frac{1}{2}m$. Since $a \equiv u \mod m$, we deduce $m_1 u \equiv m_2 \mod m$. Squaring, we obtain $m_1^2 u^2 \equiv m_2^2$. By Theorem 3.1.1, $\{m, m_1, m_2\}$ is a Markoff triple. By the Markoff equation $m^2 + m_1^2 + m_2^2 = 3mm_1m_2$, we see that $m_1^2 \equiv -m_2^2$ so that $m_1^2 u^2 \equiv -m_1^2$. By Corollary 3.2.1, any two of the three integers in a Markoff triple are relatively prime. Thus m_1 is invertible modulo m, and consequently $u^2 \equiv -1$. Therefore $u^2 + 1 = vm$ for some natural integer v. Finally, since $a = 2m + u$, $3m - a = m - u$. Take the determinant in (12.1): $1 = (2m + u)(m - u) - mc = 2m^2 - um - u^2 - mc$, hence $mc = 2m^2 - um - u^2 - 1 = 2m^2 - um - mv$, and therefore $c = 2m - u - v$.

The congruence $m_1 u \equiv m_2 \mod m$ determines u uniquely, since m_1 is invertible modulo m; and the equation $u^2 + 1 = vm$ determines v uniquely. □

In the next result, for any word h on $\mathbb{P}$, we denote by $p(h)$, the number equal to the continuant polynomial $p(n_1, \ldots, n_k)$ if $h = n_1 \cdots n_k$.

Theorem 12.3.5 *(Frobenius [F], (6) p. 612) Let* $w = ahb$ *be a proper lower Christoffel word, let* $q = \chi(h)$*, which is a palindrome, and let* m, u, *and* v *as in Theorem 12.3.1. Then* $m = p(2q2)$, $u = p(2q) = p(q2)$, $v = p(q)$.

For example, let $w = aabab$; then $h = aba$, $q = 112211$, $m = p(21122112)$, $u = p(2112211)$, and $v = p(112211)$.

Proof. We know that $m = \mu(w)_{12}$ by Theorem 3.1.1 and Lemma 3.1.2. Since $\mu(a) = P(1)^2$ and $\mu(b) = P(2)^2$ (see the proof of Theorem 8.1.2), we have by (5.1) and Theorem 12.3.1

$$\begin{pmatrix} 2m + u & m \\ 2m - u - v & m - u \end{pmatrix} = \begin{pmatrix} p(11q22) & p(11q2) \\ p(1q22) & p(1q2) \end{pmatrix}$$

We have $m = p(11q2) = p(2q2)$ by (5.3).

Moreover, $m - u = p(1q2)$, hence $p(11q2) = p(1q2) + u$; since by the dual recursion formula $p(11q2) = p(1q2) + p(q2)$, we infer that $u = p(q2) = p(2q)$, since q is a palindrome.

Finally, using the left and right recursion formulae, $2m - u - v = p(1q22)$ so that $v = 2m - u - p(1q22) = 2p(11q2) - p(q2) - p(1q22) = 2p(1q2) + 2p(q2) - p(q2) - 2p(1q2) - p(1q) = p(q11) - p(q1)$ (since $p(q2) = p(q11)$ and $p(1q) = p(q1)) = p(q)$. □

12.4 Markoff quadratic forms

Recall that the Markoff quadratic form f_w is defined by $f_w(x, y) = mx^2 + (s-p)xy - ry^2$, where $\mu(w) = \begin{pmatrix} p & m \\ r & s \end{pmatrix}$ and μ has been defined in Section 3.1. In particular, $3m = p + s$. By Corollary 12.3.2, one also has $f_w(x, y) = mx^2 - (m + 2u)xy - (2m - u - v)y^2$. The Markoff forms corresponding to the 13 lower Christoffel words in Section 12.1 are respectively

$$f_a = x^2 - xy - y^2, f_b = 2x^2 - 4xy - 2y^2, f_{ab} = 5x^2 - 9xy - 7y^2$$

$$f_{a^2b} = 13x^2 - 23xy - 19y^2, f_{ab^2} = 29x^2 - 53xy - 41y^2, f_{a^3b} = 34x^2 - 60xy - 50y^2$$

$$f_{a^4b} = 89x^2 - 157xy - 131y^2, f_{ab^3} = 169x^2 - 309xy - 239y^2$$

$$f_{a^2bab} = 194x^2 - 344xy - 284y^2, f_{a^5b} = 233x^2 - 411xy - 343y^2$$

$$f_{abab^2} = 433x^2 - 791xy - 613y^2, f_{a^6b} = 610x^2 - 1076xy - 898y^2$$

$$f_{ab^4} = 985x^2 - 1801xy - 1393y^2$$

In the previous list, some quadratic forms have coefficients that are not relatively prime. This is elucidated by the next result.

Proposition 12.4.1 *(variant of Frobenius [F], Section 6 p. 606) Let w be a lower Christoffel word and m the associated Markoff number. Then the coefficients of the quadratic form f_w (resp. $\frac{1}{2}f_w$) are relatively prime integers if m is odd (resp. even).*

It is shown in Exercise 12.4.1 that m is even if and only if $|w|_a \equiv 0 \mod 3$.

Proof. We have $s = 3m - p$ so that $s - p = 3m - 2p$ and $1 = \det(\mu(w)) = p(3m - p) - mr$. Thus $p^2 + 1 = m(3p - r)$. Hence p and m are relatively prime. If l is some prime number dividing $s - p$ and m, then l divides $2p$; since l cannot divide p (because p and m are relatively prime), $l = 2$. This can happen only if m is even. In particular, $s - p$ and m are relatively prime if m is odd.

Conversely, if $m = 2n$ is even, then $s - p$ is even and p is odd (because p and m are relatively prime). Modulo 4, we have $2n(3p - r) = p^2 + 1 \equiv 2$ so that n and $3p - r$ are odd, and r is even; moreover, the gcd of m and $s - p$ is 2 (since $4|m \Rightarrow 4|2n(3p - r)$, but $2n(3p - r) \equiv 2 \mod 4$), which ends the proof. □

Note that the examples of quadratic forms given in Markoff's article [M1], pp. 405–406, are not the same as those given here (but are of course equivalent forms).

They are obtained as follows: let $w = aub$ be a lower Christoffel word, distinct from a, b; define $M_w = P(2)\,\mu(au)P(2)$. Since $\mu(w) = \mu(au)\mu(b) = \mu(au)P(2)^2$, $\mu(w)$ and M_w are conjugate in $\mathrm{GL}_2(\mathbb{Z})$ by the matrix $P(2)$ of determinant -1, but $\mu(w)$ and $\mu(\tilde{w})$ are conjugate in $\mathrm{SL}_2(\mathbb{Z})$. It follows that M_w and $\mu(\tilde{w})$ are conjugate by a matrix of determinant -1 so that, by Proposition 6.1.2, $f_{\mu(\tilde{w})}$ and $-f_{M_w}$ are equivalent quadratic forms (see (6.3) for the definition of the latter notation). The matrix $\mu(\tilde{w})$ is the transpose of the matrix $\mu(w)$ (since the matrices $\mu(a)$ and $\mu(b)$ are symmetric); hence, by (6.3) and the definition of f_w, one has $f_w = f_{\mu(\tilde{w})}$. In conclusion, f_w and $-f_{M_w}$ are equivalent quadratic forms.

For example, let $w = ab$. Then $M_{ab} = P(2)P(1)^2P(2) = \begin{pmatrix} 13 & 5 \\ 5 & 2 \end{pmatrix}$ and the quadratic form is $f_{M_{ab}} = 5x^2 - 11xy - 5y^2$, which is a multiple of the quadratic form denoted f_2 by Markoff [M1], p. 405.

As another example, let $w = aabab$; then $\tilde{w} = babaa$ and $M_{aabab} = P(2)\mu(aaba)$ $P(2) = \begin{pmatrix} 507 & 196 \\ 194 & 75 \end{pmatrix}$; the quadratic form is $f_{M_{aabab}} = 194x^2 - 432xy - 196$, which is a multiple of the form denoted f_8 in [M1], p. 406.

Other variants of the original Markoff forms, equivalent to them, are the following ones. Taking the notations of Theorem 12.3.1, the form is $mx^2 + (3m - 2u)$ $xy + (v - 3u)y^2$ ([F], p. 598, [D1], Theorem 80 p. 106, [Ca2], (3) p. 31, [CF], Theorem 6 p. 10); it is indeed equivalent to f_w, since it is equal to $f_w(x+2y, y) = m(x+2y)^2 - (m+2u)(x+2y)y - (2m-u-v)y^2 = mx^2 + (4m-m-2u)xy + (4m-2m-4u-2m+u+v)y^2$.

One may find in [Re8], Theorem 17.1 and Corollary 17.1, the determination of the set of integral points where the Markoff quadratic form f_w attains "small values" and, in particular, its mimimum.

12.4.1 Exercises

Exercise 12.4.1 *Show that for a lower Christoffel word w, the Markoff number* $(\mu(w))_{12}$ *is even if and ony if* $|w|_a \equiv 0 \mod 3$. *Hint: use Theorem 3.1.1 and consider the matrices modulo 2.*

Exercise 12.4.2 *Two real binary quadratic forms* $f(x,y)$ *and* $g(x,y)$ *are called* $\mathrm{SL}_2(\mathbb{Z})$-equivalent *if for some matrix M in* $\mathrm{SL}_2(\mathbb{Z})$ *one has* $g = f \cdot M$. *By opposition, the equivalence considered at the beginning of Section 6.1 is called* $\mathrm{GL}_2(\mathbb{Z})$-equivalence. *In the literature, one also says* proper *(resp.* improper*) equivalence, for* $\mathrm{SL}_2(\mathbb{Z})$*-equivalence (resp.* $\mathrm{GL}_2(\mathbb{Z})$*-equivalence). Show that* $f(-y,x)$ *is* SL_2*-equivalent to* $f(x,y)$. *Show that* $f(y,x)$ *and* $f(-x,y)$ *are both* GL_2*-equivalent to* $f(x,y)$. *Show that each* GL_2*-class is either a union of two disjoint* SL_2*-classes or equal to a single* SL_2*-class; show that the latter case is equivalent to the fact that* $f(y,x)$ *(equivalently* $f(-x,y)$*) is* $\mathrm{SL}_2(\mathbb{Z})$*-equivalent to f.*

Exercise 12.4.3 *Show that the Markoff form* f_w *is* $\mathrm{SL}_2(\mathbb{Z})$*-equivalent to* $-f_w$ *(Frobenius [F], p. 458; see also [CF], Theorem 2 (B), p. 20). Hint: w and* $\tilde{w}$ *are conjugate; hence* $A = \mu(w)$ *and* $B = \mu(\widetilde{w})$ *are conjugate by a matrix of determinant 1;*

by Proposition 6.1.2, f_A and f_B are $SL_2(\mathbb{Z})$*-equivalent; A* and *B are also transpose of each other; hence with the notations of (6.3),* $f_A = cx^2 + (d - a)xy - by^2$, *then* $f_B = bx^2 + (d - a)xy - cy^2$. *Hence* $-f_B(x,y) = f_A(-y,x)$ *which is* $SL_2(\mathbb{Z})$*-equivalent to* f_A*; finally, by definition,* $f_w = f_B$.

Exercise 12.4.4 *Show that the Markoff form* $f_{ab}(x,y) = 5x^2-9xy-7y^2$ *is* SL_2*-equivalent neither to* $f_{ab}(y,x)$ *nor to* $f_{ab}(-x,y)$*. Hence, its* GL_2*-class is the union of two disjoint* SL_2*-classes.*

Exercise 12.4.5 *For the notion of* cycles of forms, *see the solution of Exercise 12.4.4. Let w be a lower Christoffel word of length n; show that the cycle containing* $f_w(x,y)$ *is labelled by the word* $u \in \{l,r\}^*$ *determined by* $u = r^{i_1} l^{i_2} \cdots r^{i_{2n-1}} l^{i_{2n}}$, *with* $\chi(\tilde{w}) = i_1 i_2 \cdots i_{2n-1} i_{2n}$. *Hint:* $f_w = f_{\mu(\tilde{w})}$*; one has* $\mu(\tilde{w}) = R^{i_1} L^{i_2} \cdots R^{i_{2n-1}} L^{i_{2n}}$*, a consequence of* $\mu(a) = RL$ *and* $\mu(b) = R^2L^2$*. Apply Proposition 6.1.2.*

13
Historical notes

Andrei Andreyevich Markoff was born in 1856. In 1879 and 1880, he published two articles in the German journal *Mathematische Annalen*; they were written in French, and Markoff wrote his own name with two final *f*'s in the French way. He is also known, and certainly better known, for the theory of Markov processes, with the English way of writing his name: with a final *v*.

His two articles [M1, M2] are the content of his master thesis in St Petersburg, written before his doctoral thesis. The main results of these articles are formulated in terms of minima of quadratic forms and Markoff numbers, the main theorems being the results of Chapter 11 in the present book, together with the characterization of Markoff numbers by the Markoff equation (Chapter 3).

After Markoff's article, there has been a lot of work by many mathematicians on the theory of Markoff. This theory was formulated either in terms of quadratic forms as Markoff had done (Theorem 11.3.1 and Corollary 11.3.2 in this book) or in terms of approximations of real numbers (Theorem 10.1.1 and Corollary 10.1.2 in this book).

Markoff's theorem on approximation of real numbers, as formulated in Chapter 10, does not appear in Markoff's work. The first result in the spirit of Corollary 10.1.2 seems to appear in Hurwitz's article [H1] (p. 284), given without proof: Hurwitz claims that it may be deduced easily from the arguments of Markoff (*'deren Beweise man leicht den Markoff'schen Betrachtungen entnimmt'*). The article begins by the famous theorem of Hurwitz (Corollary 10.1.3 in this book).

The link between Markoff's work and that of Christoffel was noted by Frobenius [F]. He notes that the expression for Markoff numbers in terms of the Euler continuant polynomials may be explained by the use of the words introduced by Elwin Bruno Christoffel in [C1, C2]; see [F], pp. 622–224. This link was apparently forgotten until recently but was also known to Caroline Series [Se2], p. 21.

It seems likely that Markoff did not know the 1875 article [C1] of Christoffel (written in Latin). The latter introduces the words which, following Berstel [B2], we call Christoffel words. These words were considered soon after, and independently, by Smith [S]. The link between Christoffel words and continued fractions, as is given in the present book in Section 19.1, was developed by Christoffel in [C2]. Note that there is a double link between Christoffel words and continued fractions: one is given by Christoffel [C2] and studied many times since then (Christoffel words encode the continued fraction expansions of all rational numbers), and the other one is deeper, given by Markoff's theory (Christoffel words encode the continued fraction expansions of certain quadratic numbers having certain extremal properties).

Markoff's theorem on quadratic forms (as given here in Chapter 8) has been revisited many times and sometimes reproved: Frobenius 1913 [F]; Bachmann 1923 [Ba]; Remak 1924 [R]; Dickson 1930 [D1], Chapter VII (he gives a complete proof using

From Christoffel Words to Markoff Numbers. Second Edition. Christophe Reutenauer, Oxford University Press.
 DOI: 10.1093/9780197907535.003.0013

the Gauss theory of chains of quadratic forms, which is somewhat similar to Markoff's proof); Cassels 1949 and 1957 [Ca1, Ca2];[1] and Cusick and Flahive 1989 [CF].

Markoff's theorem on continued fractions (Chapter 7 in the present book) is considered in Heawood 1922 [He], Perron 1921 [P1], Perron 1921 [P2] (between both articles, Perron seems to have learnt the existence of Markoff's work), Cassels 1949 and 1957 [Ca1, Ca2], Cusick and Flahive 1989 [CF], Bombieri 2007 [B], and Aigner 2013 [A].

[1] He gives quadratic-form arguments quite different from Markoff's.

PART II
THE THEORY OF CHRISTOFFEL WORDS

14
Palindromes and periods

The geometrical definition of Christoffel words shows that such a word is of the form amb for some palindrome m. These palindromes are special palindromes on a two-letter alphabet, called central words, which have many characterizations. In particular, they may be computed by iterated palindromization, as follows from a theorem of Aldo de Luca. This combinatorial operation is obtained by morphisms, using Justin's formula. They are also characterized by the existence of two relatively primes periods, p and q, and length equal to $p+q-2$ (Mignosi and de Luca). Christoffel words are a product of two palindromes (Chuan). A word m is central if and only if it is a palindrome and amb is a product of two palindromes (Mignosi and de Luca). The minimal automaton of the set of suffixes of a central word is constructed (Sciortino and Zamboni). The tree of Christoffel words is a specialization of the tree of Christoffel pairs. There is a natural bijection from the set of Christoffel words onto the monoid $SL_2(\mathbb{N})$.

14.1 Central words

A word m on the alphabet $\{a<b\}$ is called *central* if amb is a lower Christoffel word. Equivalently, bma is an upper Christoffel word. We know that a central word is a palindrome (Theorem 2.3.1). In other words, a central word is a cutting word as defined in Section 2.3.

For later use, note that a^n and b^n are central words, since $a^{n+1}b, ab^{n+1}$ are Christoffel words. Note also that if m is central, so is $E(m)$; indeed, w is a Christoffel word if and only if so is $E(w)$.

We shall characterize central words in several ways. Recall that $\tilde{p}$ denotes the reversal of the word p.

Proposition 14.1.1 *(de Luca [dL1], Lemma 5 pp. 59–60) Given a word w in a free monoid, let $w = ps$ where s is its longest palindromic suffix. Then $w\tilde{p}=ps\tilde{p}$ is the unique shortest palindrome having w as a prefix.*

The word $ps\tilde{p}$ is called the *right palindromic closure* of w, denoted by $w^{(+)}$. For example, $(abbab)^{(+)} = ab\,bab\,ba$, with $p = ab$, $s = bab$.

Proof. Clearly $ps\tilde{p}$ is a palindrome having w as a prefix. Let wq be a palindrome having w as a prefix and of minimum length. Then $|q|\leq|p|$. Since $psq=wq$ is a palindrome, we have $psq=\widetilde{psq}=\tilde{q}s\tilde{p}$; since $|\tilde{q}|\leq|p|$, $\tilde{q}$ is a prefix of p: $p=\tilde{q}u$. Since $wq=psq=\tilde{q}usq$ is a palindrome, so is us, and $w=ps=\tilde{q}us$ has us as a suffix.

From Christoffel Words to Markoff Numbers. Second Edition. Christophe Reutenauer, Oxford University Press.
 DOI: 10.1093/9780197907535.003.0014

By maximality of s, $u = 1$, so that $q = \tilde{p}$ and $wq = w\tilde{p}$. Thus $w\tilde{p}$ is the unique shortest palindrome having w as a prefix. □

Again, following de Luca [dL1], Proposition 8, we define the mapping Pal from a free monoid into itself, called the *iterated palindromization*, as follows: $\mathrm{Pal}(1) = 1$ and for each word u and each letter x, $\mathrm{Pal}(ux) = (\mathrm{Pal}(u)x)^{(+)}$.

Theorem 14.1.2 *(de Luca [dL1]) The function* Pal *is a bijection from* $\{a, b\}^*$ *onto the set of central words on the alphabet* $\{a, b\}$.

If m is a central word and $m = \mathrm{Pal}(u)$, then u is called the *directive word* of m. It exists and is unique by the theorem.

Theorem 14.1.3 (Justin's formula *[J], Lemma 2.1) Denote by G the endomorphism of a free monoid sending some letter a onto a and any other letter b onto ab. Then, for any word v, one has* $\mathrm{Pal}(av) = G(\mathrm{Pal}(v))a$.

Lemma 14.1.4

(i) *A non-empty word is in the image of G if and only if it begins with a and has no factor of the form bc for any letters* $b, c \neq a$.
(ii) *For any words x* and *y, y is a suffix of x if and only if G(y) is a suffix of G(x).*
(iii) $G(ua)^{(+)} = G(u^{(+)}a)$ *and* $G(ub)^{(+)} = G((ub)^{(+)}a)$.

Proof.

(i) The image of G is the submonoid generated by a and the words ab for b in the set of letters distinct from a. Thus, there can be no factor bc ($b, c \neq a$) in any word in the image. Conversely, let w be empty or satisfy the condition in (i). If w is empty or of length 1 (so equal to a), it is in the image of G. Otherwise, it begins with a, which is followed by some letter. If the next letter is a, then $w = au$ and u satisfies the condition, so that u is in the image by induction, and w is too. If the next letter is $b \neq a$, then either $w = ab$ is in the image, or $w = abu$ and u must begin with a (otherwise, w has a forbidden factor bc). Moreover, u has no factor of the form bc; by induction, u is in the image and therefore w is too.
(ii) Note that if x and y are distinct letters, then $G(y)$ is not a suffix of $G(x)$. Thus, (ii) follows by induction on the lengths of the words x and y.
(iii) It follows from Lemma 4.1.4[1] that for any word v, $G(va)$ is a palindrome if and only if v is a palindrome. Hence, if v is a palindromic suffix of u, then $G(va)$ is a palindromic suffix of $G(ua)$, and it is non-empty. Conversely, if w is a non-empty palindromic suffix of $G(ua)$, w ends by a so that it begins with a and therefore, by (i), w is in the image of G; moreover, w ends by a, hence $w = G(va)$. Since $G(va)$ is a suffix of $G(ua)$, va is a suffix of ua by (ii),

[1] This lemma was stated and proved for a two-letter alphabet, but its extension to any alphabet has the same proof.

and v is a suffix of u; next, v is a palindrome, since $G(va)$ is. We thus see that the longest palindromic suffix of $G(ua)$ is $G(sa)$, where s is the longest palindromic suffix of u. Write $u = ps$. Then by Proposition 14.1.1, $u^{(+)} = ps\tilde{p}$ and $G(ua)^{(+)} = (G(p)G(sa))^{(+)} = G(p)G(sa)\widetilde{G(p)} = G(p)G(s)a\widetilde{G(p)} = G(p)G(s)\widetilde{G(p)}$ $a = G(p)G(s)G(\tilde{p})a$ (by (4.1)) $= G(ps\tilde{p})a = G(u^{(+)})a = G(u^{(+)}a)$. This proves the first formula.

In order to prove the second formula, note that $G(ub)^{(+)} = (G(u)ab)^{(+)}$. We claim (proof below) that this latter word is equal to $(G(u)aba)^{(+)}$. This is equal to $G(uba)^{(+)}$. By replacing u by ub in the first formula, we see that $G(uba)^{(+)} = G((ub)^{(+)}a)$. The second formula follows.

The claim follows from two observations:

(a) With the notations of Proposition 14.1.1, if x is the last letter of p, then $w^{(+)} = (wx)^{(+)}$; this follows since $w^{(+)} = ps\tilde{p} = psx\ldots$ is necessarily the shortest palindrome having psx as prefix;

(b) Let s_1 be the longest palindromic suffix of $G(u)ab$, and write $G(u)ab = p_1s_1$; then s_1 ends with b, so begins with b, and p_1 is non-empty with a as the last letter. Indeed, this is clear if $s_1 = b$, and if s_1 is longer than b, then it begins with a b lying in $G(u)$, which by (i) must be preceded by an a.

□

Proof of Theorem 14.1.3. The formula is true when v is the empty word. Suppose now that $v = wx$ for some letter x. Then $\text{Pal}(av) = \text{Pal}(awx) = (\text{Pal}(aw)x)^{(+)}$, which is equal by induction to $(G(\text{Pal}(w))ax)^{(+)}$. If $x = a$, then $\text{Pal}(av) = (G(\text{Pal}(w))aa)^{(+)} = (G(\text{Pal}(w)a)a)^{(+)}$, which by the previous lemma is equal to $G((\text{Pal}(w)a)^{(+)}a) = G(\text{Pal}(wa)a) = G(\text{Pal}(v)a)$, and the induction step is complete in this case. If $x = b \neq a$, then $\text{Pal}(av) = (G(\text{Pal}(w))ab)^{(+)} = G(\text{Pal}(w)b)^{(+)}$; by the lemma, this is $G((\text{Pal}(w)b)^{(+)}a) = G(\text{Pal}(wb)a) = G(\text{Pal}(v)a)$, which ends the induction step in this case and completes the proof. □

Lemma 14.1.5 *Let $\rho : \{a, b\}^* \to \text{End}(\{a, b\}^*)$ be the monoid homomorphism sending a onto G and b onto $\tilde{D}$. Then, for any word w in $\{a, b\}^*$, one has $\rho(w)(ab) = a\text{Pal}(w)b$; in particular, assuming that $a < b$, $a\text{Pal}(w)b$ is a lower Christoffel word and $\text{Pal}(w)$ is a central word.*

Proof. The proof of the formula is by induction on the length of w. If $w = 1$, then $\rho(1)(ab) = ab = a\text{Pal}(1)b$. Suppose that $w = av$. Then $a\text{Pal}(w)b = a\text{Pal}(av)b = aG(\text{Pal}(v))ab$ (by Theorem 14.1.3) $= G(a\text{Pal}(v)b) = G(\rho(v)(ab))$ (by induction) $= G \circ \rho(v)(ab) = \rho(a) \circ \rho(v)(ab) = \rho(av)(ab) = \rho(w)(ab)$. Suppose now that $w = bv$. Then $a\text{Pal}(w)b = a\text{Pal}(bv)b = aD(\text{Pal}(v))bb$ (by Justin's formula and the exchange of a and b, noting that $D = EGE$) $= ab\tilde{D}(\text{Pal}(v))b$ (since $D(u) = b\tilde{D}(u)b^{-1}$, Exercise 2.6.1) $= \tilde{D}(a\text{Pal}(v)b) = \tilde{D}(\rho(v)(ab))$ (by induction) $= \tilde{D} \circ \rho(v)(ab) = \rho(b) \circ \rho(v)(ab) = \rho(bv)(ab) = \rho(w)(ab)$.

The final assertion follows from Corollary 2.6.2 and the definition of central words. □

Proof of Theorem 14.1.2. By the previous lemma, $\mathrm{Pal}(w)$ is a central word for any word w. The function Pal is surjective. Indeed, if m is a central word, then amb is a lower Christoffel word, and by Theorem 2.6.1, we may find a word w such that $\rho(w)(ab) = amb$; then, by the previous lemma, $m = \mathrm{Pal}(w)$.

The function Pal is injective. Indeed, if $\mathrm{Pal}(w) = m$, then, by the previous lemma $amb = \rho(w)(ab)$ is a lower Christoffel word. It has a unique standard factorization $amb = uv$, and the node (u, v) in the tree of Christoffel pairs satisfies $\mathcal{N}(w) = (u, v)$; indeed, let $\mathcal{N}(w) = (u', v')$. Then by Theorem 2.6.1, $\rho(w)(ab) = u'v'$, hence $uv = u'v'$, and by uniqueness of the standard factorization, $u = u'$ and $v = v'$. By Section 2.5, w depends only on (u, v); hence Pal is injective. □

Another, more algorithmic way of proving injectivity of Pal is a consequence of the following proposition.

Proposition 14.1.6. *If* $m = \mathrm{Pal}(w)$, $w = a_1 \cdots a_n$, *then the set of palindromic prefixes of* m *is* $\{\mathrm{Pal}(a_1 \cdots a_i), i = 0, \ldots, n\}$, *and for any* $i = 0, \ldots, n-1$, $\mathrm{Pal}(a_1 \cdots a_i)a_{i+1}$ *is a prefix of* m *whose right palindromic closure is* $\mathrm{Pal}(a_1 \cdots a_{i+1})$.

Proof. This follows from the definitions. □

This implies that in order to recover w, one looks for the proper palindromic prefixes of m and the letter immediately following them: w is obtained by concatenating these letters. For example, let $w = abab$; then $\mathrm{Pal}(a) = a^{(+)} = a$, $\mathrm{Pal}(ab) = ab^{(+)} = aba$, $\mathrm{Pal}(aba) = abaa^{(+)} = abaaba$, $\mathrm{Pal}(abab) = abaabab^{(+)} = abaababaaba = m$, and $m = \mathbf{ab}a\mathbf{a}ab\mathbf{b}aaba$: w is the word of bold letters, and the proper palindromic prefixes are the prefixes ending before a bold letter.

The *tree of central words* is the tree obtained from the tree of Christoffel pairs by replacing each node (u, v) with the word m, where $uv = amb$. Equivalently, the node $\mathcal{N}(w)$ coded by the word $w \in \{a, b\}^*$ (with a meaning "left" and b meaning "right") is the central word $\mathrm{Pal}(w)$; see Figure 14.1. For example, the node coded by abb in Figure 2.10 is (a^2bab, ab), the standard factorization of the lower Christoffel word $a^2babab = a(ababa)b$, and the corresponding node in Figure 14.1 is $ababa$; one indeed has $a^{(+)} = a$, $(ab)^{(+)} = aba$ and $(abab)^{(+)} = ababa$, so that $\mathrm{Pal}(abb) = ababa$.

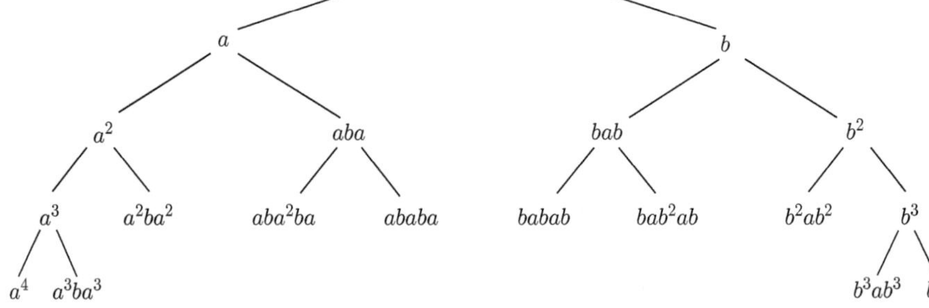

Figure 14.1 The tree of central words

Corollary 14.1.7 *The palindromic prefixes or suffixes of a central word m are central and are the nodes on the path from the root to m in the tree of central words.*

Corollary 14.1.8 *The mapping* $u \mapsto a\mathrm{Pal}(u)b$ *is a bijection from* $\{a < b\}^*$ *onto the set of lower Christoffel words.*

14.1.1 Exercises

Exercise 14.1.1 *Show that the number of central words of length n is* $\varphi(n+2)$, *where* φ *is the Euler function.*

Exercise 14.1.2 *Show that* $\mathrm{Pal}(a^n) = a^n$, $\mathrm{Pal}(a^n b) = a^n b a^n$, *and* $\mathrm{Pal}(ba^n) = b(ab)^n$.

Exercise 14.1.3 *Show that if u is a central word, then so are* $G(u)a$ *and* $b\tilde{D}(u)$ *(use Corollary 2.6.2).*

Exercise 14.1.4 *Let w be a word and a a letter. Show that* $|w^{(+)}| \leq 2|w| - 1$. *Deduce that* $|\mathrm{Pal}(wa)| \leq 2|\mathrm{Pal}(w)| + 1$.

Exercise 14.1.5 *Let A be an alphabet and a some letter in A. Let* L_a *(resp.* R_a*) be the endomorphism of* A^* *sending a onto a and any other letter b onto ab (resp. ba). Define* $R_u = R_{a_1} \circ \cdots \circ R_{a_n}$ *if* $u = a_1 \cdots a_n$, $a_i \in A$, *and* $L_u = L_{a_1} \circ \cdots \circ L_{a_n}$. *Recall that* $\mathcal{R}$ *denotes the reversal mapping, that is, the anti-automorphism which is the identity on the alphabet. Show that* $R_u = \mathcal{R} \circ L_u \circ \mathcal{R}$ *for any word u, and that* $\mathcal{R} \circ \mathrm{Pal} = \mathrm{Pal}$.

Using Theorem 14.1.3, prove that

$$\mathrm{Pal}(uv) = L_u(\mathrm{Pal}(v))\mathrm{Pal}(u)$$

for any words u, v (Justin's formula *[J], Lemma 2.1). Using the two relations above, show that*

$$\mathrm{Pal}(uv) = \mathrm{Pal}(u)R_u(\mathrm{Pal}(v))$$

Show that for $w \in \{a, b\}^*$, $ba\,\mathrm{Pal}(w) = R_w(ba)$ *and* $\mathrm{Pal}(w)ab = L_w(ab)$.

Exercise 14.1.6 *Let w be a word on any alphabet and a a letter. Show that if a does not appear in w, then* $\mathrm{Pal}(wa) = \mathrm{Pal}(w)a\,\mathrm{Pal}(w)$.

Suppose now that a appears in w and let $w = uav$, *where a does not appear in v. Show that the longest palindromic prefix p of* $\mathrm{Pal}(w)$ *which is followed by a (that is, such that pa is also a prefix of* $\mathrm{Pal}(w)$*) is* $p = \mathrm{Pal}(u)$ *(use the algothmic point of view after the proof of Theorem 14.1.2). Let* $\mathrm{Pal}(w)a = xy$, *where y is the longest palindromic suffix of* $\mathrm{Pal}(w)a$, *so that* $\mathrm{Pal}(wa) = xy\tilde{x}$; *show that* $y = ay'a$ *and that* y' *is the longest palindromic suffix of* $\mathrm{Pal}(w)$ *which is preceded by an a; show that*

y' is the longest palindromic prefix of Pal(w) *which is followed by an a; deduce that* $y' = p$. *Deduce that* Pal$(w) = xap$, *and for another formula of Justin, see [J] (1):*

$$\mathrm{Pal}(wa) = \mathrm{Pal}(w)\mathrm{Pal}(u)^{-1}\mathrm{Pal}(w),$$

where one computes in the free group on $F(A)$ on A.

Exercise 14.1.7 *Show that the tree of central words may be constructed directly as follows: the nodes on the left and right extreme branches are the powers of a and b; any other node is equal to ubav, where u is the node after the first north-west step, and v is the node after the first north-east step, on the path from the current node to the root.*

Exercise 14.1.8 *Define a monoid homomorphism f from $\{a, b\}^*$ into the symmetric group S_3: $f(a) = (1, 2)$ and $f(b) = (2, 3)$. Show that the central word* Pal(w) *is of even length if and only if $f(w)$ fixes 2. Hint: define a right action of the free monoid $\{a, b\}^*$ onto the set $\{(1, 1), (1, 0), (0, 1)\}$ by:*

$$(1,1) \cdot a = (1,0), (1,0) \cdot a = (1,1), (0,1) \cdot a = (0,1)$$
$$(1,1) \cdot b = (0,1), (1,0) \cdot b = (1,0), (0,1) \cdot b = (1,1)$$

Show that the Christoffel pair $(u, v) = \mathcal{N}(w)$ (with the notations of Section 2.6) satisfies $(|u|, |v|) = (i, j)$ (where the equality has to be taken modulo 2) if and only if $(1, 1) \cdot w = (i, j)$ (use the rule for constructing the tree of Christoffel pairs described in Figure 2.9); deduce that the Christoffel word $uv = a\,\mathrm{Pal}(w)b$ is of even length if and only if $(1, 1) \cdot w = (1, 1)$. Identify $(1, 1), (1, 0), (0, 1)$ with 2, 1, 3, respectively ([Re5], Corollary 15.3).

Exercise 14.1.9 *For any alphabet A, define, for $a \in A$, a function γ_a on A^* by $\gamma_a(a^{i_0}b_1a^{i_1}b_2 \cdots a^{i_{\ell-1}}b_\ell a^{i_\ell}) = a^{i_0+1}b_1a^{i_1+1}b_2 \cdots a^{i_{\ell-1}+1}b_\ell a^{i_\ell+1}$, where the b_i are letters distinct from a, and $\ell \geq 0$. Using Justin's formula, show that* Pal$(aw) = \gamma_a($Pal$(w))$. *Noting that γ_a is injective, let θ_a be the inverse of γ_a on A^*; it is a partial function. Show that a word u is central if and only if the following algorithm stops at $u = 1$ and that it outputs the word w such that $u =$* Pal(w):

- *let $w = 1$;*
- *if u begins with the letter a, replace u by $\theta_a(u)$ and replace w by wa. Iterate this step until $u = 1$.*

For $A = \{a, b\}$, show that, viewing each word on A as a path, as in Section 2.2, the previous formula has a geometric interpretation: Pal(aw) *is obtained from* Pal(w) *by increasing the length of each horizontal step by 1 (including empty steps), and similarly for* Pal(bw)

In his article ([Bf], Section 2.2), Ian Banfield associates with each lower Christoffel word a Laurent polynomial Δ_w with coefficients in $\mathbb{N}$; he shows that this polynomial is log-concave; *that is, the sequence of its coefficients is log-concave. Then he shows* *that the Alexander polynomial of each* two-bridge knot *is of the form* Δ_w *for some lower Christoffel word* w *and is therefore log-concave. A log-concave sequence is a sequence* (a_i) *of non-negative real numbers such that* $a_{i-1}a_{i+1} \le a_i^2$ *for any* i*; such a sequence is necessarily* unimodal, *that is,* $a_0 \le a_1 \le \ldots \le a_k \ge a_{k+1} \ge \ldots \ge a_n$ *for some* k.

In [D'AdLDL], D'Aniello, de Luca, and De Luca define, using the Sturmian morphisms G and $\tilde{D}$, the *derivative* of a Christoffel word w, denoted by $\partial(w)$, which is itself a Christoffel word. The derivative is related to the directive word and to the expansion into continued fractions of the slope. The smallest n such that $\partial^n(w)$ is a letter is the *depth* of w; it is equal to the *height* of the directive word. See also [Se2], p. 21.

14.2 Palindromic factorization

Christoffel words are uniquely products of two palindromes, as stated in the next result.

Theorem 14.2.1 *(Chuan [Ch1], Theorem 3.1; see also Borel and the author [BR], Proposition 6.1) Each proper lower Christoffel word* $w = uv$ *with its standard factorization has a unique factorization* $w = v'u'$*, where* u' *and* v' *are palindromes. The associated upper Christoffel word is then* $\tilde{w} = u'v'$*. Moreover, one has* $|u| = |u'|, |v| = |v'|$*, and* u' *(resp.* v'*) begins and ends with* b *(resp. with* a*). Moreover, if* u *(resp.* v*) is proper, then* $u = amb$ *and* $u' = bmb$ *(resp.* $v = anb$ *and* $v' = ana$*).*

The factorization $w = v'u'$ is called the *palindromic factorization* of w. As an example, consider $w = (aab)(aabaabab) = uv$, its standard factorization. Then its palindromic factorization is $w = (aabaabaa)(bab)$, obtained by exchanging the two factors u and v and replacing the first letter of u and the last letter of v. The palindromic factorization has a geometric interpretation: one cuts the path determined by the Christoffel word at the point that is the furthest from the diagonal. See, for example, Figure 2.3, where this point is $A + (6, 2)$.

Lemma 14.2.2 *(de Luca and Mignosi [dLM], Lemma 1) If a radical word is the product of two palindromes, then it is uniquely so.*

Proof. Write the letters of the word w at the vertices of a regular polygon with $|w|$ vertices. Suppose that this figure has two distinct axial symmetries. Then their composition is a non-trivial rotation preserving the figure. Thus, the word is a power of a shorter word. This is not possible if w is a radical word. Thus, there can exist at most one symmetry preserving the figure.

The existence of a factorization into two palindromes implies the existence of such a symmetry: its axis is obtained by joining, in the figure, the centres of the two palindromes (the centre of a palindrome is either a letter or the space between two letters);

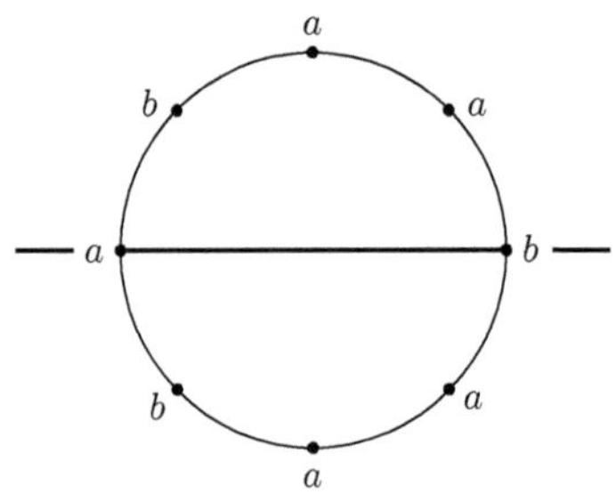

Figure 14.2 Palindromic factorization $(aabaa)(bab)$ and its axial symmetry

see Figure 14.2, where the word with its palindromic factorization is $(aabaa)(bab)$. Thus, we see that there exists only one such factorization. □

A combinatorial proof of the previous lemma is given in the proof of Lemma 1 of [dLM].

Proof of Theorem 14.2.1. Suppose that u and v are proper Christoffel words. Then $w = apb$, $u = amb$, $v = anb$, where p, m, and n are palindromes. Then $w = ambanb$, $p = mban$, $p = \tilde{p} = nabm$, and so $w = anabmb = v'u'$, with $v' = ana$, $u' = bmb$. One has, since m and n are palindromes, $\tilde{w} = bmbana = u'v'$.

If u is not proper, then $u = a$; hence $w = a^n b$ (see Figure 2.10), $n \geq 1$, $v = a^{n-1}b$, $v' = a^n$, $u' = b$. One has $\tilde{w} = ba^n = u'v'$.

The case where v is not proper is similar.

For uniqueness, note that a Christoffel word is radical; see Section 2.2; then apply the previous lemma. □

14.2.1 Exercises

Exercise 14.2.1 *Let $w = uv$ be a Christoffel word with its standard factorization. Show that the longest common prefix of v and w (resp. suffix of u and w) is of length $|v| - 1$ (resp. length $|u| - 1$). Hint: use Theorem 14.2.1; this gives another solution for Exercise 2.5.3.*

Exercise 14.2.2

(i) *Show that a word w is a product of two palindromes if and only if w* and $\tilde{w}$ *are conjugate.*
Consider now such a word.

(ii) *Show that Cw is a product of palindromes, where C is the conjugator defined in Chapter 1.*

(iii) *Show that w either has a conjugate that is a palindrome or has a conjugate of the form $aub\tilde{u}$ for some letters a* and b *and some word u (write w regularly on a circle, and show that this figure has an axial symmetry; see the proof* *of Lemma 14.2.2; there are two cases: the axis of symmetry passes through two letters or passes through at most one letter).*

Exercise 14.2.3 *Let w be a lower Christoffel word with palindromic factorization $w = v'u'$. Recalling that G and $\tilde{D}$ preserve lower Christoffel words (Corollary 2.6.2), show that the palindromic factorization of $G(w)$ (resp. $\tilde{D}(w)$) is $(G(v')a)(a^{-1}G(u'))$ (resp. $(\tilde{D}(v')b^{-1})(b\tilde{D}(u'))$). Use Lemma 4.1.4.*

Exercise 14.2.4 *Define the* tree of palindromic factorizations *by replacing in the tree of Christoffel pairs each pair (u, v) by the pair (v', u'), where $v'u'$ is the palindromic factorization of the lower Christoffel word uv (in particular, $uv = v'u'$). Show that this tree may be constructed directly by the following rules:*

- *the root is (a, b);*
- *the left (resp. right) child of a pair (v', u') is the pair (ama, u') (resp. (v', bmb)), where $v'u' = amb$.*

Show that if (v', u') is an ancestor of (v'', u'') in this tree, then v' is a prefix of v'' and u' is a suffix of u''.

14.3 Periods

We say that a (finite) word $w = a_1 \cdots a_n$, where the a_i's are letters, has the *period* $p \in \mathbb{N}$ if whenever $i, i+p \in \{1, \ldots, n\}$, one has $a_i = a_{i+p}$. Clearly, w has the period $p > 0$ if and only if it is a prefix of the infinite word u^∞ for some word u of length p. The period is *non-trivial* if $0 < p < n$, and we say that w is *periodic* if it has a non-trivial period p. If $p > 0$ is a period of w, each factor of the form $a_i a_{i+1} \cdots a_{i+p-1}$ is called a *periodic pattern* of w ($i = 1, \ldots, n-p+1$).

A *border* of a word w is a non-trivial proper prefix of w which is also a suffix of w. If w has no border, we say that w is *unbordered*.

Lemma 14.3.1

(i) *If w is a prefix of uw, then w has the period $|u|$.*
(ii) *The word w has a non-trivial period p if and only if it has a border of length $|w| - p$.*

Proof.

(i) Since 0 is certainly a period, we may assume that u is non-empty. Then w is a prefix of uw, hence uw is a prefix of uuw, and so on; this implies that w is a prefix of u^∞, and therefore w has the period $|u|$.
(ii) Assume that w has a border m of length $|w| - p$. Then $0 < p < |w|$, $w = um = mv$, and u is of length p. Then $wv = umv = uw$, hence w is a prefix of uw. By (i), w has the non-trivial period p.

Conversely, assume that w has the non-trivial period p. Then, writing $w = a_1 \cdots a_n$, for some letters a_i, we have $a_1 \cdots a_{n-p} = a_{p+1} \cdots a_n$, which is therefore a border of w. □

Corollary 14.3.2 *A word is unbordered if and only if it is not periodic.*

Theorem 14.3.3 *(de Luca and Mignosi [dLM]) A word in $\{a, b\}^*$ is central if and only if, for some relatively prime natural numbers p and q, it has the two periods p and q and is of length $p + q - 2$.*

Lemma 14.3.4 *(de Luca [dL1], Lemma 3) If a palindrome has a palindromic suffix, then it has as period the length of the corresponding prefix.*

Proof. Let $w = us$, with w, s palindromes. We may assume that s is a proper non-trivial suffix. We have $w = \tilde{w} = \tilde{s}\tilde{u} = s\tilde{u}$. Thus s is a border of w, and by Lemma 14.3.1 (ii), w has the period $|u|$. □

Lemma 14.3.5 *Let $w = aw' \in \{a, b\}^*$ be a proper Christoffel word with standard factorization $w = uv$. Let p be the prefix of length $|v| - 1$ of w. Then $w'p$ is a palindrome and has v as periodic pattern and thus has the period $|v|$, which is non-trivial if $|w| \geq 3$.*

Proof. Suppose that u and v are both proper Christoffel words. Then, by Theorem 2.3.1, $u = amb$, $v = anb$, $w = ambanb$, where m, n, and $mban$ are palindromes. Moreover, w has, by Theorem 14.2.1, the factorization into two palindromes $w = (ana)(bmb)$. Thus $p = an$, $w' = nabmb$, and $w'p = nabmban$ is a palindrome. Now, $w'p$ has $mban$ as a suffix, and $mban$ is a palindrome, as said above. Hence, by Lemma 14.3.4, $w'p$ has as a period the length of the corresponding prefix, that is, nab, which has the same length as $v = anb$. Note that v is a proper suffix of w, hence a factor of $w'p$, which therefore has v as a periodic pattern.

If u and v are not both proper Christoffel words, then (by an observation in Section 2.5) either $w = a^n b$, $n \geq 1$, $u = a$, $v = a^{n-1}b$, $w'p = a^{n-1}ba^{n-1}$, which has v as a periodic pattern; or $w = ab^n$, $n \geq 1$, $u = ab^{n-1}$, $v = b$, $w'p = b^n$, which has v as a periodic pattern.

Finally, if $|w| \geq 3$, then $|v| < |v| + |w| - 2 = |p| + |w'| = |w'p|$, and the period is non-trivial. □

Lemma 14.3.6 *Let p* and *q be relatively prime natural numbers and $n = p + q$. The oriented graph, with vertices $0, 1, \ldots, n-1$ and arrows $i \to j$ if $j = i + q$ or $j = i - p$, is a cycle.*

Proof. Note that q and n are relatively prime. Since q generates the cyclic group $\mathbb{Z}/n\mathbb{Z}$, the graph G', with set of vertices $\mathbb{Z}/n\mathbb{Z} = \{0, 1, \ldots, n-1\}$ and arrows $i \to j$ if $j \equiv i + q \mod n$, is a cycle. This graph coincides with the graph G of the lemma. Indeed, each arrow in G is an arrow of G' because $-p \equiv q \mod n$. Moreover, for each arrow $i \to j$ in G', one has either $i + q < n$ and then $j = i + q$ so that it is an arrow in G, or $i + q \geq n$ and then, since $i + q < 2n$, $j = i + q - n = i - p$, and it is an arrow in G too. □

Corollary 14.3.7 *Let p and q be two relatively prime positive integers. The alphabet of a word m of length $p + q - 2$ having the periods p and q has at most two elements. Let $m \in \{a, b\}^*$. If $p + q - 2 = 0$, m is the empty word. If $p + q - 2 > 0$, then:*

(i) if p or $q = 1$, m is the power of a single letter;
(ii) if $p, q > 1$, m may take exactly four different values, two of them being as in (i), and the others having two letters, and each obtained from the other by exchange of these letters.

Proof. Let $n = p + q > 2$ (the case $p + q - 2 = 0$ is immediate) and $m = a_1 \cdots a_{n-2}$. Let H be the undirected graph obtained from the graph G of the previous lemma by removing the vertices 0 and $n - 1$ and forgetting the orientations of the edges. Thus the vertices of H are $1, \ldots, n - 2$. Let $1 \le i, j \le n - 2$ and suppose that there is an edge between i and j in H; we show that $a_i = a_j$. We may suppose that $i < j$; then $j = i + q$ or $i = j - p$, hence $a_i = a_j$ by the periodicity hypothesis. Thus, if i and j are connected by a path in H, then $a_i = a_j$. By the previous lemma, this graph has one or two connected components. Thus the alphabet of the word must have one or two elements.

In order to prove (i) and (ii), one notes that H has only one connected component if and only if 0 and $n - 1$ are connected in G, that is, if and only if $p = 1$ or $q = 1$. □

Lemma 14.3.8 *If $w = amb$ is a Christoffel word with standard factorization uv, then m has the relatively prime periods $|u|$ and $|v|$.*

Proof. Using Lemma 14.3.5, we see that m has the period $|v|$. Symmetrically, it has the period $|u|$. We conclude using Corollary 2.5.1. □

Proof of Theorem 14.3.3. If m is central, let w be the Christoffel word amb. By the previous lemma, we may conclude since $|m| = |u| + |v| - 2$.

Conversely, let p and q be two relatively prime natural integers and m be a word of length $p + q - 2$ having the periods p and q. We may assume that m is not the power of a single letter (because such a word is central). Then $p, q \ge 2$. By Corollary 14.3.7, m is unique, up to the exchange of a and b. By Corollary 2.5.1, there exists a lower Christoffel word $w = uv$ (standard factorization) such that $p = |u|$ and $q = |v|$. By what has been proved above, we have $w = am'b$, where m' is a central word having the periods p and q. Then m' is not the power of a letter; otherwise $w = a^n b$ or ab^n, and $|u| = 1$ or $|v| = 1$, a contradiction with $p, q \ge 2$. By uniqueness in Corollary 14.3.7, we thus deduce $m = m'$ or $m = E(m')$. Hence m is central, since E exchanges lower and upper Christoffel words. □

For later use, we also state and prove the next result, which is a theorem due to Fine and Wilf [FW].

Corollary 14.3.9 *Let p and q be two positive natural integers. If a word w of length at least $p + q - \gcd(p, q)$ has the periods p and q, then it has the period* $\gcd(p, q)$.

Proof. We may assume that p and q are not both equal to 1. Assume first that p and q are relatively prime and that w is of length $n = p + q - 1$; note that $n \ge 2$. Let H be the undirected graph obtained from the graph of Lemma 14.3.6 by removing the vertex 0 and forgetting the orientations of the edges. This graph is connected so that, arguing as in the proof of Corollary 14.3.7, w must be the power of a single letter. If w is longer, then by considering the successive factors of length n (which is ≥ 2), we see that w is also the power of a single letter and hence has period $1 = gcd(p, q)$.

In the general case, we show that w has the period $d = \gcd(p, q)$; equivalently, for each factorization $w = uxu'yu''$, where x and y are letters and u' is a word of length $d-1$, one has $x = y$. It is enough to show that for each factorization $w = u_0a_1u_1a_2 \cdots a_ku_k$, where $a_1, \ldots, a_k$ are letters and $u_0, \ldots, u_k$ are words such that $|u_0|, |u_k| \leq d-1$ and $|u_i| = d-1$ for $i = 1, \ldots, k-1$, one has $a_1 = a_2 = \cdots = a_k$. Now, one has $p + q - d \leq |w| \leq d - 1 + kd$, since $|a_iu_i| \leq d$. We deduce that the length k of the word $w' = a_1 \cdots a_k$ satisfies $k \geq \frac{p}{d} + \frac{q}{d} - 1 - \frac{d-1}{d}$, and therefore (since $\frac{p}{d} + \frac{q}{d} - 1$ is an integer) $k \geq \frac{p}{d} + \frac{q}{d} - 1$. Now w' clearly has the periods $\frac{p}{d}$ and $\frac{q}{d}$ so that, by the previous gcd = 1 case, w' is the power of a single letter. □

Theorem 14.3.10 *(de Luca and Mignosi [dLM], Proposition 8) A word m on the alphabet $\{a, b\}$ is central if and only if it is a palindrome and if amb is a product of two palindromes.*

Equivalently, $w = amb$ is a Christoffel word if and only if it is a product of two palindromes and if m is a palindrome.

Proof. If m is central, then it is a palindrome; moreover, amb is a Christoffel word which has a palindromic factorization by Theorem 14.2.1.

Conversely, suppose that m is a palindrome such that $amb = uv$, where u and v are palindromes. Since amb is not a palindrome, u and v are non-empty. If v is of length 1, then $v = b$, $am = u$, and by Lemma 14.3.4, am has the period 1, so is a power of a and m too; thus m is central. Similarly, if u is of length 1.

Thus we may assume that the lengths of v and u are ≥ 2. Then $u = au'a$, $v = bv'b$, and u', v' are palindromes. Then $amb = au'abv'b$, $m = u'abv'$. Then by the same lemma, m has the periods $p = |abv'|$ and $q = |u'ab|$, and we have $|m| = p + q - 2$. If p and q are relatively prime, m is central by Theorem 14.3.3.

By contradiction, suppose that p and q are not relatively prime; then $p, q \geq 2$, and by Corollary 14.3.9, m has the period $d = \gcd(p, q) \geq 2$. Thus m is a prefix of w^∞ for some word w of length d. Then, since $m = u'abv'$ and $q = |u'ab|$ is a multiple of d, we have $u'ab = w^i$. Thus $w = w'ab$, and therefore $m = (w'ab)^j w'$, since d divides $|m| + 2 = p + q$. Note that $j > 0$, since $j = 0$ would imply that $m = w'$ and then $d - 2 = |m| = p + q - 2$, $d = p + q$, which cannot be true, since $p, q \geq 2$. But a word of the form $w'ab \cdots abw'$ cannot be a palindrome, since $ab \neq ba$. □

14.3.1 Exercises

Exercise 14.3.1 *(Gabriele Fici) Show that each non-trivial prefix (resp. suffix) of a lower Christoffel word is the power with some rational exponent ≥ 1 of some lower Christoffel word (if u is a word and r a rational number of the form $r = N/|w|$, then*

w^r is by definition the word $w^q p$, $N = |w|q + r$ (Euclidean division by $|w|$), p the prefix of length r of w). Hint: use the symmetric version of Lemma 14.3.5.

Exercise 14.3.2 *Show that a word w in $\{a, b\}^*$ is central if and only if w is either a power of a letter or belongs to $P \cap PbaP$, where P is the set of palindromes in $\{a, b\}^*$. Hint: use Exercise 14.1.7 for one direction and Theorem 14.3.10 (exchanging a and b) for the other.*

Let w be a central word, $w = pbaq = qabp$, $p, q \in P$; prove that p and q *are unique. Hint: use the uniqueness of the palindromic factorization of a Christoffel word. See de Luca and Mignosi [dLM].*

Exercise 14.3.3 *Let m be a central word, $m = pbaq = qabp$, p and q palindromes (see Exercise 14.3.2); prove that* $\min(|p|, |q|) + 2$ *is the smallest period of w. Hint: use Lemma 14.3.8. See de Luca and Mignosi [dLM].*

14.4 Suffix automaton of a central word

A *finite deterministic automaton* (we say automaton for simplicity) on an alphabet A is a finite directed graph whose vertices are called *states*, with a distinguished state called the *initial state*, with a distinguished set of states called the *set of terminal states*, such that each edge has a label which is a letter in A; moreover, it is *determinisc*, that is, for each vertex q and each letter a, there is at most one edge labelled a outgoing from q.

Each path gets a label, as follows; if the path is empty, it is the empty word; if the path is an edge labelled a, its label is a; if the path is the concatenation of two paths, then its label is the concatenation of the labels of the two paths. A path is called *successful* if it starts form the initial state and reaches some terminal state.

A word is *recognized* by the automaton if it is the label of a successful path. A *language* is a subset of A^*. The language *recognized by an automaton* is the set of words that it recognizes. Such a language is called *recognizable*. Among the automata recognizing a given language, there is one with the minimum number of states. It is shown that this *minimal automaton* is unique, up to isomorphism (that is, renaming of states). See Exercise 14.4.5.

Following Sciortino and Zamboni, we construct the minimal automaton of the set S_w of suffixes of the central word $w \in A^*, A = \{a < b\}$. Its set of states is the set of prefixes of w; the initial state is the empty word 1; a state is terminal if and only if it is a palindrome; a terminal state p is said to be of *type* c ($c \in A$) if pc is a prefix of w. For any state p and letter c such that pc is a prefix of w, there is an edge $p \overset{c}{\to} pc$ labelled c; such an edge is called *direct*. Moreover, for each terminal (that is, palindromic) state p and letter c such that pc is a prefix of w, let d be the other letter, and search, in the prefix order, for the smallest terminal state q of type d which is $> p$; if such a q exists, there is an edge $p \overset{d}{\to} qd$; such an edge is called a *bypass*.

Denote by $\mathcal{A}_w$ this automaton. Observe that in this automaton, the label of each edge depends only on the vertex it reaches: it is equal to the last letter of the latter. Observe also that for each path $1 \xrightarrow{x} p$ in this automaton, $|p| \geq |x|$.

Theorem 14.4.1 *(Sciortino and Zamboni [SZ], Proposition 7) Let w be a central word. The automaton $\mathcal{A}_w$ is the minimal automaton of the set S_w of suffixes of w.*

Lemma 14.4.2 *Let $g \in \{a, b\}^*$, $n \in \mathbb{N}$, and consider the central words $w = \mathrm{Pal}(gba^n b)$, $u = \mathrm{Pal}(gba^n)$, and $z = \mathrm{Pal}(g)$.*

(i) For some word v, $w = ubv$, $u = zbv$, u is the longest palindromic proper prefix of w, and z is the longest palindromic prefix of u such that zb is a prefix of u.
(ii) The automaton $\mathcal{A}_u$ is obtained from $\mathcal{A}_w$ by removing the states in the latter which are not prefixes of u, and the following edges of $\mathcal{A}_w$:

- *the direct edges which reach a state that is not a prefix of u;*
- *the bypasses $p \xrightarrow{b} ub$, where p is a palindromic prefix of w, such that $z <_{pref} p <_{pref} u$ for the prefix order.*

Proof. Observe that, by Proposition 14.1.6, the palindromic prefixes of w (resp. the palindromic prefixes of w followed in w by letter $c = a$ or b) are the words $\mathrm{Pal}(p)$, p prefix of $gba^n b$ (resp. p prefix of $gba^n b$ and followed by c in $gba^n b$). This observation is used several times in the following proof.

(i) From this it follows that u is the longest palindromic proper prefix of w and that z is the longest palindromic prefix of u such that zb is a prefix of u.
We now use Exercise 14.1.5, based on Justin's formula (Theorem 14.1.3): $\mathrm{Pal}(fh) = \mathrm{Pal}(f)R_f(\mathrm{Pal}(h))$, where $f \mapsto R_f$ is the monoid homomorphism $A^* \to \mathrm{End}(A^*)$ sending a to $R_a = (a, ba)$ and b to $R_b = (ab, b)$. Thus $\mathrm{Pal}(gba^n b) = \mathrm{Pal}(gba^n)R_{gba^n}(b) = ubv$, if we denote $bv = R_{gba^n}(b)$, noting that the latter word begins by b (indeed, by Proposition 14.1.6, $u = \mathrm{Pal}(gba^n)$ is followed by b in $w = \mathrm{Pal}(gba^n b)$. Moreover, $\mathrm{Pal}(gba^n) = \mathrm{Pal}(g)R_g(\mathrm{Pal}(ba^n)) = zR_g(\mathrm{Pal}(ba^n))$. We verify that $R_g(\mathrm{Pal}(ba^n)) = R_{gba^n}(b)$, which will end the proof of (i).
Indeed, $\mathrm{Pal}(ba^n) = b(ab)^n$; see Exercise 14.1.2. Moreover $R_{gba^n}(b) = R_{gb}R_a^n(b) = R_{gb}(ba^n) = R_gR_b(ba^n) = R_g(b(ab)^n) = R_g(\mathrm{Pal}(ba^n))$.
(ii) The assertion about the direct edges follows from the definition of these automata. Note that a bypass of $\mathcal{A}_w$, not in $\mathcal{A}_u$, necessarily reaches ub (since u is the longest palindromic prefix of w, and since u is followed by b in w). By construction of the automaton, its origin must be a state of the form $\mathrm{Pal}(gba^i)$, $i = 0, \ldots, n-1$. Hence it is a palindromic prefix p of w such that $z = \mathrm{Pal}(g) <_{pref} p <_{pref} u = \mathrm{Pal}(gba^n)$.

□

Proof of Theorem 14.4.1. The automaton is deterministic; indeed, consider some vertex: if it is not terminal, there is one direct outgoing edge; if it is w, there is no outgoing edge; and if it is terminal $\neq w$, there is one direct outgoing edge labelled c, and at most another outgoing edge, which is a bypass with the other letter as label.

We now show that $\mathcal{A}_w$ recognizes S_w. We have $w = \mathrm{Pal}(f)$ for some word $f \in A^*$. We leave to the reader the cases where f is a power of a or b or if $f = a^n b$ or $b^n a$ (Exercise 14.4.1). So we may assume that $f = gba^n b$, applying E if necessary. Take the notations of Lemma 14.4.2: $w = ubv$, $u = zbv$. We use this lemma constantly later. Observe that since u is a palindromic prefix of the palindrome w, u is a suffix of w.

By induction on the length, $\mathcal{A}_u$ recognizes S_u (u is a central word). Note that by the deterministic property, for each word s, a path labelled s is uniquely determined by its first state.

Let s be a word recognized by $\mathcal{A}_w$. Then there is a succesful path $1 \to p$ in $\mathcal{A}_w$, where p is a palindromic prefix of w. If p is a prefix of u, then the path $1 \to p$ is in $\mathcal{A}_u$, and therefore s is a suffix of u, thus of w. Otherwise, $p = w$; if $s = w$, s is a suffix of w; otherwise, the path $1 \to w$ must use one of the bypasses, which are in $\mathcal{A}_w \setminus \mathcal{A}_u$, described in Lemma 14.4.2; thus there is a decomposition of the path $1 \overset{x}{\to} p \overset{b}{\to} ub \overset{v}{\to} w$. Indeed, the label of the third part of the path is v, since this path uses only direct edges because u is the longest proper palindromic prefix of w, and since $w = ubv$. Therefore, $s = xbv$ is a suffix of $w = ubv$, since x is a suffix of u, being recognized by $\mathcal{A}_u$.

Conversely, let s be a suffix of w. If $|s| \leq |u|$, then s is a suffix of u by the previous observation; thus s is recognized by $\mathcal{A}_u$, hence by $\mathcal{A}_w$ (by Lemma 14.4.2). If, on the other hand, $|s| > |u|$, then since $u = zbv$ is a suffix of $w = ubv$, $|s| > |bv|$ and one has $s = xbv$, with x a suffix of u, with $|x| > |z|$; then, by induction, x is recognized by $\mathcal{A}_u$: there is in $\mathcal{A}_u$ a path labelled x from 1 to p, with p palindromic (since it is a terminal state), a prefix of u, with $|p| \geq |x| > |z|$; since u is a prefix of w, z and p are prefixes of w, and therefore $z <_{pref} p \leq_{pref} u$. The previous path is also a path in $\mathcal{A}_w$. Moreover, there is an edge labelled b from p to ub in $\mathcal{A}_w$: indeed, if $|p| < |u|$, this edge is a bypass and it follows from Lemma 14.4.2, and if $p = u$, it is a direct edge since u and ub are prefixes of w. Finally, since there is a direct path $ub \overset{v}{\to} w$, there is a path from 1 to w labelled $xbv = s$; hence s is recognized by $\mathcal{A}_w$.

Note that an automaton recognizing S_w has at least $|w| + 1$ states. Indeed, it has a successful path labelled w, since $w \in S_w$; this path must have $|w| + 1$ distinct states; otherwise it contains some closed path, which may be iterated, and the language contains infinitely many words; however, S_w is finite. Thus $\mathcal{A}_w$ is minimal, since it has $|w| + 1$ states. □

14.4.1 Exercises

Exercise 14.4.1 *Prove Theorem 14.4.1 in the case where* $w = \mathrm{Pal}(a^n)$ *and* $w = \mathrm{Pal}(a^n b)$. *Use Exercise 14.1.2.*

The following exercises give the basic theory of finite deterministic automata and their minimization.

Exercise 14.4.2 *Let L be a language and u be a word. Define*

$$L \circ u = \{w \mid uw \in L\}$$

In other words, $L \circ u$ is the set of suffixes w of words of L whose corresponding prefix is u ($L \circ u$ has also the self-explanatory notation $u^{-1}L$). Show that

$$L \circ (uv) = (L \circ u) \circ v, L \circ 1 = L$$

In other words, we have a right action *of the free monoid on the set of languages. We call $L \circ u$ a* translate *of L.*

Exercise 14.4.3 *Consider a finite deterministic automaton on the alphabet A with the set of states Q. It is called* complete *if, for any state q and any letter a, there is an outgoing edge labelled by this letter; denote by $\delta(q, a)$ the final state of this edge so that one has the edge $q \overset{a}{\to} \delta(q, a)$. Show that the mapping $\delta : Q \times A \to Q$ extends uniquely to a right action $Q \times A^* \to Q$, also denoted δ, of A^* on Q; that is,*

$$\delta(q, 1) = q, \delta(q, uv) = \delta(\delta(q, u), v)$$

for any words u, v and any state q. One may therefore adopt the notation

$$q \cdot u = \delta(q, u)$$

Hint: $\delta(q, u)$ is the final state of the unique path outgoing from q and with label u; uniqueness follows from the deterministic property.

Show that a finite complete deterministic automaton (disregarding initial and terminal states) is equivalent to a right action of the free monoid on a set.

Exercise 14.4.4 *Show that if a finite deterministic automaton is not complete, then one may add*

- *a new non-terminal state s;*
- *an edge $q \overset{a}{\to} s$ for each state q and letter a such that there is no edge outgoing from q and labelled a;*
- *and edges $s \overset{a}{\to} s$ for any letter a;*

and one obtains a finite complete deterministic automaton recognizing the same language. The state s is called a sink state.

Exercise 14.4.5 *Consider a finite complete deterministic automaton, with initial state q_0, and let L be the language recognized by it.*

(i) *Show that the set of translates of L is finite. Hint: $L \circ u$ is the language recognized by the automaton with new initial state $q_0 \cdot u$.*

(ii) Show that if the set of translates of L is finite, then L is recognizable: L is recognized by the automaton $\mathcal{A}_L$, where the set of states of $\mathcal{A}_L$ is the set Q_L of translates of L, the right action is the natural action defined in Exercise 14.4.2, the initial state is L, and the set of terminal states is the set of translates containing the empty word.

(iii) Suppose that the automaton is accessible, *that is, $Q = q_0 \cdot A^*$. Show that the mapping $Q \to Q_L$, sending each state $q = q_0 \cdot w$ onto $L \circ w$, is well defined, is surjective, and preserves the action.*

Exercise 14.4.6 *A finite deterministic (resp. a finite complete deterministic) automaton, recognizing the language L is called* minimal *if it has the minimal number of states among all finite deterministic (resp. finite complete deterministic) automata recognizing L.*

Deduce that, in this case, its number of states is equal to the number of non-empty translates (resp. of translates) of L.

14.5 The Tree of Christoffel words

A variant of the tree of Christoffel pairs (Figure 2.10) is obtained by replacing each node (u, v) with $w = uv$; see Figure 14.3. This tree is called the *tree of Christoffel words.* The two trees are equivalent, because we recover (u, v) from w by taking the standard factorization of w. Note that the nodes of this tree are exactly all proper lower Christoffel words.

Theorem 14.5.1 *The construction of the tree of Christoffel words is obtained directly as follows: let w be any node, distinct from the root ab. Consider the path from w towards the root.*

(i) Suppose that w is not on the two extreme branches of the tree; then this path has north-west steps and north-east steps. Let u be the node after the first north-west step and v be the node after the first north-east step. Then w = uv.

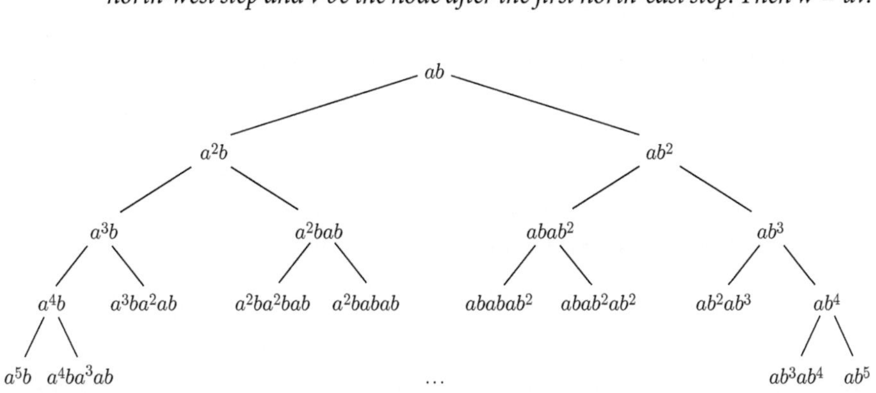

Figure 14.3 The tree of (lower) Christoffel words

(ii) *If w is on the left (respectively right) extreme branch, then no north-west (respectively north-east) step exists. Choose $u = a$ and v as in (i) (respectively u as in (i) and $v = b$).*

For example, in Figure 14.3, let $w = a^2babab$; then $u = a^2bab$ and $v = ab$. This allows one to directly construct the Christoffel tree by starting with the root ab.

Proof. The validity of this construction follows easily by induction, using the construction rules of the tree of Christoffel pairs; see Figure 2.9. □

Let us return for a moment to the tree of Christoffel pairs, by describing the commutative image of each pair in this tree and deriving some consequences. For this, denote by $SL_2(\mathbb{N})$ the set of 2-by-2 matrices of determinant 1 over the natural numbers; it is a monoid under matrix multiplication. Let ν be the monoid homomorphism from the free monoid $\{a, b\}^*$ into the multiplicative monoid $SL_2(\mathbb{N})$ defined by

$$\nu(a) = R = \begin{pmatrix} 1 & 1 \\ 0 & 1 \end{pmatrix}, \quad \nu(b) = L = \begin{pmatrix} 1 & 0 \\ 1 & 1 \end{pmatrix}$$

Proposition 14.5.2 *Let (u, v) be a node in the tree of Christoffel pairs, and let the path from the root to this node be coded by the word $m \in \{a, b\}^*$ (recall that a means "left" and b means "right"). Let $\nu(m) = \begin{pmatrix} p & q \\ r & s \end{pmatrix}$. Then the commutative images of the words u, v, and uv are respectively the tuples (p, r), (q, s), and $(p + q, r + s)$.*

For example, the node (a^2bab, ab) corresponds to the path coded by abb (meaning left-right-right); see Figure 2.10. One has

$$\begin{aligned} \nu(abb) &= \begin{pmatrix} 1 & 1 \\ 0 & 1 \end{pmatrix}\begin{pmatrix} 1 & 0 \\ 1 & 1 \end{pmatrix}\begin{pmatrix} 1 & 0 \\ 1 & 1 \end{pmatrix} \\ &= \begin{pmatrix} 1 & 1 \\ 0 & 1 \end{pmatrix}\begin{pmatrix} 1 & 0 \\ 2 & 1 \end{pmatrix} = \begin{pmatrix} 3 & 1 \\ 2 & 1 \end{pmatrix} \end{aligned}$$

and indeed the commutative images of a^2bab and ab are $(3, 2)$ and $(1, 1)$.

Proof. The pair (u, v), viewed as an endomorphism of the free monoid $\{a, b\}^*$, is equal to $\rho(m)$, by Theorem 2.6.1 and with its notations. Recall that Ab denotes the abelianization $\mathrm{End}(\{a, b\}^*) \to \mathbb{Z}^{2\times 2}$; see Chapter 1. Then, recalling that $G = (a, ab)$ and $\tilde{D} = (ab, b)$, $Ab(G) = \nu(a)$ and $Ab(\tilde{D}) = \nu(b)$. Therefore, $Ab \circ \rho = \nu$. It follows that the abelianization of (u, v) is $Ab \circ \rho(m) = \nu(m)$, and the proposition follows, noting that the abelianization of the endomorphism (u, v) is the matrix $\begin{pmatrix} |u|_a & |v|_a \\ |u|_b & |v|_b \end{pmatrix}$. □

Corollary 14.5.3 *The mapping associating the matrix $\begin{pmatrix} |u|_a & |v|_a \\ |u|_b & |v|_b \end{pmatrix}$ with a proper lower Christoffel word w (where $w = uv$ is the standard factorization of w) is a bijection from the set of proper lower Christoffel words onto $SL_2(\mathbb{N})$.*

Proof. The mapping is well defined: by the previous result, this matrix is in $SL_2(\mathbb{N})$.

We now show that the mapping is bijective. Let $\begin{pmatrix} p & q \\ r & s \end{pmatrix}$ be an element of $SL_2(\mathbb{N})$. The integers $p + q$ and $r + s$ are relatively prime, since $(p + q)s - (r + s)q = 1$. Let w be the lower Christoffel word of slope $\frac{r+s}{p+q}$. Since $p + q, r + s \neq 0$, this slope is $\neq 0$ or ∞, and therefore w has length ≥ 2; let $w = uv$ be its standard factorization. We show below that the commutative images of u and v are (p, r) and (q, s) respectively. This implies bijectivity, since there is exactly one proper lower Christoffel word for any given positive rational slope.

Consider in the discrete plane a point A and the points $M = A+(p, r)$, $N = A+(q, s)$, $B = A+(p + q, r+s)$. Then the slope r/p of $\overrightarrow{AM}$ is smaller than the slope s/q of $\overrightarrow{AN}$ (since $ps - qr = 1$), as in Figure 2.7. As follows from Section 2.1, the parallelogram $AMBN$ is primitive, hence does not contain any integer point except the four vertices. Therefore, the discretization from below of the segment AB is the concatenation of that of AM and that of MB. This shows (by Theorem 2.4.1) that the standard factorization of the first is the product of the two others. Hence $w = uv$. □

Corollary 14.5.4 *The multiplicative submonoid* $SL_2(\mathbb{N})$ *of* $SL_2(\mathbb{Z}))$ *is a free monoid, freely generated by the matrices* $\nu(a) = \begin{pmatrix} 1 & 1 \\ 0 & 1 \end{pmatrix}$ *and* $\nu(b) = \begin{pmatrix} 1 & 0 \\ 1 & 1 \end{pmatrix}$.

Proof. It is enough to show that ν is a bijection. This is true, since, as follows from the proof of Proposition 14.5.2, ν is the composition of the bijection $m \mapsto \rho(m)(ab)$ (it is a bijection by Theorem 2.6.1 and Section 2.5), followed by the bijection of the previous corollary. □

Corollary 14.5.5 *Let* $n \geq 2$*. The number of 2-by-2 matrices* $\begin{pmatrix} a & b \\ c & d \end{pmatrix}$ *of determinant 1 over the natural numbers of total sum* $a + b + c + d = n$ *is* $\varphi(n)$*, where* φ *is the Euler function.*

Proof. The number of lower Christoffel words of length $n \geq 2$ is $\varphi(n)$ (Exercise 2.2.1). By Corollary 14.5.3, there is a bijection between them and the set described in the corollary, which therefore follows. □

Let $\frac{q}{p}$ be some positive reduced fraction. Define the fraction $\frac{q^*}{p^*}$, where q^* (resp. p^*) is the inverse modulo $p + q$ of p (resp. of q), taken in $\{0, 1, \ldots, p + q - 1\}$. Since modulo $p + q$, $p \equiv -q$, we have $p^* \equiv -q^*$, and therefore $p^* + q^* = p + q$. Moreover, p^* and q^* are relatively prime. The mapping $\frac{q}{p} \mapsto \frac{q^*}{p^*}$ is an involution.

Corollary 14.5.6 *(Berstel and de Luca, [BdL], Theorem 5.1) If* $w = uv$ *is a proper lower Christoffel word of slope* $\frac{q}{p}$*, with its standard factorization, then the length of* u *is* q^* *and that of* v *is* p^*.

Recall that $|w|_a = p$, $|w|_b = q$.

Lemma 14.5.7 *Let* $\begin{pmatrix} p & q \\ r & s \end{pmatrix} \in SL_2(\mathbb{Z})$. *Then, modulo* $p + q + r + s$, $p + q$ *is the inverse of* $q + s$ *and* $p + r$ *is the inverse of* $r + s$.

Proof. Since $ps = 1 + qr$, one has $(p + q)(q + s) = pq + ps + q^2 + qs = 1 + pq + qr + q^2 + qs = 1 + q(p + q + r + s)$, which proves the first statement. The second follows by transposing. □

Proof of Corollary 14.5.6. By Section 2.2, $q = |u|_b + |v|_b, p = |u|_a + |v|_a$. Therefore, by Corollary 14.5.3 and the previous lemma, $p^* = |v|_a + |v|_b = |v|, q^* = |u|_a + |u|_b = |u|$. □

In [Bo], Borel constructs, for each Möbius transformation f, a *transducer* that maps a Christoffel word of slope r onto the Christoffel word of slope $f(r)$. In [Lu], Laurier gives algorithm that computes the Christoffel word of slope $r + s$, directly from the Christoffel words of slope r and s, and an algorithm computing the Christoffel word of an integral multiple of the slope.

14.5.1 Exercises

Exercise 14.5.1 *Consider the tree of Christoffel words and two nodes $u = amb = xy$ (standard factorization) and v. Show that if u is an ancestor of v, then u is a factor of v, x is a prefix of v, and y is a suffix of v. Hint: use Exercice 2.5.1. Show that if v is in the left (resp. right) subtree under u, then ama is a prefix of v (resp. bmb is a suffix of v). Hint: use Exercice 14.2.4.*

Exercise 14.5.2

(i) *Suppose that x is an ancestor of w in the tree of Christoffel words. Show that each circular factor of x is a circular factor of w. Hint: one may assume that w is a child of x. If it is a left child, then $w = uv$, standard factorization, $v = x$. Use Exercise 2.5.3 to show that each conjugate of v is a circular factor of w.*
(ii) *Let (x, y) be a standard pair. Recall that for $n \geq 1$, $x^n y$ and xy^n are lower Christoffel words, with standard factorization $x \cdot x^{n-1}y$ and $xy^{n-1} \cdot y$, respectively (Exercise 2.5.2). Show that if $x^n y$ (resp. xy^n) is an ancestor of w in the tree of Christoffel words, then each circular factor of x^n (resp. y^n) is a circular factor of w. Hint: using Exercise 2.5.3, show that each conjugate of y^n is a circular factor of xy^n.*

Exercise 14.5.3 *Show that if u is an ancestor of v in the tree of Christoffel words, then $G(u)$ (resp. $\tilde{D}(u)$) is an ancestor of $G(v)$ (resp. $\tilde{D}(v)$). Hint: use Theorem 2.6.1.*

Exercise 14.5.4 *Show that each ancestor, in the tree of central words, of a central word is a factor of this word. Hint: Corollary 14.1.7.*

Show that each palindromic factor of a central word is a central factor of some ancestor of this word in the tree of central words. Hint: consider the functions

$G'(w) = G(w)a$ and $D'(w) = b\tilde{D}(w)$; show that G', D' preserve the property of being an ancestor by using Exercise 14.5.3. Then use the bijections in Exercise 4.1.2 (ii) and the similar bijections for $\tilde{D}$. Use Exercise 4.1.1.

Exercise 14.5.5 *Let $n \geq 1$. Consider the following subtree T of the tree of Christoffel words: it is the complete binary tree whose internal nodes are the lower Christoffel words of length $\leq n$. Show that T has $\varphi(2) + \cdots + \varphi(n)$ internal nodes, and one leaf more. Show that the set of leaves of T is the set of lower Christoffel words w, of length $\geq n+1$, whose standard factorization $w = uv$ satisfies $|u|, |v| \leq n$. Deduce that this latter set has $\varphi(1) + \varphi(2) + \cdots + \varphi(n)$ elements. See [Re5] (proof of Proposition 5.2).*

Exercise 14.5.6 *Let $m \in \{a, b\}^*$ and let w be the lower Christoffel word $a\mathrm{Pal}(m)b$. Show that the length of w is equal to the sum of the entries in the matrix $\nu(m)$. Use Theorem 2.6.1, Lemma 14.1.5, and Proposition 14.5.2.*

Exercise 14.5.7 *Let $v \in \{a, b\}^*$ and let w be the lower Christoffel word $a\mathrm{Pal}(v)b$.*

(i) Show that $\mu(a\mathrm{Pal}(v)b)_{12} = (2,1)\mu(\mathrm{Pal}(v))\,{}^t(2,1)$.
(ii) Define a homomorphism of $\{a, b\}^*$ into $\mathrm{SL}_2(\mathbb{Z})$ by

$$\omega(a) = \begin{pmatrix} 2 & 1 \\ 1 & 1 \end{pmatrix}, \quad \omega(b) = \begin{pmatrix} 1 & 1 \\ 1 & 2 \end{pmatrix}$$

Let $P = P(1)$. Show that for any word x, one has

$$\mu(x) = P^{-1}(\omega G(x))P$$

(iii) Define an endomomorphism θ of the free monoid $\{a, b\}^*$ by $\theta(a) = ab, \theta(b) = ba$ (θ is called the Thue-Morse homomorphism*). Show that $\nu\theta = \omega$, and that $S\nu\theta\,\mathrm{Pal}(av) = S\omega\,\mathrm{Pal}(av)$, where S sends any matrix onto the sum of its entries. By using Theorem 14.1.3, deduce that $S\nu\theta\,\mathrm{Pal}(av) = (1,1)\omega G\,\mathrm{Pal}(v)\,{}^t(3,2)$.*
(iv) Let $x \in \{a, b\}^*$. Show that $(1,1)\omega G(x)\,{}^t(3,2) = (2,1)\mu(x)\,{}^t(2,1)$.
(v) Deduce from Exercise 14.5.6 that the Markoff number $\mu(w)_{12}$ is equal to the length of the lower Christoffel word $a(\mathrm{Pal}\,\theta\,\mathrm{Pal}(av))b$.

Note: The latter word is equal to $a(\mathrm{Pal}\,\mathrm{Antipal}(av))b$; hence the function $v \mapsto 2 + |\mathrm{Pal}\,\mathrm{Antipal}(av)|$ is a surjection from the free monoid onto the set of Markoff numbers. The injectivity of this function is equivalent to the Markoff number injectivity conjecture (the author and Vuillon [RV]). The word function Antipal *is defined similarly to* Pal, *by replacing palindromes with* antipalindromes, *that is, words u such that $\tilde{u} = \theta(u)$; it is shown by de Luca and De Luca that* Antipal $= \theta\mathrm{Pal}$ *[dLDL2], Theorem 7.1.*

(vi) Show that the set of Markoff numbers $\neq 1$ is equal to the set of $S\omega(u)$ for all central words u, equivalently to the set of $S\nu(\mathrm{Antipal}(v))$ for all words $v \in \{a, b\}^*$.

15
Conjugates and factors

Christoffel words are naturally cyclic objects. They may be defined by the Cayley graphs of finite cyclic groups. Lexicographic properties of the set of conjugates of Christoffel words are studied. Pirillo's characterization of conjugates of Christoffel words is given. The set of circular factors of length k of a given Christoffel word may be partitioned according to a composition of k: the theorem of Clough and Myerson asserts that the number of classes is $k + 1$. In particular, the number of circular factors of length k is $k + 1$, and this characterizes conjugates of Christoffel words (theorem of Borel and the author), a characterization reminiscent of the similar property of Sturmian sequences. The multiplicities of factors take only three values, which is a variant of the three-distance theorem. Another characterization of conjugates of Christoffel words, similar to that of Droubay and Pirillo for Sturmian sequences, rests on the count of palindromic factors. The monoid of Sturmian morphisms is generated by three of them.

15.1 Cayley graphs

A simple way to define and generate the lower Christoffel word of slope $\frac{q}{p}$ is given by the following result. The alphabet is $\{a < b\}$. Recall that $x \bmod n$ denotes, as usual, the unique $x \in \{0, \dots, n-1\}$ such that $x \equiv n \bmod n$.

Theorem 15.1.1 *The lower Christoffel word $w = a_0 \cdots a_{n-1}$ of slope $\frac{q}{p}$, with p and q relatively prime and $n = p + q$, is completely defined as follows: for any i in $\{0, \dots, n-1\}$, $a_i = a$ if $qi \bmod n < q(i+1) \bmod n$, and $a_i = b$ if $qi \bmod n > q(i+1) \bmod n$.*

Note that q and n are relatively prime so that q generates the cyclic group $\mathbb{Z}/n\mathbb{Z} = \{0, \dots, n-1\}$; drawing the *Cayley graph* of this group with respect to the generator q allows us to visualize the theorem: the word w appears on the edges of the graph, and there is a letter a when the vertices of the edge $i \to j$ increase (that is, $i < j$), and a letter b otherwise; we call this labelled graph the *Cayley graph of* w; see Figure 15.1. Note that the underlying graph is the same as the graph considered in Lemma 14.3.6.

Proof. We label each integral point on the path defined by w (see Section 2.2) by the vertical distance from this point to the diagonal of the defining rectangle for w. For an example, see Figure 15.2. Since the equation of the diagonal is $y = \frac{q}{p}x$, each label is a fraction $\frac{i}{p}$ for some natural integer i. It follows also that if $\frac{i}{p}$ and $\frac{j}{p}$ are the labels on two successive points on the path, then $j = i + q$ if the step between these two points is horizontal, and $j = i - p$ if the step is vertical. Note that if a point has the

From Christoffel Words to Markoff Numbers. Second Edition. Christophe Reutenauer, Oxford University Press.
 DOI: 10.1093/9780197907535.003.0015

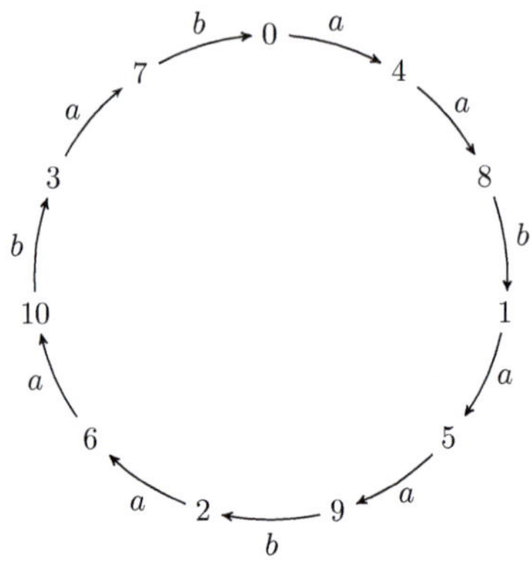

Figure 15.1 Cayley graph of $\mathbb{Z}/11\mathbb{Z}$, with generator 4, generating the lower Christoffel word *aabaabaabab* of slope $\frac{4}{7}$

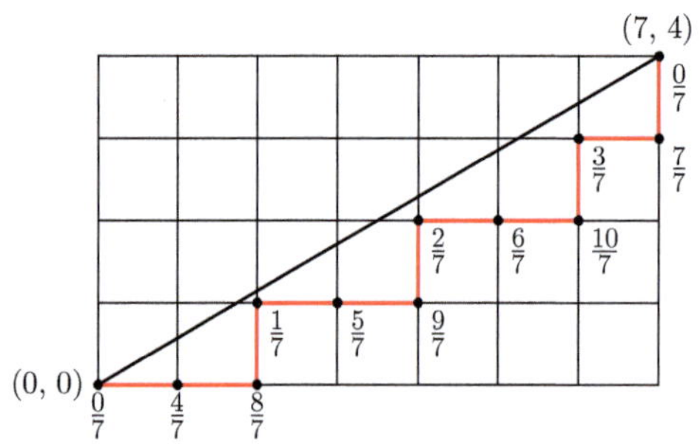

Figure 15.2 Vertical distances from points on the path to the diagonal

label $\frac{i}{p}$ with $i < p$, then the next step is horizontal, since otherwise the path crosses the diagonal, and the next point must therefore have the label $\frac{i+q}{p}$; likewise, if a point has label $\frac{i}{p}$ with $i \geq p$, then the next step must be vertical, since otherwise the polygon delimited by the path and the diagonal has an inner integral point, and the next point must therefore have the label $\frac{i-p}{p}$.

Except for the first and the last labels, which are equal, the numerators are all distinct; otherwise, there is some integral segment PQ parallel to the diagonal in the interior of the rectangle, contradicting the relative primality of p and q. The possible numerators i are therefore exactly the numbers $i = 0, \ldots, p + q - 1$.

Thus, we may take as labels i instead of $\frac{i}{p}$; these n labels, excluding the last, form the set $\{0, \ldots, n-1\}$, and if i and j are two consecutive labels, we have $j \equiv i + q \bmod n$, and $i < j$ if the step is horizontal, $i > j$ if the step is vertical. □

The previous theorem may be restated as follows: consider the mapping ρ from $\mathbb{Z}/n\mathbb{Z} = \{0, \ldots, n-1\}$ into itself defined by $\rho(i) \equiv i + q \bmod n$; then $a_i = a$ if $\rho^i(0) \in \{0, \ldots, p-1\}$ and $a_i = b$ if $\rho^i(0) \in \{p, \ldots, n-1\}$. We leave the verification of this property to the reader (using the theorem).

The previous proof, together with the geometric definition of the standard factorization, also implies the following result, since the factorization is obtained by cutting the path at the closest point (see the proofs of Corollary 2.1.1 and Theorem 2.4.1).

Corollary 15.1.2 *Let $w = uv$ be a proper lower Christoffel word with its standard factorization. Then u (resp. v) is the label, in the Cayley graph of w, of the simple path from 0 to 1 (resp. from 1 to 0).*

For example, $w = (aab)(aabaabab)$ and aab is the label of the path from 0 to 1 in Figure 15.1.

One also has the following result.

Corollary 15.1.3 *Let $w = v'u'$ be a proper lower Christoffel word of length n, with its palindromic factorization. Then v' (resp. u') is the label, in the Cayley graph of w, of the simple path from 0 to $n-1$ (resp. from $n-1$ to 0). In particular, the upper Christoffel word $\tilde{w} = u'v'$ is the label of the simple closed path from $n-1$ to itself.*

For example, $w = (aabaabaa)(bab)$ and $aabaabaa$ is the label of the path from 0 to 10 in Figure 15.1. We leave the proof to the reader (Exercise 15.1.4).

15.1.1 Exercises

Exercise 15.1.1 *Take the letters 0 and 1 instead of a and b (respectively), with $0 < 1$. Consider the lower Christoffel word $w = a_0a_1 \cdots a_{n-1}$ of slope q/p, with p and q relatively prime, and define $S = q/(p+q)$, which is the Slope of w (see Exercise 5.2.1). Show that w is completely defined by*

$$a_i = \lfloor (i+1)S \rfloor - \lfloor iS \rfloor$$

Exercise 15.1.2 *Let p and q be natural numbers, and $n = p + q$. Consider the permutation σ of $\mathbb{Z}/n\mathbb{Z} = \{0, 1, \dots, n-1\}$ defined by the equality of words $\sigma(0)\sigma(1) \cdots \sigma(p+q-1) = q(q+1)\dots(q+p-1)01\dots(q-1)$. Show that in $\mathbb{Z}/n\mathbb{Z}$, it is the translation $x \mapsto x + q$. Show that σ has only one orbit if and only if p and q are relatively prime. Show that in this latter case, its unique cycle may be identified with the Cayley graph of $\mathbb{Z}/n\mathbb{Z}$ with respect to the generator q. Define $A = \{0, 1, \dots, p-1\}$, $B = \{p, p+1, \dots, p+q-1\}$; show that the lower Christoffel word $w = a_0a_1 \cdots a_{n-1}$ of slope q/p is determined by*

$$a_i = \begin{cases} a & \text{if } \sigma^i(0) \in A \\ b & \text{if } \sigma^i(0) \in B \end{cases}$$

Note: the permutation σ, associated with the pair (p, q) is a particular case of what is called a discrete interval exchange, *which is associated with a composition, whose parts are the lengths of the successive intervals (see [SmP, FZ, BeR]); they correspond to perfectly clustering words (see Section 18.4). For a composition (a, b, c) of length 3,*

Pak and Redlich [PR] show that the associated discrete interval exchange is circular (that is, an n-cycle) if and only if $a+b, b+c$ are relatively prime, and that among all discrete exchanges of three intervals, the proportion of the circular ones tends to $\pi^2/6$ when n goes to infinity.

Exercise 15.1.3 *Let $A = \{a < b\}$; let p and q be relatively prime natural integers and $n = p + q$. Define the additive monoid homomorphism $A^* \to \mathbb{N}$, $\mu(a) = q, \mu(b) = -p$. Let f be a word of length $n-1$ such that for any prefix u of f, $\mu(u) \in \{0, 1, \ldots, n-1\}$. Show that fb is the lower Christoffel word of slope q/p. Hint: show that for each prefix u of length i of f, one has $\mu(u) = iq \bmod n$.*

Exercise 15.1.4 *Prove Corollary 15.1.3. Hint: prove that the label of the simple path $0 \to n-1$ (resp. $n-1 \to 0$) is a palindrome by applying to the path the edge-reversing involution $i \mapsto n-1-i$.*

Exercise 15.1.5 *Consider the permutation σ and the word w of Exercise 15.1.2, assuming that σ has only one cycle. Show that σ has the cycle form $(0, i_1, \ldots, i_{n-2}, p)$. Consider the words $f = i_1 \cdots i_{n-2}$ and $g = i_2 \cdots i_{n-2}$. Given a word $u = b_1 \cdots b_k$ of length k, with distinct letters, say that two letters b_i and b_j are symmetric in u if $i + j = k + 1$ (symmetry with respect to the center); extend this definition to the elements of an interval in $\mathbb{N}$, by identifying it with the word of its successive elements. Show that:*

(i) *for any two letters in f which are symmetric, they lie both in $A \setminus 0$ or both in $B \setminus p$ and are symmetric in this interval. In particular, this shows that the word $a_1 \cdots a_{n-2}$ is a palindrome;*
(ii) *for any two letters in g which are symmetric, their sum is n;*
(iii) *$\sigma^{-1} = \omega \circ \sigma \circ \omega$, where ω is the involution of $\{0, 1, \ldots, n-1\}$ defined by $\omega(i) = n-1-i$. The latter result is generalized to discrete interval exchanges in [SmP], (4.4).*

15.2 Conjugates

The set of conjugates of a Christoffel word has a remarkable structure with respect to the lexicographical ordering.

Theorem 15.2.1 *(see [MRS, JZ, BR]) Let $w = w_0 <_{lex} w_1 <_{lex} \cdots <_{lex} w_{n-1}$ be the conjugates of some proper lower Christoffel word $w = amb \in \{a < b\}^*$ of length n. Then for each $i = 1, \ldots, n-1$, one has $w_{i-1} = uabv$, $w_i = ubav$ for some words u, v, and $vu = m$. One has $w_i = C^{q^*}(w_{i-1}) = C^{iq^*}(w_0)$ and $|ua| = |ub| \equiv ip^* \bmod n$, where the slope of w is $\frac{q}{p}$ and p^*, q^* are the multiplicative inverses of p, q modulo n respectively.*

The theorem is illustrated by the following array, which represents the conjugates of $w = aabaabab$ in lexicographical order, with $m = abaaba$. In each word (except the last), the factor ab (which in the following word is changed into ba) is underlined.

The fact that $vu = m$ is seen, for example, here for $i = 0$: $u = aaba, v = ab$, and $abaaba = ab \cdot aaba$.

$$\begin{array}{c} aaba\underline{ab}ab \\ a\underline{ab}abaab \\ abaaba\underline{ab} \\ aba\underline{ab}aba \\ \underline{ab}abaaba \\ baaba\underline{ab}a \\ ba\underline{ab}abaa \\ babaabaa \end{array} \tag{15.1}$$

Proof. Recall that the set of vertices of the Cayley graph of w is $\{0, 1, \ldots, n-1\} = \mathbb{Z}/n\mathbb{Z}$ and that the label of an edge $k \to l$ is a if and only if $k < l$.

The conjugates of w are the labels of the simple closed paths of the Cayley graph of w. Let p_i denote such a path starting and ending at $i \in \{0, \ldots, n-1\}$. Let u_i be the label of p_i. By definition of the Cayley graph, $C^i(w) = u_{iq \bmod n}$, which implies that $u_i = C^{iq^*}(w)$. We show that $u_i = w_i$.

Let $i \in \{1, \ldots, n\}$. The vertices of path p_{i-1} form the sequence $i - 1 = i_0, i_1, i_2, \ldots, i_n = i - 1$, and by definition of the Cayley graph, $i_{j+1} \equiv i_j + q \bmod n$. The sequence of vertices of the path p_i is therefore $i_0 + 1, i_1 + 1, i_2 + 1, \ldots, i_0 + 1$, taken modulo n. It follows that the labels of these two paths are the same, except for the two successive edges $i_{j-1} \to i_j \to i_{j+1}$ with $i_j = n - 1$. As part of the path p_{i-1}, they are respectively labelled a and b, and adding 1, they become the successive edges $i_{j-1} + 1 \to 0 \to i_{j+1} + 1$ of the path p_i, respectively labelled b and a. Thus $u_{i-1} = uabv$, $u_i = ubav$.

Now, vu is the label of the simple path from i_{j+1} to i_{j-1}. The labels of the edges $n - 1 \to i_{j+1}$ and $i_{j-1} \to n - 1$ are respectivey b and a. Hence $bvua$ is the label of the simple closed path from $n - 1$ to iself, which is the upper Christoffel word $\tilde{w}$ (Corollary 15.1.3); the latter is equal to bma, which shows that $vu = m$.

Note that the length of ua is j, since ua is the label of the subpath $i_0 \to i_j$ of p_{i-1}. Now, since $i_j \equiv i_0 + jq$, we have $n - 1 \equiv i - 1 + jq$; since $qp^* \equiv -1$ (because $p \equiv -q$), we obtain $j \equiv ip^*$. This ends the proof, since clearly $u_{i-1} <_{lex} u_i$ so that the sequence $u_0, u_1, \ldots, u_{n-1}$ is equal to the sequence $w_0, w_1, \ldots, w_{n-1}$. □

The next result is about the prefixes of the conjugates of a Christoffel word.

Lemma 15.2.2 *Let $w = w_0 <_{lex} w_1 <_{lex} \cdots <_{lex} w_{n-1}$ be the conjugates of some lower Christoffel word $w \in \{a < b\}^*$, $k \in \{0, \ldots, n-1\}$; and p_i be the prefix of length k of w_i. Then:*

(i) the set $\{p_i, i = 0, \ldots, n-1\}$ has $k + 1$ elements;
(ii) this set is equal to the set of prefixes of length k of the words $C^j(w)$, $j = n - k, \ldots, n$;
(iii) if $k \geq 1$, then for some h, one has $\forall i = 0, \ldots, n-1$, $|p_i|_a = h$ or $h+1$; moreover, if $i \leq j$, then $|p_i|_a \geq |p_j|_a$.

Note that (i) is particular case of Theorem 15.3.1 due to Clough and Myerson.

Proof. In what follows, we use Theorem 15.2.1 and its notations. Observe that, for $i = 0, \ldots, n-2$, p_{i+1} is either equal to p_i, or is obtained by replacing some factor ab within it by ba, or by replacing the final letter a by b. In the first case, $p_i = p_{i+1}$, and in the two others, $p_i <_{lex} p_{i+1}$. Moreover, the replacements occur in all possible positions (see, for example, (15.1)) so that (i) follows. It follows also that the set in (i) is equal to the union of $\{p_0\}$ together with the set of p_i, $i \geq 1$, such that the factorization $w_i = ubav$ satisfies $|ub| \leq k$. We have $w_i = C^j(w)$ with $j \equiv iq^* \mod n$, $j \in \{1, \ldots, n-1\}$. Hence $i \equiv jq$. Thus $|ub| \equiv ip^* \equiv jqp^* \equiv -j \equiv n-j$. Thus the condition $|ub| \leq k$ is equivalent to $n - j \leq k$, that is, $j \in \{n-k, \ldots, n-1\}$. This proves (ii), taking into account that p_0 is the prefix of length k of $w = C^n(w)$.

Let $|p_0|_a = h + 1$ (note that p_0 begins by a, hence $h \geq 0$). Thus (iii) follows from the previous observation. □

We say that a set of words on the alphabet $\{a, b\}$ is *balanced* if, for any elements u and v of the same length in this set, the number of a's in u and v differ in absolute value by at most 1: $||u|_a - |v|_a| \leq 1$. Note that the condition with b replacing a is equivalent, since the word has only two letters. We say that a word is *balanced* if the set of its factors is balanced.

The following result is a consequence of Lemma 15.2.2 (iii), since the set of circular factors of a word is equal to the set of prefixes of its conjugates.

Corollary 15.2.3 *The set of circular factors of a Christoffel word is balanced.*

The following result strengthens Theorem 8.2.3.

Lemma 15.2.4 *Let C be the conjugation class of some Christoffel word. Then C is closed under the reversal mapping, and this mapping is decreasing for the alphabetical order.*

Proof. Since the reversal is an anti-automorphism, the set of reversals of a conjugation class is a conjugation class. This implies the first statement, by Theorem 8.2.3.

Let n be the length of the lower Christoffel word w, which is also the cardinality of its conjugation class C. We use Figure 8.2: the path from A to B, which discretizes from below the segment AB, is coded by w; the same holds for the path from B to C. Each conjugate u of w codes the path from some point E on the path from A to B, to the point F, the latter being defined by the condition $\overrightarrow{EF} = \overrightarrow{AB}$. We obtain exactly all conjugates of w by taking all points E on the path from A to B, excluding B.

The vertical distance from E to the segment AB is in the set $I = \{\frac{i}{p}, i = 0, \ldots, n-1\}$, where $p = |w|_a$ (for this, see the proof of Theorem 15.1.1). The mapping from C into I, given by mapping u onto the vertical distance from E to the segment AB, is a bijection by Section 15.1, which is increasing by Corollary 8.2.5 (with C lexicographically and I naturally ordered).

Symmetrically, by considering the discretization from above of the segment AB and the vertical distances, we obtain a decreasing bijection from C (which is, by Theorem 8.2.3, the conjugation class of the upper Christoffel word $\tilde{w}$ associated with

w) onto the set I. The composition of the first mapping followed by the inverse of the second is then a decreasing mapping from C into itself. It is enough to show that this mapping is the reversal mapping. But this is seen by taking symmetry with respect to the centre of the defining rectangle of w: this symmetry exchanges lower and upper discretization and preserves vertical distances. □

Theorem 15.2.5 *(Pirillo [Pi]) Let $w = amb \in \{a, b, \ldots\}^*$, where a and b are distinct letters. Then amb and bma are conjugate if and only if w is a lower Christoffel word on the alphabet $\{a < b\}$.*

Proof. If w is a lower Christoffel word on the alphabet $\{a < b\}$, then amb and bma are conjugate by Theorem 2.3.1 and Theorem 8.2.3.

Conversely, suppose that amb and bma are conjugate. We know that a Christoffel word of length ≥ 2 begins with a if and only if it is a lower Christoffel word. Thus it is enough to show that w is a Christoffel word on the alphabet $\{a, b\}^*$; equivalently, that m is a central word.

Note that mab and mba are conjugate (since so are amb and mba, bma and mab). Thus $mab = uv$, $mba = vu$ for some words u and v. Then u and v are both non-empty because $a \neq b$.

If u is of length 1, then $u = a$ and therefore $mb = v$, $mab = uv = amb$; thus $ma = am$ and m is a prefix of am. By Lemma 14.3.1 (i), m has the period 1 and thus m is a power of a, hence a central word. Similarly, if v is of length 1, m is a central word.

Suppose now that u and v are both of length at least 2. Then $mabu = uvu = umba$. Thus m is a prefix of um, and by Lemma 14.3.1 (i), m has the period $|u|$. Symmetrically, m has the period $|v|$. The equalities $mab = uv$, $mba = vu$ imply that ba is a suffix of u and, since the length of m is $|u| + |v| - 2$ and that of v is ≥ 2, that u is a prefix of m; in particular, the letters a and b appear in m. Suppose that $|u|$ and $|v|$ are relatively prime. By Corollary 14.3.7, $m \in \{a, b\}^*$. We conclude by Theorem 14.3.3 that m is a central word.

On the contrary, if $d = \gcd(|u|, |v|)$ is ≥ 2, then by Corollary 14.3.9, m has the period d. Note that u and v are prefixes of m so that they also have this period, and u and v are power of some word x of length d. Thus so are uv and vu, but they have different suffixes ab and ba, so this is not possible. □

Corollary 15.2.6 *(Ellis, Ruskey, Sawada, Simpson [ERSS], Theorem 5) Let $k_1, \ldots, k_n \in \mathbb{N}$, $n \geq 2$, and let w be the word $k_1 \cdots k_n$. The following conditions are equivalent:*

(i) w is a lower Christoffel word on the alphabet $\{k_1, k_1 + 1\}$, naturally ordered;
(ii) w is conjugate to the word $(k_1 + 1)k_2 \cdots k_{n-1}(k_n - 1)$.

Proof. (i) implies (ii): since w is a lower Christoffel word on the alphabet $\{k_1 < k_1 + 1\}$, its last letter is $k_n = k_1 + 1$. By the previous theorem, w is conjugate to $k_n k_2 \cdots k_{n-1} k_1$. Hence (ii) holds.

(ii) implies (i): since two conjugate words have the same letters with the same multiplicities, the multisets $\{k_1, k_2, \ldots, k_{n-1}, k_n\}$ and $\{k_1 + 1, k_2, \ldots, k_{n-1}, k_n - 1\}$ are equal; thus we must have $\{k_1, k_n\} = \{k_1 + 1, k_n - 1\}$ as multisets. By inspection, it is seen that the only possibility is that $k_n = k_1 + 1$. Thus w is conjugate to $k_n k_2 \cdots k_{n-1} k_1$. By Theorem 15.2.5, (i) holds. □

15.2.1 Exercises

Exercise 15.2.1 *Let w be a Christoffel word. Use the notation $A(u)$ of Exercise 2.3.2. Show that the set of $A(u)$, u a conjugate of w, is an interval in $\mathbb{N}$, with smallest value $\frac{1}{2}(|w|_a - 1)(|w|_b - 1)$. Show that the mapping $u \mapsto A(u)$ from the set of conjugates of w into $\mathbb{N}$ is strictly increasing for the lexicographical order. Hint: use Theorem 15.2.1 and Exercise 2.3.1.*

Exercise 15.2.2 *Given a word w on the alphabet $\{a, b\}$, let $M(w) = \sum_{w=ubv} |bv|$, called* the moment *of w. Show that $M(w) = \sum_{w=ps} |p|_b$. Show that $M(w) = \binom{w}{b} + \binom{w}{ba} + \binom{w}{bb}$, using the subword functions (see Exercise 2.3.2). Show that $M(w) = A(w) + \binom{w}{b} + \binom{w}{bb}$ (notations of Exercise 2.3.2). Show that the set of $M(u)$, u a conjugate of w, is an interval in $\mathbb{N}$, with smallest value $\frac{1}{2}(|w| + 1)(|w|_b - 1) + 1$. See Chuan [Ch4, Ch5].*

Exercise 15.2.3 *Let w be a word of length $n \geq 1$. Let N be the number of circular factors of w.*

(i) Show that if w is a Christoffel word, then $N = \frac{n(n+3)}{2}$. Hint: use Lemma 15.2.2.
(ii) Suppose that w is a power with exponent $\ell \geq 2$ of a radical word. Show that $N \leq 1 + n^2/\ell$. Hint: it has, for $i = 1, \ldots, n$, at most n/ℓ circular factors of length i. Show that this latter number is smaller than the number in (i).

Exercise 15.2.4 *Suppose that x is an ancestor of w in the tree of Christoffel words. Deduce from Exercise 14.5.2 that for $k < |x|$, x and w have the same set of circular factors of length k. Hint: use Lemma 15.2.2.*

Exercise 15.2.5 *Show that the conjugation class of the word aabb is closed under reversal, but that the final statement of Lemma 15.2.4 does not hold.*

Exercise 15.2.6 *Let L be the set of circular factors of a word $w \in \{a, b\}^*$, and denote by L_k the set of words of length k in L. Show the equivalence between the three following conditions:*

(i) L is balanced;
(ii) for each k, the set $\{|w|_b, w \in L_k\}$ has cardinality at most 2;
(iii) for each k, the set L_k intersects at most 2 commutation classes (a commutation class *is the set of words having the same commutative image; see Chapter 1).*

See Carey and Clampitt [CC] (Proposition 1).

Exercise 15.2.7

(i) *Show that if the set of circular factors of w is balanced, so is the set of factors of the bi-infinite word* $^{\infty}w^{\infty}$.
(ii) *Show that if the set of factors of a periodic bi-infinite word s is balanced, then s satisfies the Markoff property for bi-infinite words.*
(iii) *Deduce that if w is a Christoffel word, then* $^{\infty}w^{\infty}$ *satisfies this property. See [Re2] (Theorem 6.1).*

In [HeN], Hegedüs and Nagy determine the set of all periods of all conjugates of a Christoffel word. In [La], Lapointe determines the smallest period of each conjugate of a Christoffel word. In [CS], Currie and Saari describe the set of all smallest periods of all factors of a Sturmian sequence.

In [PaR], following Morikawa [Mo] and Simpson [Sm], Paquin and the author study the *superimposition* of two Christoffel words; this is related to the *Fraenkel conjecture* and to the *coin problem.*

15.3 Circular factors

A *composition of* m is a sequence $C = (c_1, \ldots, c_k)$ of positive integers such that $c_1 + \cdots + c_k = m$; its *length* is k, and the *parts* of C are the integers c_i. The definition makes sense also when $m = 0$ and $k = 0$, and one obtains the empty composition.

We need two constructions with compositions. If C is a composition as above and u a word of length m, we obtain a unique factorization $u = u_1 \cdots u_k$ with $|u_i| = c_i$ for any i; we say that this factorization of u is the *C-factorization* of u; conversely, each factorization of u as earlier gives the composition $(|u_1|, \ldots, |u_k|)$ of $|u|$, provided the u_i are non-empty words.

For the second construction, let C be as earlier, and let $E = \{c_1, c_1 + c_2, \ldots, c_1 + \cdots + c_{k-1}\}$, which is a subset of $\{1, 2, \ldots, m-1\}$; the mapping $C \mapsto E$ is a bijection from the set of compositions of m, of length k, onto the set of subsets of $\{1, 2, \ldots, m-1\}$, of cardinality $k-1$; the reverse bijection associates with the subset $E = \{e_1 < \cdots < e_{k-1}\}$ of $\{1, 2, \ldots, m-1\}$ the composition $(e_1 - e_0, e_2 - e_1, \ldots, e_k - e_{k-1})$, where we let $e_0 = 0, e_k = m$.

Theorem 15.3.1 *(Clough and Myerson [CM1, CM2]; see Carey and Clampitt [CC])* *Let* $w = w_0 <_{lex} w_1 <_{lex} \ldots <_{lex} w_{n-1}$ *be the conjugates of some lower Christoffel word* $w \in \{a < b\}^*$ *of length n and slope* q/p. *Let* $m \in \{0, \ldots, n-1\}$ *and* $C = (c_1, \ldots, c_k)$ *be a composition of m. Define the equivalence relation* $\sim_C$ *on* $\{0, 1, \ldots, n-1\}$ *as follows: denoting by* p_i *the prefix of length m of* w_i, *consider the C-factorizations* $p_i = u_1 \cdots u_k$ *and* $p_j = v_1 \cdots v_k$ *of* p_i *and* p_j; *then* $i \sim_C j$ *if and only if the words* u_t *and* v_t *have the same commutative image for each* $t = 1, \ldots, k$. *Then:*

(i) *the equivalence relation* $\sim_C$ *has* $k+1$ *equivalence classes;*
(ii) *the classes are intervals of* $\{0, 1, \ldots, n-1\}$;

(iii) *the cardinalities of the classes, ordered from the first interval to the last, form the composition of* n *associated with the subset* $E = \{c_1 p \bmod n, (c_1 + c_2)p \bmod n, \ldots, (c_1 + \cdots + c_k)p \bmod n\}$ *of* $\{1, 2, \ldots, n-1\}$.

An example may be useful. Let $w = aabaabab$, of length $n = 8$ and slope 3/5, and $C = (1, 2, 1)$, $m = 3$; then, by Figure 15.1, the words p_i, $i = 0, \ldots, 7$, are $a|ab|a, a|ab|a, a|ba|a, a|ba|a, a|ba|b, b|aa|b, b|aa|b, b|ab|a$, where the vertical bar indicates the C-factorizations. The equivalence classes of $\sim_C$ are therefore $\{0, 1, 2, 3\}, \{4\}, \{5, 6\}, \{7\}$. Their cardinalities determine the composition $(4, 1, 2, 1)$ of 8, which is associated with the subset $E = \{4, 5, 7\}$ of $\{1, 2, 3, 4, 5, 6, 7\}$, and we have $\{1 \cdot 5 \bmod 8, 3 \cdot 5 \bmod 8, 4 \cdot 5 \bmod 8\} = \{5, 7, 4\}$, as predicted by (iii), since $p = 5$. Note that the largest elements of the classes of $\sim_C$, excluding the last class, are 3, 4, 6, which are obtained by subtracting 1 from the elements of the aforementioned set E.

Note that if C has only one part, then Condition (i) in the theorem means that the set of circular factors of w is balanced (see Exercise 15.2.6). And if $C = (1, 1, \ldots, 1)$, then it means that there are $m + 1$ circular factors of length m.

Proof. We first prove (ii). Let $h(u) = (|u|_a, |u|_b) \in \mathbb{N}^2$ denote the commutative image of $u \in \{a, b\}^*$, and denote $H(p_i) = (h(u_1), \ldots, h(u_k)) \in (\mathbb{N}^2)^k$, with the notations of the statement. Totally order $\mathbb{N}^2$ by $(s, t) < (s', t')$ if $s + t < s' + t'$, or if $s + t = s' + t'$ and $t < t'$. Then order $(\mathbb{N}^2)^k$ lexicographically from left to right.

We know by Theorem 15.2.1 that w_{i+1} is obtained from w_i by replacing some factor ab by ba. If this factor is at the right of $p_i = u_1 \cdots u_k$ in w_i, or if it is completely within some u_j, then $H(p_i) = H(p_{i+1})$. Otherwise, this factor is at the end of some u_j, and either the b is in u_{j+1} or $j = k$; in both cases, $H(p_i) < H(p_{i+1})$.

From this follows that $H(p_0) \leq \cdots \leq H(p_{n-1})$. Since $i \sim_C j$ is equivalent to $H(p_i) = H(p_j)$, it follows that the classes of $\sim_C$ are successive intervals of $\{0, 1, \ldots, n-1\}$.

Thus (ii) is proved, and we prove (iii), which is equivalent to the following statement: the largest elements of the classes (which are intervals), excluding the last class, form the subset $\{x - 1, x \in E\}$, where E is described in (iii), (a look at the example before the proof may be useful, explaining the -1). As recalled earlier, one obtains w_{i+1} from $w_i = u_1 \cdots u_k \cdots$ by replacing a factor ab by ba. If this factor ab is completely in some u_j, or is at the right of u_k, then $i \sim_C i + 1$; thus, $i, i + 1$ are in the same class of $\sim_C$, except if the letter a of the factor ab is the last letter of some u_j; therefore, this a is the last letter of $u_1 \cdots u_j$, which is of length $c_1 + \cdots + c_j$, and i is the largest element of its equivalence class. It follows from Theorem 15.2.1 (where we replace i by $i + 1$) that $c_1 + \cdots + c_j \equiv (i + 1)p^* \bmod n$; hence $i + 1 \equiv (c_1 + \cdots + c_j)p$, and finally $i = ((c_1 + \cdots + c_j)p \bmod n) - 1$. Indeed, both sides are congruent modulo n and belong to $\{0, 1, \ldots, n-1\}$ because $0 < c_1 + \cdots + c_j \leq m \leq n - 1$ and p is invertible modulo n so that $(c_1 + \cdots + c_j)p$ is not congruent to $0 \bmod n$.

Finally, (i) follows from (iii). Indeed, the composition in (iii) is of length $k + 1$, since the subset in (iii) has k elements because the k partial sums $c_1 + \cdots + c_j$ are in $\{1, \ldots, n-1\}$ and are distinct, and since p is a generator of the additive group $\mathbb{Z}/n\mathbb{Z}$. □

Let $k < n$ and w be a Christoffel word of length n. Let u be a circular factor of length k of w. We call *multiplicity of* u the number of times u appears as a circular factor of w. It is convenient to view the conjugation class of w, denoted by (w), as a *circular word*, that is, a regular n-gon whose vertices are labelled by the letters of w, together with the orientation of the polygon which is compatible with the word, read from left to right; a circular factor of w is then identified with a sequence of consecutive edges of this polygon (not more than n edges); we then say that it is a *factor* of the circular word (w). The multiplicity may be seen on the circular word associated to w; see, for example, Figure 15.1 (disregarding the numbers): aba has multiplicity 4, since it appears four times as a factor of the circular word.

Equivalently, the multiplicity of u is the number of $i = 0, \dots, n-1$ such that $u = p_i$, with the notations of Lemma 15.2.2. Note that there are n occurrences of circular factors of length k in w; hence the sum of the multiplicities of all circular factors of length k is n.

The following corollary was observed by David Clampitt and the author.

Corollary 15.3.2 *Let $k < n$ and w be a Christoffel word of length n and slope q/p. The multiplicities of the circular factors of length k of w take at most three values, and if there are three values, one is the sum of the two others.*

This result is a discrete version of the famous *three-distance theorem* (conjectured by Steinhaus); see Exercise 15.3.4 and [AB, Be] for a survey of this subject, references, and applications to Sturmian sequences.

Let us mention a related result. Bucci, De Luca, and Zamboni [BDLZ] define *reversible Sturmian factorizations*; this is a factorization of a Sturmian sequence as a product of infinitely many Christoffel words, such that the sequence obtained by reversing each of these words is also Sturmian. They show that, in this case, only two or three Christoffel words appear in the factorization and that if there are three, then one is a product of the two others; hence this product is its standard factorization (Theorem 3.1); the proof uses the three-distance theorem.

For the proof of the corollary, we have followed Liang [Li] and Shiu [Si].

Proof. In the following proof, all numbers are modulo n, and $\mathbb{Z}/nZ$ is identified with $\{0, 1, \dots, n-1\}$. We also use the usual order on the latter set.

Let $x, y \in \mathbb{Z}/n\mathbb{Z}$; the interval $[x, y]$, called the *interval from x to y*, is the subset $\{x, x+1, \dots, y\}$ if $x \leq y$, and $\{x, \dots, n-1, 0, \dots, y\}$ if $x > y$. In other words, it is a cyclic interval of $\mathbb{Z}/n\mathbb{Z}$, if one views the latter as a circular set, as in Figure 15.3, with $n = 11$: for example, the interval $[2, 6]$ is $\{2, 3, 4, 5, 6\}$, and the interval $[8, 2]$ is $\{8, 9, 10, 0, 1, 2\}$; the *length* of the interval $[x, y]$ is its cardinality minus 1; in particular, it is $y - x$ if $x \leq y$.

(i) We consider the graph $G_0 : 0 \to p \to 2p \to \cdots \to kp$; we call *legal* the vertices and edges of this graph; the set of its vertices is denoted V. We may write $V = \{0 = x_0 < x_1 < \cdots < x_k\}$. The *legal intervals* are $[x_0, x_1], [x_1, x_2], \dots, [x_k, x_{k+1}]$, where $x_{k+1} = 0$; thus $[x, y]$ is *legal* if and only if $x, y \in V$, $x \neq y$, and if it does not contain any other legal point.

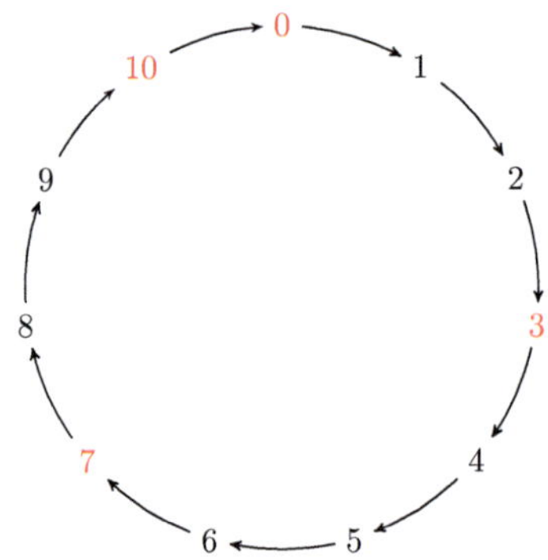

Figure 15.3 The legal points $0 \cdot 7 \equiv 0, 1 \cdot 7 \equiv 7, 2 \cdot 7 \equiv 3, 3 \cdot 7 \equiv 10$ in $\mathbb{Z}/11\mathbb{Z}$, $k = 3, p = 7$

Note that there are $k+1$ elements in V, since p and n are relatively prime, and $k < n$. Let x be a legal vertex; then $x + p$ (resp. $x - p$) is not legal only if $x = kp$ (resp. $x = 0$).

We define a graph G whose vertex set is the set of legal intervals, with edges $[x, y] \to [x', y']$ if both edges $x \to x'$ and $y \to y'$ are legal. Then $[x, y]$ and $[x', y']$ have necessarily the same length. Since G_0 is a chain, it follows that G is a union of chains: indeed, each vertex of G has at most one ingoing and one outgoing, edge.

Consider a sink in G (that is, a vertex with no outgoing edge). Then either (i) it is of the form $[kp, y]$; or (ii) of the form $[x, kp]$; or (iii) of the form $[x_k - p, x_1 - p]$. Indeed, if $[x, y]$ is a sink not of the form (i) or (ii), then $x + p, y + p$ are legal points, so that $[x + p, y + p]$, being not a legal interval must contain a third legal point; if this point is of the form $z + p$, with z legal, then $[x, y]$ contains the legal point z, a contradiction with $[x, y]$ being legal; hence $x + p, y + p$ contains 0, and no other legal point (otherwise this fourth legal point is of the form $z + p$, z legal, and we find the same contradiction). This implies that $x + p = x_k, y + p = x_1$.

We deduce that G has at most three sinks; hence G is the union of at most three chains.

We show that the legal intervals $[0, x_1]$ and $[x_k, 0]$ have the same length as, respectively, the intervals in (ii) and (i). Indeed, there is a path in G from $[0, x_1]$ until a sink; this sink cannot be $[x_k - p, x_1 - p]$, since they have not the same length (the length of $[x_k - p, x_1 - p]$ is equal to that of $[x_k, x_1]$ and the latter contains strictly $[0, x_1]$); it cannot be $[kp, y]$; otherwise this path is formed by two paths in G_0: $0 \to p \to \cdots \to kp$ and $x_1 = ip \to (i + 1)p \to \cdots \to jp = y$, with $i \geq 1$, and they are of length k so that the second one must contains the edge $kp \to (k + 1)p$, which is not legal, a contradiction. Thus there is a path from $[0, x_1]$ to $[x, kp]$, and they have the same length; similarly $[x_k, 0]$ and $[kp, y]$ have the same length.

(ii) It follows from (i) that there are at most three possible lengths of legal intervals. Moreover, if there are three, then they are the lengths of $[0, x_1]$, $[x_k, 0]$, $[x_k, x_1]$, and the largest length (that of $[x_k, x_1]$) is the sum of the two others. The lengths of the $k + 1$ legal intervals are the numbers $x_1 - x_0, x_2 - x_1, \ldots, x_k - x_{k-1}, n - x_k$. Thus the corollary follows from Theorem 15.3.1, applied to the composition $C = (1, 1, \ldots, 1)$ with k parts all equal to 1; the set E of the theorem is $\{x_1, \ldots, x_k\}$; the associated composition (as in part (iii) of the theorem) is $(x_1 - x_0, x_2 - x_1, \ldots, x_k - x_{k-1}, n - x_k)$, whose parts are the cardinalities of the classes, which are precisely the multiplicities of the circular factors of length k of w. □

The following result was obtained by Borel and the author [BR]. The implication $(i) \Rightarrow (ii)$ is a special case of Theorem 15.3.1 (case of compositions with all parts equal to 1).

Theorem 15.3.3 *The following conditions are equivalent for any word w of length $n \geq 2$ on the alphabet $\{a, b\}$:*

(i) *w is a conjugate of a Christoffel word;*
(ii) *for any $k = 0, 1, \ldots, n-1$, w has $k+1$ circular factors of length k;*
(iii) *w is a radical word and has $n-1$ circular factors of length $n-2$.*

Lemma 15.3.4 *Let w be a word of length n. The following conditions are equivalent:*

(i) *w is radical;*
(ii) *for $k = 0, 1, \ldots, n-1$, w has at least $k+1$ circular factors of length k.*

Proof. If w is not radical, then $w = u^p, p \geq 2$. Then $k = |u| < n$ and w has $\leq k$ circular factors of length k.

Conversely, denote by a_i the number of circular factors of length i, $i = 0, 1, \ldots, n$, of w. One has $1 = a_0 \leq a_1 \leq \ldots \leq a_n$, since each circular factor of length $i < n$ is a prefix of some circular factor of length $i+1$. Suppose that $a_k \leq k$ for some $k \leq n-1$. Then $k \geq 1$ and for some $l \leq k$, one has $a_{l-1} = a_l$. Consider the following directed graph: the vertices are the a_{l-1} circular factors of length $l-1$ of w, and there is an edge $u \xrightarrow{x} v$ if, for some letters x and y, $ux = yv$ is a circular factor of length l of w. Since $a_{l-1} = a_l$, each circular factor of length $l-1$ of w is a prefix of a unique circular factor of length l. Hence each vertex has exactly one outgoing edge. It follows that each strongly connected component of the graph is a simple closed path.

The sequence w^∞ is the label of some infinite path in the graph: indeed, write $w^\infty = a_1 a_2 a_3 \cdots$; then the path is the sequence of successive edges $a_{i+1} a_{i+2} \cdots a_{i+l-1} \rightarrow a_{i+2} a_{i+3} \cdots a_{i+l}$. This path is ultimately in some strongly connected component. It follows that w^∞ has a period not greater than the number of vertices, which is $a_{l-1} \leq a_k \leq k < n$; thus w^∞ has a period $< n$, and w is therefore not radical. □

Proof of Theorem 15.3.3. (i) implies (ii): we may assume that w is a lower Christoffel word and that $a < b$. Let $k = 0, \ldots, n-1$. The circular factors of length k of w are the prefixes of length k of the conjugates of w. Thus (ii) follows from Lemma 15.2.2.

(ii) implies (iii): use Lemma 15.3.4.

(iii) implies (i): Note that the circular word (w) has n occurrences of factors of length $n-2$. By hypothesis there are $n-1$ distinct such factors. Therefore (w) has exactly one factor, say m, of length $n-2$ that appears twice, whereas all other factors of length $n-2$ appear only once.

Since m has two occurrences as a factor of (w), and since $|w| = n = |m| + 2$, there are letters c, d, e, and f such that these two occurrences extend into two conjugates cmd and emf of w, which are distinct since all conjugates $C^i(w)$, $i = 0, \dots, n-1$, are distinct (w being radical). We thus have $c \neq e$ or $d \neq f$. Now, since the two words cmd and emf are conjugate, we must have $\{c, d\} = \{e, f\}$; we cannot have $c = d$, otherwise $c = d = e = f$, a contradiction. Thus $\{c, d\} = \{e, f\} = \{a, b\}$. If $c = a$, then $d = b$, and $e = b$ (since $e = a$ would imply $f = b$, and thus $e = c$ and $f = d$, a contradiction), thus $f = a$. We deduce that amb and bma are conjugate, which implies by Theorem 15.2.5 that amb is a Christoffel word and w is conjugate to a Christoffel word. The case $c = b$ is treated similarly. □

Let L be some subset of $\{a, b\}^*$. We say that $w \in L$ is *right special with respect to L* if $wa, wb \in L$. We say simply *special* if there is no ambiguity.

Corollary 15.3.5 *Let $w = amb$ be a Christoffel word of length n, and let L be the set of circular factors of w. Then the set of elements of L which are right special with respect to L is totally ordered for the suffix ordering and has exactly one element of each length $k = 0, \dots, n-2$. This set is the set of suffixes of m.*

Proof. By Theorem 15.3.3, w has $k+1$ circular factors of length k for any $k = 0, 1, \dots, n-1$. Observe that for each u in L of length $0, 1, \dots, n-2$, ua or ub is in L. It follows that for $k = 0, \dots, n-2$, there is exactly one right-special word in L. Moreover, a suffix of a right-special word is clearly right special. Now, the word m is special, since the upper Christoffel word bma is a conjugate of w so that $ma, mb \in L$. Since the length of m is $n-2$, the corollary follows. □

Corollary 15.3.6 *Let w be a Christoffel word of length n, $k \in \{0, \dots, n-1\}$, and s (resp. p) be the suffix (resp. prefix) of length k of w. Then the $k+1$ circular factors of length k of w are the factors of length k of sp.*

Note that sp is of length $2k$ and so has $k+1$ occurrences of factors of length k, which are therefore all distinct. If we take, for example, $w = aabaabaabab$ and $k = 4$, then $sp = ababaaba$, and the five factors are $abab$, $baba$, $abaa$, $baab$, and $aaba$.

Proof. Note that the factors of length k of sp are the prefixes of length k of the words $C^h(w)$, $h = 0, -1, \dots, -k$. Equivalently (since $C^n(w) = w$), they are the prefixes of length k of the words $C^j(w)$, $j = n, n-1, \dots, n-k$. Thus the lemma follows from Lemma 15.2.2 (ii). □

See Labbé and Lapointe [LL], Corollary 2.6, for another approach to the previous result. For further use, we state and prove the following lemma.

Lemma 15.3.7 *The set of circular factors of a Christoffel word is closed under reversal.*

Proof. This follows from Theorem 8.2.3, which implies that the reversal of a Christoffel word w is a conjugate of w, hence a circular factor of w. □

15.3.1 Exercises

Exercise 15.3.1 *Using the bijection given at the beginning of this section, show that the number of compositions of length k of m is equal to* $\binom{m-1}{k-1}$ *and that the number of all compositions of m is* 2^{m-1}.

Exercise 15.3.2 *Let w be some word, and suppose that for some* $k = 1, \ldots, |w| - 1$, *the circular factors of length k of w have all the same commutative image (see Section 15.1). Show that w is not radical.*

Exercise 15.3.3 *Let* $t_1 < t_2 < t_3$ *be the remainders of* $3, 6, 12$ *modulo* 10. *With* $t_0 = 0, t_4 = 10$, *show that the set* $\{t_{i+1} - t_i, i = 0, 1, 2, 3\}$ *is equal to* $\{1, 2, 3, 4\}$. *Deduce from Theorem 15.3.1 that the cardinalities of the classes modulo* $\sim_C$ *may take more than three values so that Corollary 15.3.2 has no extension in this setting.*

Exercise 15.3.4 *Consider a real number* α *and* $n \in \mathbb{P}$; *if* α *is rational, assume that n is smaller than the smallest positive denominator of* α. *Denote by* $\{x\}$ *the fractional part of any real x. Denote by* $0 = x_0 < x_1 < \cdots < x_{n-1}$ *the numbers* $\{i\alpha\}$, *reordered (with* $0 \le i \le n-1$*). Let* $x_n = 1$. *Show that the differences* $x_i - x_{i-1}, i = 1 \cdots n$ *take at most three values, and that if there are three of them, then the largest is the sum of the two others. This is the three-distance theorem of Sós [So], Surányi [Su], and Świerczkowski [Sw]. Hint: if* α *is rational, multiply everything by its denominator and use the proof of Corollary 15.3.2. If* α *is irrational, take a sequence of rational numbers whose limit is* α, *and use the pigeonhole principle (see [BeR], C. p. 4).*

Exercise 15.3.5 *Let w be a radical word of length n, which is not the conjugate of some Christoffel word, and let N be the number of circular factors of w. Show that* $N > \frac{n(n+3)}{2}$. *Hint: use Lemma 15.3.4 and Theorem 15.3.3. Deduce that for any word w, on any alphabet, w is a conjugate of a Christoffel word if and only if* $N = \frac{n(n+3)}{2}$. *Hint: use Exercise 15.2.3. Conclusion: N is* $<, =, > \frac{n(n+3)}{2}$ *if w is, respectively, not radical, conjugate to a Christoffel word, radical but not conjugate to a Christoffel word.*

Exercise 15.3.6 *Call* factor set *of length n the set of circular factors of some Christoffel word of length* $> n$. *Such a set has cardinality* $n + 1$ *(Theorem 15.3.3). It is known that factors sets may equivalently be defined as sets of factors of length n of infinite Sturmian words (see Lothaire [Lo], Chapter 2). Let the alphabet be* $\{a < b\}$. *Let F be a factor set of length* $n \ge 1$.

(i) *Show that there is a unique word* $\ell(F)$ *such that* $a\ell(F), b\ell(F) \in F$. *Show that if F is the set of circular factors of length n of the Christoffel word amb of length* $> n$, *then* $\ell(F)$ *is the prefix of length* $n - 1$ *of the central word m. Hint: use Corollary 15.3.5.*

(ii) *Show that, for the alphabetical order,* $a\ell(F)$ *is the minimum and* $b\ell(F)$ *the maximum of F. Hint: use (i) and Lemma 15.2.2. See Perrin, Restivo*

[PeR] (Proposition 2). See also Labbé and Lapointe [LL] (Lemma 4.1). An interesting result on factor sets is that each factor set of length n in $\{a,b\}^*$ is a basis of some subgroup of index n of the free group F(a, b), due to Berstel, De Felice, Perrin, Rindone, and the author, Corollary 6.2.2 in [BDFPRR].

Exercise 15.3.7 A maximal repeat *in a non-empty word w is a factor m of w such that, for some words x, x', y, y', one has $x \neq x'$, $w = xmy = x'my'$, x and x' do not end with the same letter, and y* and *y' do not begin with the same letter. In other words, m has two* occurrences *in w which cannot be extended, neither to the left nor to the the right, into occurrences of the same longer factor.*

(i) *Show that if m is a border of w, or if m is empty, then it is a maximal repeat. Deduce that a palindromic prefix of a palindrome w is a maximal repeat of w.*
(ii) *Suppose that w is a central word, and awb the corresponding lower Christoffel word. Let m be a maximal repeat, with the notations x, y, x', y' mentioned earlier. Show that if y and y' are non-empty, then m is right special with respect to the set of factors of m and also to the set of circular factors of awb; deduce by Corollary 15.3.5 that m is a suffix of w. Deduce that m is in all cases a suffix of w. Similarly, show that m is a prefix of m. Show that m is a palindrome.*
(iii) *Deduce that the maximal repeats of a central word w are the palindromic prefixes (equivalently, palindromic suffixes) of w (result due to Epifanio, Mignosi, Shallit, and Venturini [EMSV], Theorem 24).*

Exercise 15.3.8 *Let L be a totally ordered set with n elements, and C a permutation of L. A* descent *of C is an element x of L* \ max(L), *with next element y in L, such that $C(x) > C(y)$. The* descent set *of C is the set of its descents. Let w be a radical word of length n on the totally ordered alphabet $\{a < b\}$, L be the set of its conjugates totally ordered by $<_{lex}$, and let $C : L \to L$ be the conjugator. Show that the unique descent of C is the largest conjugate of w beginning by a.*

Exercise 15.3.9 *Let $w_0 <_{lex} w_1 <_{lex} \dots <_{lex} w_{n-1}$ be the n conjugates of a radical word w of length n on the totally ordered alphabet $\{a < b\}$. Show that the two following conditions are equivalent:*

(i) *w is a conjugate of some Christoffel word;*
(ii) *for some integer m, one has for any i, $0 \leq i \leq n-1$, $w_i = C^{im}(w_0)$.*

Hint: for "(i) implies (ii)", use Theorem 15.2.1; for "(ii) implies (i)", use Exercises 1.0.4, 15.3.8, and 15.1.2. See [JZ] (Theorem C).

Exercise 15.3.10 *Let $w = a_1 \cdots a_n$ be a word of length n; a subset I of $\{1, \dots, n\}$ is called* a circular attractor *of w if, for each non-empty circular factor f of w, there exist i and j such that either $1 \leq i \leq j \leq n$, $f = a_i \cdots a_j$, and I intersects the interval $\{i, i+1, \dots, j\}$ or $1 \leq i < j \leq n$, $f = a_j \cdots a_n a_1 \cdots a_i$, and I intersects $\{1, 2, \dots, i\} \cup \{j, j+1, \dots, n\}$. Informally, it means that each circular factor of w has an occurrence as a circular factor which contains some letter a_k, $k \in I$. Show that a radical word w is a conjugate* *of some Christoffel word if and only if it has a circular attractor of the form* $\{k, k+1\}$ *or* $\{n, 1\}$*. Hint: for the "only if" part, use Corollary 15.3.6; for the "if" part, use Theorem 15.3.3 and Lemma 15.3.4.*

Suppose that w *is a Christoffel word, and let* $w = v'u'$ *its palindromic factorization. Show that* $\{|v'|, |v'| + 1\}$ *is then a circular attractor, too. Use Theorem 14.2.1. See [MRRRS] (Theorem 34).*

Exercise 15.3.11 *Let* $w = a_1 \cdots a_n$ *be a word of length* n*; a subset* I *of* $\{1, \ldots, n\}$ *is called an* attractor *of* w *if, for each non-empty factor* f *of* w*, there exist* i *and* j *such that* $1 \le i \le j \le n$*,* $f = a_i \cdots a_j$*, and* I *intersects the interval* $\{i, i+1, \ldots, j\}$*. Suppose that* $w = uv = v'u'$ *is a Christoffel word with its standard and palindromic factorizations so that* $|u| = |u'|, |v| = |v'|$ *(Theorem 14.2.1). Let* $k = |v'|$*. Show that* $\{k, k+1\}$ *is an attractor of* w*. Show that, moreover, for each factor* f *of* w*,* i *and* j *as earlier are unique. Hint: suppose that some factor* f *of* w *is strictly at the rigth of* a_{k+1}*; use Lemma 14.3.5 to show that* f *has an occurence whose first letter is within* v'*; show that if* a_k *is not in this occurrence, then the length of* f *is smaller than that of* v' *and also smaller than that of* u'*. Using Exercise 15.3.10, show that some circular occurrence of* f *in* w *contains* a_k *or* a_{k+1} *and, by the length condition, that it is an occurrence in* w*. See [LR] (Theorem 4.1).*

15.4 Palindromic circular factors

The following result is a finitary version of a theorem of Droubay and Pirillo [DP], which characterizes Sturmian sequences.

Theorem 15.4.1 *([Re5], Theorem 17.1) A word* w *of length* n *on the alphabet* $\{a, b\}$ *is conjugate to a Christoffel word if and only if, for any* $k = 0, 1, \ldots, n-1$*,* w *has one (resp. two) palindromic circular factors of length* k *if* k *is even (resp. odd).*

We need two lemmas.

Lemma 15.4.2 *Let* w *be some word. If for each* $k = 0, 1, \ldots, |w|-1$*,* w *has one (resp. two) palindromic circular factors of length* k *if* k *is even (resp. odd), then* w *is radical.*

Proof. Suppose that w is not radical. Then $w = u^l$ for some radical word u and some integer $l \ge 2$.

If $|u|$ is even, then there is precisely one palindromic circular factor of w of length $|u|$ (since this length is $< |w|$); it is a conjugate of u. Replacing w by a conjugate if necessary, we may assume that u is a palindrome. Thus $u = v\tilde{v}$. Since $w = v\tilde{v}v\tilde{v}\cdots$, $v\tilde{v}$ and $\tilde{v}v$ are both circular factors of w. By the hypothesis, they must be equal so that $v = \tilde{v}$ and u is not radical, a contradiction.

Thus $|u|$ is odd. There are two palindromic circular factors of w of length $|u|$. They are both conjugate to u. By Exercise 15.4.1, u is not radical, a contradiction. □

The second lemma is a pretty result on balanced words due to Coven and Hedlund [CH], Lemma 3.06.

Lemma 15.4.3 *Let F be a set of words on the alphabet $\{a, b\}$, which is closed under taking factors. Then F is not balanced if and only if, for some palindrome m, it has the two elements ama and bmb.*

Proof, following [Lo], proof of Proposition 2.1.3. If F contains ama and bmb, it is not balanced, since $|ama|_a - |bmb|_a = 2$.

Conversely, suppose that F is not balanced. Then it has two elements u and v of the same minimal length such that $||u|_a - |v|_a| \geq 2$. By minimality, the first (resp. last) letters of u and v are distinct. We may assume that $u = amxu'$, $v = bmyv'$ with $\{x, y\} = \{a, b\}$. We must have $x = a$ and $y = b$; otherwise $u = ambu'$, $v = bmav'$, and by the inequality mentioned earlier $||u'|_a - |v'|_a| \geq 2$, $u', v' \in F$, contradicting minimality. Thus $u = amau'$, $v = bmbv'$ and in fact, $u = ama$, $v = bmb$ by minimality.

We show that m is a palindrome. There exists a prefix p of maximal length of m such that $\tilde{p}$ is a suffix of m. If $p \neq m$, then for some x and y such that $\{x, y\} = \{a, b\}$, px is a prefix of m and $y\tilde{p}$ is a suffix of m. Then apx is a proper prefix of u and $y\tilde{p}b$ is a proper suffix of v. If $x = a$ and $y = b$, we contradict the minimality. Thus $x = b$, $y = a$, and $u = apbu''$, $v = v''a\tilde{p}b$, contradicting the minimality again, with the words u'' and v''. Thus $p = m$ and m is therefore a palindrome. □

The proof of Theorem 15.4.1 follows closely, *mutatis mutandis*, the clever proof of Theorem 5 in [DP].[1]

Proof of Theorem 15.4.1.

(i) Suppose that w is a Christoffel word of length n; we may assume that $n \geq 2$; thus w contains both letters. Clearly, the number of palindromic circular factors of length 0 (resp. 1) of w is 1 (resp. 2). We prove the existence of a bijective mapping θ from the set of palindromic factors of length $k = 2, 3, \ldots, n-1$ of w into the set of palindromic factors of length $k = 0, 1, \ldots, n-3$ of w (this will prove the "only if" part). Define θ by $\theta(xmx) = m$, for any palindrome xmx, where x is a letter. If this mapping is not injective, then for some word m, ama and bmb are circular factors of w. This contradicts Lemma 15.2.3. Thus θ is injective.

We prove now that θ is surjective. Let m be a palindromic circular factor of length $k \leq n-3$ of w. There exists a letter a, say (the case for b is similar), such that am is a circular factor of w. Then either ama is a circular factor and we are done, because $\theta(ama) = m$, or amb is a circular factor. Then, by Lemma 15.3.7, bma is also a circular factor. Thus m is right special with respect to the set of circular factors of w so that by Corollary 15.3.5, it is a suffix of the right-special circular factor of length $n-2$. Thus there exists a letter x such that xm is right special so that xmx is a circular factor and $m = \theta(xmx)$.

(ii) For the reverse implication, define the following properties of a word $w \in \{a, b\}^*$ of length n and an integer k with $0 < k < n$:

[1] It would be interesting to have a simpler proof.

- $B_k(w)$: for any circular factors u and v of length at most k of w, and for any letter x, one has $||u|_x - |v|_x| \leq 1$.
- $S_k(w)$: there is a unique circular factor u of each length $\leq k-1$ of w such that ua and ub are circular factors of w (in other words, u is right special with respect to the set of circular factors of w).
- $R_k(w)$: for any circular factor u of length k of w, $\tilde{u}$ is also a circular factor of w.

We assume that w is a word of length n such that for any $k = 0, 1, ..., n-1$, w has one (resp. two) palindromic circular factors of length k if k is even (resp. odd). We prove by induction that for any k, $1 \leq k \leq n-1$, the three properties hold. First, it is easily verified that they hold for $k = 1$. We assume that they hold for some $k < n-1$ and show that they hold also for $k+1$.

$B_{k+1}(w)$: suppose that this property is false. Then, since $B_k(w)$ holds, there exists, by Lemma 15.4.3, a palindrome p such that apa and bpb are circular factors of length $k+1$ of w. By hypothesis on w, we deduce that $k+1$ is odd. Since p is a palindromic circular factor of w of length $k-1$, there is another one; call it m. There exists a letter x such that mx is a circular factor of w. Since mx is of length k, $R_k(w)$ implies that xm is a circular factor of w. Note that pa and pb are circular factors of w so that by $S_k(w)$, ma and mb are not both circular factors of w. It follows, since mx is a circular factor of w, that the unique right extension of xm as a circular factor of w is xmx. Thus we have three palindromic circular factors of length $n+1$ (namely apa, bpb, xmx, noting that $m \neq p$), contradicting the hypothesis. Thus $B_{k+1}(w)$ must hold.

$S_{k+1}(w)$: since there is a unique right-special circular factor of each length $0, 1, ..., k-1$, the number of circular factors of length k is $k+1$. By Lemma 15.4.2, w is radical, and by Lemma 15.3.4, the number of circular factors of length $k+1$ is at least $k+2$. Hence there must be at least one right-special circular factor of length k. Suppose that property $S_{k+1}(w)$ does not hold. Since $k < n$, we deduce that there exist two distinct right-special circular factors of length k of w: say, u and v. We may write $u = u'as$, $v = v'bs$, where s is the longest common suffix of u and v. Then u and v being right special, $u'asa$ and $v'bsb$ are circular factors of length $k+1$ of w. Thus asa and bsb are circular factors of length $\leq k+1$ of w, which contradicts property $B_{k+1}(w)$.

$R_{k+1}(w)$: since $S_{k+1}(w)$ holds, the number of circular factors of length $k+1$ is $k+2$. If $k+1$ is even, the number of non-palindromic circular factors of length $k+1$ is $k+1$, and if $k+1$ is odd, it is k; in both cases, this number is even. So the number of circular factors of length $k+1$ such that their reversal is not a circular factor is even. If $R_{k+1}(w)$ does not hold, we must therefore have at least two words of length $k+1$ whose reversal is not a circular factor; call them u_1 and u_2. Since $k+1 \geq 2$, we may write $u_1 = xmx'$, $u_2 = ypy'$ for some letters x, x', y, y'. By $R_k(w)$, the word $x'\tilde{m}$ is a circular factor. Since $\tilde{u}_1 = x'\tilde{m}x$ is not a circular factor, we deduce $u_3 = x'\tilde{m}\hat{x}$ is a circular factor, where $\hat{a} = b$ and $\hat{b} = a$. Similarly, $u_4 = y'\tilde{p}\hat{y}$ is a circular factor.

By $R_k(w)$ again, $\tilde{m}x$ is a circular factor. Thus $\tilde{m}$ is right special (since $\tilde{m}x$ and $\tilde{m}\hat{x}$ are circular factors). Similarly, $\tilde{p}$ is right special. From $S_k(w)$ it follows that they are equal: $\tilde{m} = \tilde{p}$, $m = p$. Thus $xmx' = u_1 \neq u_2 = ymy'$. If $x' = y'$, then $x \neq y$, that is, $x = \hat{y}$, and we see that $u_4 = x'\tilde{m}x = \tilde{u}_1$, a contradiction, since $\tilde{u}_1$ is not a circular factor.

Thus we must have $x' \neq y'$, which means $y' = \hat{x}'$. We have $u_1 = xmx'$, $u_2 = ym\hat{x}'$, $u_3 = x'\tilde{m}\hat{x}$, $u_4 = \hat{x}'\tilde{m}\hat{y}$. We have four cases, and we indicate in each case two words from u_1, u_2, u_3, and u_4 which contradict $B_{k+1}(w)$:

- $x = y$ and $x = x'$: $u_1 = xmx$, $u_4 = \hat{x}\tilde{m}\hat{x}$.
- $x = y$ and $x = \hat{x}'$: $u_2 = xmx$, $u_3 = \hat{x}\tilde{m}\hat{x}$.
- $x = \hat{y}$ and $x = x'$: $u_1 = xmx$, $u_2 = \hat{x}m\hat{x}$.
- $x = \hat{y}$ and $x = \hat{x}'$: $u_3 = \hat{x}\tilde{m}\hat{x}$, $u_4 = x\tilde{m}x$.

Thus, finally, S_{n-1} holds: there is a unique right-special circular factor for each $k = 0, 1, \ldots, n-2$. This implies that the number of circular factors of length k is $k+1$, for each $k = 0, 1, \ldots, n-1$. From this and Theorem 15.3.3, it follows that w is conjugate to a Christoffel word. □

15.4.1 Exercises

Exercise 15.4.1 *Show that if a word of odd length has two palindromic conjugates, then it is not radical.*

Exercise 15.4.2 *Show that a word $w \in \{a < b\}^*$ with set of factors F is balanced if and only if it has the following property: let v be any element of F; if $v = aub$, one has $u \geq_{lex} \tilde{u}$; if $v = bua$, one has $u \leq_{lex} \tilde{u}$.*

Hint: if w is not balanced, use Lemma 15.4.3 to find ama, $bmb \in F$, with m palindrome; if ama is at the left of bmb in w, then there exists $v = aub \in F$ such that ma is a prefix of u and bm a suffix of u; then $u <_{lex} \tilde{u}$. The case where ama is at the right is similar. Conversely, if $v = aub \in F$ with $u <_{lex} \tilde{u}$, find in v the two factors apa and $b\tilde{p}b$. See [Re7] (Lemma 5.1).

Exercise 15.4.3 *Denote by* pal(w) *(resp.* pal$_x(w)$*) the number of distinct palindromic factors (resp. ending by the letter x) of the word w. Show that* pal$(w) = p(\tilde{w})$ *and that* pal$_x(w) =$ pal$_x(\tilde{w})$.

Say that a factor is unioccurrent *in w if it appears exactly once in w. Show that w has at most one palindromic suffix which is unioccurrent, which must be the longest palindromic suffix.*

Show that if a and b are letters, then pal$_a(wb) =$ pal$_a(w) + 1$ *if $a = b$ and wb has a palindromic suffix which is unioccurrent, and* pal$_a(wb) =$ pal$_a(w)$ *otherwise. Deduce that* pal$_a(w) \leq |w|_a$ *and that* pal$(w) \leq |w| + 1$.

A word w is full *if* pal$(w) = |w| + 1$.[2] *Show that w is full if and only if for each prefix p of w, the longest palindromic suffix of p is unioccurrent in p. Show that each factor of a full word is full (do it first for prefixes, then for suffixes, and then for factors). Show that w is full if and only for each letter x,* pal$_x(w) = |w|_x$.

[2] In [GJWZ], full words are called *rich* words, but we prefer the terminology "full" since it contains the idea of saturation: if something is full, it cannot be more full, but a rich person may become richer.

See Droubay, Justin, Pirillo [DJP] (Section 2). Show that aabbabaa is not full, using the unioccurrent property discussed earlier.

15.5 Sturmian morphisms again

By Theorem 8.2.3, lower and upper Christoffel words of the same slope are conjugate. Thus an endomorphism of the free monoid $\{a, b\}^*$ is Sturmian if and only if it maps the set of conjugates of Christoffel words into itself.

Theorem 15.5.1 *([BdLR], Theorem A.1) The monoid of Sturmian morphisms is generated by $E, G, D, \tilde{G}, \tilde{D}$ and also by $E, G, \tilde{G}$ or $E, D, \tilde{D}$.*

This theorem is a variant of a result due to Mignosi and Séébold [MS], Théorème 1; see also [Lo], Theorem 2.3.7: they consider the monoid of morphisms that send each Sturmian sequence onto a Sturmian sequence and prove that this monoid is generated by the morphisms $E, G, \tilde{G}$. It follows that both monoids coincide.

We use several times the following fact, a consequence of Corollary 15.2.3: if a word w is the conjugate of some Christoffel word, then the words aa and bb cannot both be factors of w.

Lemma 15.5.2 *If φ is a Sturmian morphism, different from the identity and from E, then $\varphi(a)$ and $\varphi(b)$ begin or end with the same letter.*

This result has some similarity with a result of Nielsen on the automorphims of the free group on two generators (see Theorem 20.1.2). Also, it will appear that Sturmian morphisms are the automorphisms of this free group that map $\{a, b\}^*$ onto itself (see Corollary 20.6.1).

Proof. Replacing, if necessary, φ by $E\varphi$, we may assume that $\varphi(a)$ begins with a. If $\varphi(b)$ begins with a, we are done. Assume that $\varphi(b)$ begins with b. There are two cases, from which will follow that $\varphi(a)$ and $\varphi(b)$ end with the same letter.

(i) $\varphi(a)$ ends with b, and we suppose by contradiction that $\varphi(b)$ ends with a. Then $\varphi(ab) = a \cdots bb \cdots a$ is conjugate of a Christoffel word, and so are all its conjugates. Among them is the word $\cdots bb \cdots aa$, a contradiction.

(ii) $\varphi(a)$ ends with a, and we suppose by contradiction that $\varphi(b)$ ends with b. Then aa is a factor of $\varphi(aab) = a \cdots aa \cdots ab \cdots b$ so that bb is not a factor of $\varphi(a)$ nor of $\varphi(b)$ (otherwise bb is a factor of $\varphi(aab)$, a contradiction). Using $\varphi(abb)$, we see similarly that aa is not a factor of $\varphi(a)$ nor of $\varphi(b)$. Thus we must have $\varphi(a) = (ab)^i a$, $\varphi(b) = b(ab)^j$. Hence $\varphi(ab) = (ab)^{i+j+1}$. This word is a conjugate of a Christoffel word only if it is radical, that is, $i+j = 0$, and we conclude that φ is the identity, a contradiction.

□

Lemma 15.5.3 *If φ is a Sturmian morphism, different from the identity and from E, then for some $H \in \{G, D, \tilde{G}, \tilde{D}\}$ and some endomorphism ψ of $\{a, b\}^*$, one has $\varphi = H\psi$ or $\varphi = EH\psi$.*

We use several times Lemma 14.1.4 (i) and its symmetric versions.

Proof. Replacing φ by $E\varphi$ if necessary, we may assume that bb is not a factor of $\varphi(ab)$ so that it is neither a factor of $\varphi(a)$ nor of $\varphi(b)$. By Lemma 15.5.2, these two words begin or end with the same letter. There are four cases to consider.

(i) $\varphi(a)$ and $\varphi(b)$ begin with a: then they belong to the submonoid $\{a, ab\}^*$, and since $G = (a, ab)$, we see that there exists ψ such that $\varphi = G\psi$.

(ii) $\varphi(a)$ and $\varphi(b)$ begin with b: if one of them contains aa, then they must both end with a (otherwise $\varphi(ab)$ or $\varphi(ba)$ contains both aa and bb, a contradiction); then they are both in $\{a, ba\}^*$, and since $\tilde{G} = (a, ba)$, we see that there exists ψ such that $\varphi = \tilde{G}\psi$. If none of them contains aa, then they are in $\{ba, b\}^*$, and we see that there exists ψ such that $\varphi = D\psi$, since $D = (ba, b)$.

The other cases are symmetric, and we omit them. □

Lemma 15.5.4 *Let $H \in \{G, D, \tilde{G}, \tilde{D}\}$. If $H(w)$ is a conjugate of a Christoffel word, then w is as well.*

Proof. We prove the lemma for $H = G$ and $H = \tilde{G}$, the other cases being similar.

If $G(w)$ is a lower Christoffel word, then w is as well by Exercise 2.6.3.

Now suppose that $G(w)$ is a conjugate of a lower Christoffel word: $G(w) = vu$, uv is a lower Christoffel word, and u and v are non-empty (otherwise it is the previous case). Then u begins with a. Since $G = (a, ab)$, v begins with a, and bb is not a factor of vu, hence not a factor of u and v. Thus u and v are in the image of G: $u = G(x), v = G(y)$, hence $G(w) = G(yx)$, $w = yx$, and $G(xy) = uv$, a lower Christoffel word so that xy is a lower Christoffel word, by what we have seen earlier. Thus w is a conjugate of the lower Christoffel word $\widetilde{xy}$.

For $\tilde{G}$, the identity $\widetilde{G(\tilde{w})} = \tilde{G}(w)$ (Exercise 2.6.1) shows that one is reduced to the previous case, knowing that if $\tilde{w}$ is a conjugate of a lower Christoffel word, then w is as well. □

Lemma 15.5.5 *A Sturmian morphism is non-erasing.*

Proof. Indeed, otherwise a or b is sent to 1. In the first case, $\varphi(abb)$ is the square $\varphi(b)^2$ and cannot be conjugate to a Christoffel word, which is radical (Section 2.2). The second case is similar, arguing with $\varphi(aab)$. □

Proof of Theorem 15.5.1. The automorphism E is a Sturmian morphism: it indeed sends lower Christoffel words onto upper ones, and these are conjugate to lower ones by Theorem 8.2.3. The four other morphisms are Sturmian morphisms by Corollary

2.6.2 and Corollary 2.6.3. We show now that they generate the whole monoid of Sturmian morphisms.

Let φ be a Sturmian morphism. The proof is by induction on $N(\varphi) = |\varphi(a)|+|\varphi(b)|$. If this number is 2, which is the smallest possible value by Lemma 15.5.5, then φ is the identity or equal to E.

We assume now that $\varphi \neq id, E$. Then by Lemma 15.5.3, for some $H \in \{G, D, \tilde{G}, \tilde{D}\}$ and some endomorphism ψ of $\{a,b\}^*$, one has $\varphi = H\psi$ or $\varphi = EH\psi$. By Lemma 15.5.4, ψ is Sturmian. One has $N(\varphi) > N(\psi)$. Indeed, note that multiplication by E does not change this quantity. Moreover, $|H(u)| \leq |u|$ implies that u is the power of some letter depending on H (by inspection of the four cases for H); hence, $\psi(ab)$ must be shorter than $\varphi(ab) = H(\psi(ab))$, except if $\psi(ab)$ is a power of some letter, but this is not possible, since $\psi(a)$ and $\psi(b)$ are non-empty (by Lemma 15.5.5, since ψ is a Sturmian morphism), and they cannot be a power of the same letter (since $\psi(ab)$ is the conjugate of a Christoffel word, hence is a radical word). We thus may conclude by induction that ψ is in the monoid generated by $E, G, D, \tilde{G}, \tilde{D}$, hence so if φ.

The final assertion follows from $D = EGE, \tilde{D} = E\tilde{G}E$ (see Exercise 2.6.1). □

The previous proof implies a stronger statement.

Corollary 15.5.6 *([BdLR], Theorem A.1) If an endomorphism of the free monoid $\{a, b\}^*$ sends the three words ab, aab, and abb onto conjugates of Christoffel words, then it is a Sturmian morphism.*

Note that it follows from a result of Berstel and Séébold ([BS], Théorème 3.1) that an endomorphism of this free monoid is a Sturmian morphism if and only if it sends the word $ba^2ba^2baba^2bab$ onto a balanced word.

15.5.1 Exercises

Exercise 15.5.1 *Show that Sturmian morphisms send full words onto full words (Exercise 15.4.3). Hint: using Exercise 4.1.2, show that G' preserves fullness, and then G too, using Exercise 15.4.3. See [GJWZ] (Proposition 6.1 and its proof).*

Exercise 15.5.2 *Notations of Exercise 4.1.2. Define the* defect $d(w)$ *of a word w by $d(w) = |w| + 1 - \mathrm{pal}(w)$ (following [BHNR]). More generally, define for each letter a, $d_a(w) = |w|_a - \mathrm{pal}_a(w)$. Show that for $w \in A^*$, $d(w) = \sum_{a \in A} d_a(w)$. Show that a word is full if and only if $d(w) = 0$, if and only if for each $a \in A$, $d_a(w) = 0$. Using Exercise 4.1.2, show that*

$$\begin{pmatrix} d_a(G'(u)) \\ d_b(G'(u)) \end{pmatrix} = \mathrm{Ab}(G) \begin{pmatrix} d_a(u) \\ d_b(u) \end{pmatrix} + \begin{pmatrix} 1 \\ 0 \end{pmatrix}$$

Find similar statements for $D, \tilde{G}, \tilde{D}$.

16
Standard words

Among conjugates of Christoffel words, standard words, studied by Aldo de Luca, play a special role. A look at the historical articles of Christoffel [C1] and Smith [S] shows that they actually consider these words, and not Christoffel words.

Standard words are obtained from the latter by applying the conjugator, that is, taking the first letter and putting it at the end. There are several ways to obtain them: by palindromicity properties (theorems of de Luca and Pirillo), by the tree of standard pairs (Berstel and de Luca), and by standard sequences (de Luca and Mignosi), which are non-commutative versions of continued fractions. Prefixes of standard words are characterized by the fact that their minimal automaton has the minimal number of states (Sciortino and Zamboni). Squares of standard words have a remarkable factorization whose factors are six possible squares (Peltomäki and Whiteland), and this is a characterizing property (Peltomäki and Saarela).

16.1 Standard words

Recall that the conjugator C is defined in Section 2.2. A *standard word* is a word of the form $C(w)$ for some Christoffel word w (lower or upper). Assume that the alphabet is $A = \{a, b\}$, with $a < b$. Then a, b are standard words, and all other standard words are called *proper*. Since proper Christoffel words are exactly the words of the form apb or bpa for some central word p (see Section 14.1), it follows that proper standard words are exactly the words of the form pab or pba for some central word p.

Theorem 16.1.1 *The following conditions are equivalent for a word $w \in \{a, b\}^*$, $w \neq a, b$:*

- *(i) w is a standard word;*
- *(ii) for some palindromes p, q, r, $w = pq$, and $w = rab$ or rba;*
- *(iii) for some word m, $w = mab$ or mba, and these two words are conjugate.*

Condition (ii) is due to de Luca and Mignosi ([dLM], Proposition 8), and condition (iii) is due to Pirillo [Pi]. Condition (ii) is called *Property R* in [dL1], since it goes back to Robinson [Ro] (see also [P]).

Proof. (i) implies (ii): each Christoffel word is a product of two palindromes by Theorem 14.2.1; this property is preserved under conjugation by Exercise 14.2.2; hence $w = pq$ for some palindromes p and q. Next, a Christoffel word of length ≥ 2

From Christoffel Words to Markoff Numbers. Second Edition. Christophe Reutenauer, Oxford University Press.
 DOI: 10.1093/9780197907535.003.0016

is of the form arb or bra for some palindrome r by Theorem 2.3.1; thus we have $w = C(arb) = rba$ or $w = C(bra) = rab$.

(ii) implies (iii): by symmetry, we may assume that $w = rab$; since $\tilde{w} = \tilde{q}\tilde{p} = qp$, w and $\tilde{w}$ are in the same conjugation class; since $\tilde{w} = ba\tilde{r} = bar$, this class also contains rba so that rab and rba are conjugate.

(iii) implies (i): we may assume that $a < b$. By hypothesis, bma and amb are conjugate; by Theorem 15.2.5, amb is a lower Christoffel word; thus, bma is an upper Christoffel word by Theorem 2.3.1; since w is the image under C of one of these words, it is a standard word. □

Recall that the mapping $Pal : A^* \to A^*$ is an injective mapping whose image is the set of central words. Hence, the first sentence of the following result follows.

Theorem 16.1.2 *The mapping* $u \mapsto \mathrm{Pal}(u)ab$ *is a bijection from* A^* *onto the set of standard words ending with* ab. *One has* $\mathrm{Pal}(u)ab = \sigma(u)(ab)$, *where* σ *is the monoid homomorphism* $A^* \to \mathrm{End}(A^*)$ *sending* a *onto* G *and* b *onto* D. *In particular, the set of standard words ending with* ab *is the smallest subset of* A^*, *closed under* G *and* D, *and containing* ab.

The endomorphisms G and D are defined in Section 2.6, where the endomorphisms $\tilde{G}$, $\tilde{D}$, and ρ, used later, are also defined. Recall that $\rho(a) = G, \rho(b) = \tilde{D}$.

Lemma 16.1.3 *Let* $u \in A^*$, $u \notin a^* \cup b^*$. *Then* $\rho(u) = (amb, apb)$, *for some words* m and p *and* $\sigma(u) = (mba, pab)$.

Proof. Since u is not the power of a letter, $\rho(u)$ is of the indicated form by Exercise 2.6.7; moreover, we may write $u = a^n bv$ or $u = b^n av$, $n \geq 1$. We consider only the case $u = a^n bv$, the other being similar.

Suppose that v is empty. By Exercise 2.6.7, $G^n = \rho(a^n) = (a, a^n b)$, hence $\rho(u) = \rho(a^n)\rho(b) = \rho(a)^n\tilde{D} = (a, a^n b)(ab, b) = (a^{n+1}b, a^n b)$; moreover, one has $\sigma(u) = \sigma(a)^n \sigma(b) = G^n D = (a, a^n b)(ba, b) = (a^n ba, a^n b)$, which settles this case with $m = a^n$, $p = a^{n-1}$.

If v is non-empty, then $v = v'x$, $x \in A$; then $u = wx$, $w = a^n bv'$ is not the power of a letter, and by Exercise 2.6.7, $\rho(w) = (afb, agb)$. By induction, we have $\sigma(w) = (fba, gab)$. We have $\rho(u) = \rho(w)\rho(x)$ and $\sigma(u) = \sigma(w)\sigma(x)$. If $x = a$, then $\rho(u) = (afb, agb)(a, ab) = (afb, afbagb)$ and $\sigma(u) = (fba, gab)(a, ab) = (fba, fbagab)$, which settles this case, with $m = f, p = fbag$. If $x = b$, then $\rho(u) = (afb, agb)(ab, b) = (afbagb, agb)$ and $\sigma(u) = (fba, gab)(ba, b) = (gabfba, gab)$, which settles this case, with $m = fbag = gabf$ (since by Theorem 2.6.1, afb, agb and $afbagb$ are lower Christoffel words so that by Theorem 2.3.1, the words f, g and $fbag$ are palindromes, implying that $fbag = \widetilde{fbag} = \tilde{g}ab\tilde{f} = gabf$), $p = g$. □

Proof of Theorem 16.1.2. It is enough to prove the formula in the statement. By Lemma 14.1.5, we have, for any word $u \in A^*$: $\rho(u)(ab) = a\mathrm{Pal}(u)b$. If u is not the power of a letter, we obtain, by Lemma 16.1.3 $\rho(u) = (amb, apb)$ and

$\sigma(u) = (mba, pab)$. Thus $a\text{Pal}(u)b = ambapb$, $\text{Pal}(u) = mbap$, and $\text{Pal}(u)ab = mbapab = \sigma(u)(ab)$ by the same lemma.

If $u = a^n$, then $\text{Pal}(u) = a^n$, and by Exercise 2.6.7, $\sigma(u)(ab) = G^n(ab) = a^{n+1}b = \text{Pal}(u)ab$. If $u = b^n$, then $\text{Pal}(u) = b^n$ and by Exercise 16.1.1, $\sigma(u)(ab) = D^n(ab) = (b^n a, b)(ab) = b^n ab = \text{Pal}(u)ab$. □

Theorem 16.1.4 *(Berstel and de Luca [BdL]) Consider an infinite complete planar rooted binary tree, whose nodes are coded by the free monoid $\{a < b\}^*$, as explained in Section 2.5. Let $\sigma(x) = (u, v)$ be the node coded by $x \in \{a, b\}^*$. Then this tree is constructed by the local rule described in Figure 16.1, together with the condition that its root is (a, b). Moreover, the set of components of its nodes is equal to the set of standard words.*

This tree is called the *tree of standard pairs*; it appeared in [BdL], Figure 3; see also [Lo], Figure 2.5. An equivalent concept is the *Rauzy rules*; see [Rz], pp. 167–168 (see also [Lo], Problem 2.2.2).

Proof. If $x = 1$, then the corresponding node is the root, which is indeed $\sigma(1) = (a, b)$, the identity mapping. Suppose that the result is true for x, that is, the node corresponding to x is $\sigma(x) = (u, v)$. Then the node corresponding to xa is, by the local rule, (u, uv); and $\sigma(xa) = \sigma(x) \circ \sigma(a) = (u, v) \circ G = (u, v) \circ (a, ab) = (u, uv)$, which settles this case. For the node coded ub, the proof is quite similar. This proves the first assertion.

The node corresponding to x is $\sigma(x) = (u, v)$. Hence its two components are u and v. If x is not the power of a letter, these two words are standard words by Lemma 16.1.3, since the words m and p in this lemma are central words, by Theorem 2.6.1 and the definition of central words in Section 14.1. If $x = a^n$, then $\sigma(x) = G^n = (a, ab^n)$ by

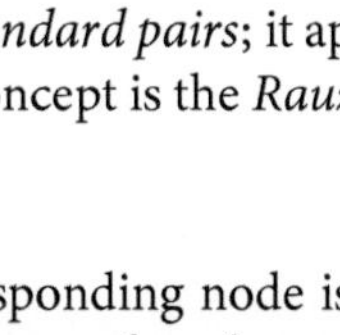

Figure 16.1 The rule for constructing the tree of standard pairs

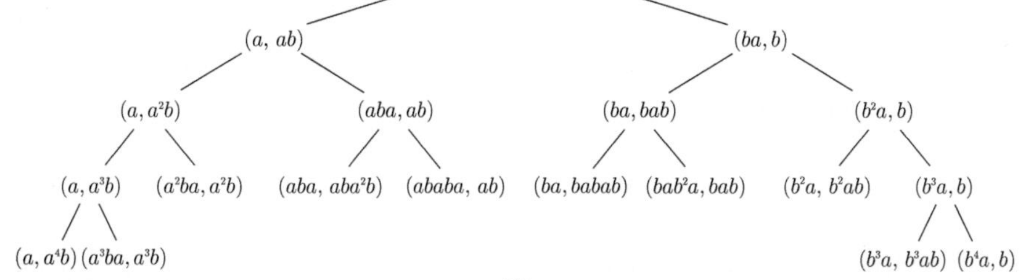

Figure 16.2 The tree of standard pairs

Exercise 2.6.7, whose two components are standard words. If $x = b^n$, $\sigma(x) = D^n = (b^n a, b)$ by Exercise 16.1.1, and one concludes similarly.

We show conversely that each standard word is the component of some standard pair. This is clear for the standard words a and b. For the others, it is enough, by Lemma 16.1.3, to show that for each central word π, there is some word x such that $\rho(x) = (a\pi b, *)$, and some word y such that $\rho(y) = (*, a\pi b)$. Equivalently, by Theorem 2.6.1, $w = a\pi b$ is the left component of some Christoffel pair and also the right component of some Christoffel pair. Now, w has a standard factorization $w = uv$ by Theorem 2.4.1 so that (u, v) is a Christoffel pair by definition; and then, by the rule of construction of the tree of Christoffel pairs (Figure 2.9), (u, w) and (w, v) are Christoffel pairs. □

Let $[d_0, d_1, d_2, ...]$ be a finite or infinite continued fraction. Following de Luca and Mignosi ([dLM] p. 365), we associate with it a *standard sequence* $s_{-1}, s_0, s_1, s_2, ...$ of words on the alphabet $A = \{a < b\}$, which is finite or infinite. It is defined by

$$s_{-1} = a, \quad s_0 = b, \quad s_n = s_{n-1}^{d_{n-1}} s_{n-2}$$

for $n \geq 1$. We say that $[d_0, d_1, d_2, ...]$ is the *directive continued fraction* of this standard sequence.

Lemma 16.1.5

(i) *Viewing pairs as endomorphisms of A^*, one has*

$$(s_n, s_{n-1}) = (s_0, s_{-1}) \circ (a^{d_0} b, a) \circ \cdots \circ (a^{d_{n-1}} b, a)$$

for any $n \geq 0$.

(ii) *For $n \geq 0$, the slope of s_n is $[d_0, ..., d_{n-1}]$.*

Recall that the slope of w is $|w|_b / |w|_a$.

Proof.

(i) For $n = 0$, this is clear. Let $n \geq 0$; taking the equality in (i) as the induction hypothesis, we multiply it on the right by the endomorphism $(a^{d_n} b, a)$. The left-hand side is

$$(s_n, s_{n-1}) \circ (a^{d_n} b, a) = (s_n^{d_n} s_{n-1}, s_n) = (s_{n+1}, s_n)$$

The right-hand side is

$$(s_0, s_{-1}) \circ (a^{d_0} b, a) \circ \cdots \circ (a^{d_{n-1}} b, a) \circ (a^{d_n} b, a)$$

which allows us to conclude.

(ii) Apply the abelianization map Ab (see Section 16.1) to the identity in (i). Recall that

$$\mathrm{Ab}(u,v) = \begin{pmatrix} |u|_a & |v|_a \\ |u|_b & |v|_b \end{pmatrix}$$

In particular, $\mathrm{Ab}(a^n b, a) = P(n) = \begin{pmatrix} n & 1 \\ 1 & 0 \end{pmatrix}$. Since Ab is a multiplicative homomorphism of End(A^*) into $\mathbb{Z}^{2\times 2}$, we obtain

$$\begin{pmatrix} |s_n|_a & |s_{n-1}|_a \\ |s_n|_b & |s_{n-1}|_b \end{pmatrix} = P(0)P(d_0)\cdots P(d_{n-1})$$

Since multiplication on the left by $P(0)$ interchanges the first and second rows, it follows from (5.5) and (5.7) that the slope $|s_n|_b/|s_n|_a$ of s_n is equal to $[d_0, \ldots, d_{n-1}]$. □

Recall that each rational number has exactly two representations as a finite continued fraction, whose difference of lengths is 1 (see Section 5.2); in particular, there is a representation of odd length and one of even length. Among others, the next result gives, for positive rational numbers, an interpretation of both representations.

Note that for each given positive rational number r, there are exactly two standard words having this slope; they are the two words C(w), for w equal to the lower and upper Christoffel word of slope r; the first ends with ba, and the second with ab.

Theorem 16.1.6 *Each member of a standard sequence is a standard word, and each standard word appears in some standard sequence. Precisely, if r is a positive rational number, then the standard word of slope r ending with ab (resp. by ba) is the last element of the standard sequence having as directive continued fraction the continued fraction of r of even length (resp. of odd length).*

The two cases are mentioned by Smith ([S], Section 8, pp. 6–7).

A small but illuminating example: let $r = 3$; then r has the two continued fractions [3] and [2, 1], of odd and even length, respectively; the first gives the standard sequence a, b, and b^3a, and the second the standard sequence a, b, b^2a, and b^2ab; b^3a and b^2ab are the two standard words of slope 3, images under C of the lower and upper Christoffel words ab^3 and b^3a.

Lemma 16.1.7 *The set of standard words is closed under the endomorphisms G, D, and E.*

Proof. We show that if w is a standard word, then so are $G(w)$ and $D(w)$. This is clear if $w = a$ or b. If w is not a letter, then $w = uab$ or uba, where u is a central word. By Exercise 14.1.3, $G(u)a = m$ and $b\tilde{D}(u)$ are central words. The latter word is equal to $D(u)b = p$ by Exercise 2.6.1. It follows that $G(uab) = G(u)aab = mab$, $G(uba) = G(u)aba = mba$, $D(uab) = D(u)bab = pab$, and $D(uba) = D(u)bba = pba$ are all standard words.

The assertion for E follows from the fact that the set of central words is closed under E; equivalently, the set of Christoffel words is closed under E (see Section 2.2). □

Proof of Theorem 16.1.6. The letters a and b are standard words. Moreover, s_n and s_{n-1} are images of a and b under the endomorphism (s_n, s_{n-1}). We have, by Exercise 16.1.1, $ED^d = (b, a)(b^d a, b) = (a^d b, a)$. Hence, by Lemma 16.1.5, the endomorphism (s_n, s_{n-1}) is a product of E's and D's. It follows from Lemma 16.1.7 that each member of a standard sequence is a standard word.

Now let r be any positive rational number. By Lemma 16.1.5, the two standard sequences corresponding to the two continued fractions representing r terminate with a standard word of slope r. Moreover, in a standard sequence, s_n ends with a if n is odd, and with b if n is even, as is seen by induction. The theorem follows, since there are exactly two standard words of any given positive slope. □

16.1.1 Exercises

Exercise 16.1.1 *Show that for $n \geq 1$, $D^n = (b^n a, b)$.*

Exercise 16.1.2 *Let w be a standard word of the form $C(f)$ for some Christoffel word f; let $v'u'$ be the palindromic factorization of the latter, and $i = |v'|$; show that $\{i-1, i\}$ is an attractor of w. Hint: use Exercise 15.3.11 and follow the lines of its hint. See [MRRRS] (Theorem 22).*

Exercise 16.1.3 *(scales in music) (i) Represent a tone by letter a and a semi-tone by letter b. Show that the sequence of tones and semi-tones from a C to the next C in the scale of C major is a standard word of length* 7.

(ii) Represent a white key on the piano by a and a black key by b. Show that the sequence of white and black keys from C to the next C, the latter excluded, is a standard word of length 12.

More general considerations lead in music theory to the study of "well-formed scales", with references to Christoffel words: Clampitt [Cl]; Clampitt and Noll [CN]; Clampitt and Shafer [ClS]; Castrillón López and M. Domínguez Romero [CLDR1, CLDR2]; Domínguez Romero, Noll and Castrillón López [DNC]; Noll [No].

Similar studies exist in musical rhythms: see Toussaint ([T], Chapter 19), who calls Euclidean rythm, *which is essentially a Christoffel word.*

16.2 Prefixes of standard words

Recall that a language is a subset of a free monoid A^*. Automata were defined in Section 14.4. We denote by S_w the language of suffixes of the word w. Recall that a translate of a language L is a language of the form $L \circ u = \{v \mid uv \in L\}$.

Theorem 16.2.1 *(Sciortino and Zamboni [SZ]) Let $w \in \{a < b\}^*$. The following conditions are equivalent:*

(i) w is a prefix of some standard word;
(ii) w is a prefix of some central word;

(iii) *each word u, such that au, bu are factors of w, is a prefix of w;*
(iv) *for each factor u of w, let $w = xuy$ with x of minimal length; then, for any other factorization $w = x'uy'$, x is a suffix of x';*
(v) *the non-empty translates of S_w are the $S_w \circ p$, p a prefix of w;*
(vi) *the number of non-empty translates of S_w is $|w| + 1$;*
(vii) *the minimal deterministic automaton recognizing S_w has $|w| + 1$ states.*

An eighth equivalent condition is stated in Corollary 18.5.6. Note that in [Fd] (Corollary 1), Anna Frid proves that each prefix of any standard word is a product of reversals of standard words; this result is a non-commutative version of the so-called *Ostrowski representation* of integers (see also Corollary 7.6 in [BRe]).

Call *Property SZ* the property stated in (vii). We begin by proving the equivalence of (v), (vi), and (vii).

Lemma 16.2.2 *The conditions (v), (vi), and (vii) in Theorem 16.2.1 are equivalent.*

Proof. We claim that the languages $S_w \circ p$, with p a prefix of w, are non-empty and distinct. The equivalence of (v) and (vi) follows, since w has $|w|+1$ prefixes. The equivalence of (vi) and (vii) follows from the fact that the states of the minimal deterministic automaton of a language are the non-empty translates of L (see Exercise 14.4.6).

It remains to prove the claim. Let $w = ps$. Since $ps \in S_w$, s in $S_w \circ p$, and it is its unique element of maximal length. These elements s, for different distinct prefixes p of w, are distinct. This proves the claim. □

Proof of Theorem 16.2.1. (i) implies (ii): if the standard word is a or b, it is also a central word, and we may therefore assume that the standard word is of the form mab or mba, for some central word m; we consider only the second case, the other being obtained by applying E. Then amb is a lower Christoffel word; therefore, by Exercise 2.6.8, $(amb)^3$ is a prefix, hence $ambamb$ is a proper prefix of some lower Christoffel word; hence mba is a prefix of a central word, and so is w, too.

(ii) implies (i): follows from the definition of standard words.

(ii) implies (iii): let w be prefix of the central word m. If au and bu are factors of w, then they are factors of m. Hence u is a left-special word with respect to the set of circular factors of the lower Christoffel word amb. By the symmetric version of Corollary 15.3.5, u is therefore a prefix of m, hence of w, since u is shorter than w.

(iii) implies (iv) (induction on the length of x): if $x = 1$, the conclusion is clear. Suppose that x is non-empty; its last letter is a, say; then the last letter of x' cannot be b: otherwise au and bu are factors of w, and by hypothesis, u is a prefix of w, contradicting the minimality of x; hence the last letter of x' is a; then we conclude by induction, replacing u by au.

(iv) implies (v): consider a non-empty translate of S_w. It is of the form $S_w \circ u$; it contains an element v, hence $uv \in S_w$; this implies that u is a factor of w. Let x be as in (iv). We show that $S_w \circ u = S_w \circ (xu)$. Let v be in the latter language; then $xuv \in S_w$, hence $uv \in S_w$ and $v \in S_w \circ u$. Conversely, let $v \in S_w \circ u$; then $uv \in S_w$; this implies that $w = x'uv$ for some word x'; by hypothesis, x' has x as a suffix, hence $x' = yx$; then $w = yxuv$, which implies that $xuv \in S_w$ and $v \in S_w \circ (xu)$.

Since xu is a prefix of w, each translate of S_w is of the required form; this proves (v), since a language $S_w \circ p$ is non-empty (if $w = ps$, it contains s).

(v) equivalent to (vi) and (vii): Lemma 16.2.2.

The proof will be completed later after some preparation. □

If p is a state of a deterministic automaton, and x a word, we denote by $p \cdot x$ the unique state q, if it exists, such that there is a path $p \overset{x}{\to} q$ (with label x); if this state does not exist, we write $p \cdot x = \emptyset$.

We denote by $\mathcal{A}_w$ the minimal deterministic automaton of S_w. If w satisfies Property SZ, the states of the minimal automaton may naturally be identified with the prefixes of w: the initial state is 1, the empty word, and each prefix is identified with the state $1 \cdot p$. Note that a word x is recognized if and only if $1 \cdot x$ is a terminal state.

Lemma 16.2.3 *(Sciortino and Zamboni [SZ], Corollary 2 and Proposition 7) Let u be a prefix of w.*

(i) If w has Property SZ, then so has u.

(ii) Precisely, $\mathcal{A}_u$ is obtained by removing from $\mathcal{A}_w$ the states which are not prefixes of u, and the edges pointing towards them.

(iii) Suppose that w is a central word so that w has Property SZ by Theorem 14.4.1. The automaton $\mathcal{A}_u$ has the following edges: let $p \in A^*, c \in A$ such that pc is a prefix of u. There is an edge

$$p \overset{c}{\to} pc$$

(such an edge is called direct*); and if p is a palindrome, let $d \neq c$ be the other letter, and let q be the shortest palindromic prefix of u which is longer than p and which is followed by d in u; if such a q exists, there is an edge*

$$p \overset{d}{\to} qd$$

(such an edge is called a bypass*).*

Proof. (i) By induction, we may assume that $w = uc$ for some letter c. Define an automaton $\mathcal{A}$ as follows: remove the state w from $\mathcal{A}_w$ and all edges pointing to it in this automaton; then declare terminal in $\mathcal{A}$ each state $p \neq w$ of $\mathcal{A}_w$ such that there is an edge $p \overset{c}{\to} q$ for some terminal state q in $\mathcal{A}_w$.

We show that $\mathcal{A}$ recognizes S_u, which will imply (i), since $\mathcal{A}$ has $|u| + 1$ states, and since an automaton recognizing S_u cannot have fewer states (see the beginning of the proof of Lemma 16.2.2). Consider a word x recognized by $\mathcal{A}$; then there is a path $1 \to p$ in $\mathcal{A}$, labelled x, for some terminal state p of $\mathcal{A}$; by construction, this path is also a path in $\mathcal{A}_w$, and there is an edge $p \overset{c}{\to} q$ for some state q which is terminal in $\mathcal{A}_w$; by composing the path and the edge we obtain a successful path in $\mathcal{A}_w$ labelled xc; hence $xc \in S_w = S_{uc}$, and therefore $x \in S_u$.

Conversely, let $x \in S_u$. Then $xc \in S_w$, and there is a successful path in $\mathcal{A}_w$ labelled xc, this path decomposes as $1 \overset{x}{\to} p \overset{c}{\to} q$, where q is a terminal state in $\mathcal{A}_w$; no state on the path $1 \xrightarrow{x} p$ is equal to w, since there is no outgoing edge from w; hence this path also exists in $\mathcal{A}$; moreover, p is a terminal state in $\mathcal{A}$ by construction; hence x is recognized by $\mathcal{A}$.

(ii) This follows from what precedes, by removing iteratively the last letter.

(iii) This follows from (ii) and Theorem 14.4.1, and the construction of $\mathcal{A}_w$ indicated before this theorem. □

Lemma 16.2.4 *(Sciortino and Zamboni [SZ]) Let c be a letter. Assume that uc has Property SZ and that in $\mathcal{A}_u$ one has $1 \cdot xc = 1 \cdot yc$. Then $x \in S_u$ if and only if $y \in S_u$.*

Proof. By Lemma 16.2.3 (ii), one has $1 \cdot xc = 1 \cdot yc$ in $\mathcal{A}_{uc}$. Suppose that $x \in S_u$; hence $xc \in S_{uc}$ and $1 \cdot xc$ is terminal in $\mathcal{A}_{uc}$. Thus by hypothesis $1 \cdot yc$ is terminal in this automaton, and $yc \in S_{uc}$. Finally $y \in S_u$. All these implications are equivalences, so that the lemma is proved. □

Lemma 16.2.5 *(Sciortino and Zamboni [SZ]) Let u be a non-palindromic prefix of a central word, and u' be the longest palindromic prefix of u. Then u' is not a suffix of u.*

Proof. By Proposition 14.1.6, there exists a word v and a letter c such that u is a prefix of $\mathrm{Pal}(vc)$ and such that $u' = \mathrm{Pal}(v)$.

By Exercise 14.1.4, $|\mathrm{Pal}(vc)| \leq 2|\mathrm{Pal}(v)| + 1 = 2|u'| + 1$. Since u is not a palindrome, and hence a proper prefix of $\mathrm{Pal}(vc)$, we deduce $|u| \leq 2|u'|$. Consequently, if u' is a suffix of u, then the two occurrences of u', as prefix and suffix of u, either overlap, or $u = u'u'$; in both cases, u is a palindrome, a contradiction. □

End of proof of Theorem 16.2.1. (vii) implies (ii) (induction on the length of w): the case where w is empty is evident. Suppose that $w = uc$, $u \in \{a, b\}^*$, $c = a$ or b. Then u has Property SZ by Lemma 16.2.3 (i). It follows by induction that u is a prefix of a central word. If u is a palindrome, then it is itself a central word by Corollary 14.1.7. Hence $u = \mathrm{Pal}(v)$ for some word v by Theorem 14.1.2. By this theorem, $\mathrm{Pal}(vc)$ is a central word; one has $\mathrm{Pal}(vc) = (\mathrm{Pal}(v)c)^{(+)} = (uc)^{(+)}$, which has $uc = w$ as a prefix so that w is the prefix of a central word.

Thus we may assume that u is not a palindrome. It is a prefix of a central word, necessarily a proper prefix; hence, for some letter x, ux is a prefix of the same central word. If we assume by contradiction that w is not a prefix of a central word, then $uc = w \neq ux$, and we see that ua and ub have both Property SZ: one of them because it is a prefix of some central word (which has Property SZ by Theorem 14.4.1) and the other being w.

We may assume that $u = a^n bv$, $n \geq 1$ (since if u has only one letter, say $u = a^n$, then $w = a^{n+1}$ or $w = a^n b$, and w is clearly a prefix of a central word). By Lemma 16.2.3 (iii) (which applies since u is a prefix of a central word), since we have in $\mathcal{A}_u$ the edge $1 \xrightarrow{a} a$, there is a bypassing edge $1 \xrightarrow{b} a^n b$. Thus $1 \cdot b = 1 \cdot a^n b$; since $1 \in S_u$, we have $a^n \in S_u$ by Lemma 16.2.4. Observe that we therefore have two palindromic prefixes of u (namely 1 and a^n), which are also suffixes of u, one followed by a in u and the other by b.

Let u' be given in Lemma 16.2.5: then u' is a palindromic prefix of u, and u' is not a suffix of u. This shows that there exist palindromic prefixes of u which are not suffixes of u. We choose such a word of smallest length, say, q; it will be longer than n, since a^i, $i = 0, \dots, n$, are prefixes and suffixes. Note that then all palindromic prefixes of u, shorter than q, are also suffixes of u.

Let c be the letter following q in u (it exists since $q \neq u$ because u is not a palindrome). We claim that for some proper prefix p of q, we have $1 \cdot pc = 1 \cdot qc$, and p is a suffix of u. This contradicts Lemma 16.2.4.

We prove the claim. Among the palindromic prefixes p of u, shorter than q, some of them are followed by a in the word u, and others by b, and both cases occur; indeed, this was observed above. Let d be such that $\{a, b\} = \{c, d\}$. We consider the longest palindromic prefix p, shorter than q, which is followed by d in u. Then by Lemma 16.2.3 (iii), there is a bypass $p \overset{c}{\to} qc$; we thus have $p \cdot c = qc$, hence $1 \cdot pc = 1 \cdot qc$, and p is a suffix of u. □

In general, the minimal deterministic automaton of the language of suffixes of a word w has at least $|w| + 1$ states, and at most $2|w| - 1$ ([BBEHCS]; see also [CrH]). Some equivalences in Theorem 16.2.1 were extended to arbitrary alphabets by Fici [Fi1], Theorem 5.1.

16.2.1 Exercises

Exercise 16.2.1 *Show that if, in the automaton $\mathcal{A}_u$, $1 \cdot xc = 1 \cdot yd \neq \emptyset$ for some letters c and d, then $c = d$. Hint: consider a word v such that there is a path from $1 \cdot xc$ to a terminal state, labelled v; show that xcv and ydv are suffixes of u.*

16.3 Factorization into squares

A *square* is a word which is of the form vv for some word v.

Let r and s be two natural integers, with $r \geq 1$. Following Saari ([Sa], Theorem 16), we define a sequence of six words $S_i(r, s)$, $i = 1, \dots, 6$, depending on r and s, as follows (we write S_i for simplicity):

$$\begin{aligned} S_1 &= a, S_2 = a^{r-1}ba, S_3 = a^r ba, S_4 = a^r b \\ S_5 &= (a^r b)^s a^{r+1} b, S_6 = (a^r b)^{s+1} a^{r+1} b \end{aligned} \tag{16.1}$$

Likewise, we define the six words $T_i(r, s)$ by exchanging a and b in $S_i(r, s)$.

Theorem 16.3.1 *(Peltomäki and Whiteland, Proposition 23 in [PW]) Let w be a standard word.*

(i) *There exist words* $v_1, \ldots, v_k$ *such that*

$$w = v_1 \cdots v_k$$

and

$$w^2 = v_1^2 \cdots v_k^2$$

and that, for some r and s, *the words* v_i *are either all of the form* $S_j(r,s)$ *or all of the form* $T_j(r,s)$.

(ii) *For any* $i = 1, \ldots, k$, v_i^2 *is the shortest suffix of* $v_1^2 \cdots v_i^2$ *which is a non-trivial square.*

The theorem leads to the following definition. We say that a set of words W has *Property S* (resp. *Property T*) if each $w \in W$ satisfies the conclusion in part (i) of the theorem with the same r and s; that is, for some r and s, each $w \in W$ satisfies: $w^2 = v_1^2 \cdots v_k^2$, $w = v_1 \cdots v_k$; moreover, the words v_i are all of the form $S_j(r,s)$ (resp. $T_j(r,s)$).

The main ingredient of the proof is the following lemma.

Lemma 16.3.2 *Let* $r, s \in \mathbb{N}, r \geq 1$, *and* $\varphi = G$ *or* D. *Then the sets* $\{\varphi(S_i(r,s)), i = 1, \ldots, 6\}$ *and* $\{\varphi(T_i(r,s)), i = 1, \ldots, 6\}$ *have Property S or Property T.*

Proof. It is enough to prove the result for the first set, since the automorphism E, which exchanges a and b, exchanges the sets S_i and T_i, and conjugates G and D (Exercise 2.6.1).

(i) Consider the set $\{G(S_i(r,s)), i = 1, \ldots, 6\}$. Recall that $G(a) = a, G(b) = ab$. We have $G(S_5(r,s)) = G((a^rb)^sa^{r+1}b) = (a^rab)^sa^{r+1}ab = (a^{r+1}b)^sa^{r+2}b = S_5(r+1,s)$. Similarly, $G(S_6(r,s)) = S_6(r+1,s)$. An even easier computation shows that $G(S_i(r,s)) = S_i(r+1,s)$ for $i = 1,2,3,4$. Hence the set has Property S.

(ii) Consider the set $\{D(S_i(r,s)), i = 1, \ldots, 6\}$. We show that, for $i = 1, \ldots, 6$, each word $w = D(S_i(r,s))$ in this set is a product of words of the form $T_i(1, r-1)$, such that w^2 is the corresponding product of the squares of these words. This will imply that this set has property T.

For short, write T_i for $T_i(1, r-1)$. Thus we have

$$T_1 = b, T_2 = ab, T_3 = bab, T_4 = ba$$

$$T_5 = (ba)^{r-1}b^2a, T_6 = (ba)^rb^2a$$

Recall that $D = (ba, b)$. We have $D(S_1(r,s)) = D(a) = ba = T_4$; $D(S_2(r,s)) = D(a^{r-1}ba) = (ba)^{r-1}bba = T_5$; $D(S_3(r,s)) = D(a^rba) = (ba)^rbba = T_6$. Thus, for $i = 1,2,3$, $w = D(S_i(r,s))$ is equal to some T_j so that the condition for w^2 is clearly satisfied.

Moreover, $D(S_4(r,s)) = D(a^r b) = (ba)^r b$. Clearly

$$((ba)^r b)^2 = (ba)^r bb(ab)^r = \begin{cases} (baba)^{r/2} bb(abab)^{r/2} & \text{if } r \text{ is even} \\ (baba)^{(r-1)/2} babbab(abab)^{(r-1)/2} & \text{if } r \text{ is odd} \end{cases}$$

$$= \begin{cases} (T_4^2)^{r/2} T_1^2 (T_2^2)^{r/2} \\ (T_4^2)^{(r-1)/2} T_3^2 (T_2^2)^{(r-1)/2} \end{cases}$$

Next, we have, removing the exponents 2 in the previous expressions,

$$\begin{cases} T_4^{r/2} T_1 T_2^{r/2} = (ba)^{r/2} b(ab)^{r/2} = (ba)^r b \\ T_4^{(r-1)/2} T_3 T_2^{(r-1)/2} = (ba)^{(r-1)/2} bab(ab)^{(r-1)/2} = (ba)^r b \end{cases}$$

This settles the case $i = 4$. For $i = 5$, we have $w = D(S_5(r,s)) = D((a^r b)^s a^{r+1} b) = ((ba)^r b)^s (ba)^{r+1} b = u^{s+1} ab$, where $u = (ba)^r b$. Suppose that r *and* s are even. We show that

$$w^2 = ((T_4^2)^{r/2} T_1^2 (T_2^2)^{r/2})^{s/2} T_4^2 (T_5^2)^{s/2} (T_4^2)^{r/2-1} T_3^2 (T_2^2)^{r/2} \tag{16.2}$$

In order to prove this equality, note the following equalities (left to the reader), computing in the free group:

$$T_4^r T_1^2 T_2^r = u^2, T_5 = (ba)^{-1} uba, baT_4^{r-2} T_3 = u, T_3 T_2^r = uab, bau = uab$$

Thus the right-hand side of (16.2) is equal to

$$(u^2)^{s/2} (ba)^2 (ba)^{-1} u^s baT_4^{r-2} T_3^2 T_2^r = u^s bau^s uuab = u^s uabu^{s-1} uuab = ww$$

Now, removing the exponents 2 on the T_j in (16.2) gives

$$((T_4)^{r/2} T_1 (T_2)^{r/2})^{s/2} T_4 (T_5)^{s/2} (T_4)^{r/2-1} T_3 (T_2)^{r/2}$$

This is equal to

$$((ba)^{r/2} b(ab)^{r/2})^{s/2} ba((ba)^{r-1} b^2 a)^{s/2} (ba)^{r/2-1} bab(ab)^{r/2}$$
$$= u^{s/2} (ba(ba)^{r-1} b)^{s/2} ba(ba)^{r/2-1} ba(ba)^{r/2-1} bab = u^{s+1} ab = w$$

This settles the case $i = 5$ when r and s are even; the three subcases where at least one of r and s is odd are similar (Exercise 16.3.3).

Now, the case $i = 6$ follows from the case $i = 5$, since $S_6(r,s) = S_5(r, s+1)$, and since the $T_j = T_j(1, r-1)$ do not depend on s. □

Proof of Theorem 16.3.1. Note that (ii) follows from Exercise 16.3.2: each word $S_i(r,s)^2$ and $T_i(r,s)^2$ has no non-trivial proper suffix which is a square.

We may assume that w ends in ab (exchanging a and b if necessary). Since, by Theorem 16.1.2, each such standard word, except ab (for which the theorem is evident), is the image under G or D of another such standard word, it is enough to prove that if w satisfies (i), then so does $\varphi(w)$, where $\varphi = G$ or D. By Lemma 16.3.2, for each $i = 1, \ldots, k$, $\varphi(v_i)^2 = \prod_{1 \le j \le n_i} y_{ij}^2$, $\varphi(v_i) = \prod_{1 \le j \le n_i} y_{ij}$, and for some p and q, the words y_{ij} either all of the form $S_h(p, q)$ or all of the form $T_h(p, q)$. Thus we obtain the theorem, since

$$\varphi(w)^2 = \prod_{1 \le i \le k} \varphi(v_i)^2 = \prod_{1 \le i \le k} \prod_{1 \le j \le n_i} y_{ij}^2,$$

and

$$\varphi(w) = \prod_{1 \le i \le k} \varphi(v_i) = \prod_{1 \le i \le k} \prod_{1 \le j \le n_i} y_{ij}.$$

□

Theorem 16.3.1 has a converse.

Theorem 16.3.3 *(Peltomäki and Saarela [PS]) Let w be a word on the alphabet $\{a, b\}$ such that $|w|_a$ and $|w|_b$ are relatively prime and that*

$$w = v_1 \cdots v_k$$

and

$$w^2 = v_1^2 \cdots v_k^2$$

where, for some r and s, the words v_i are either all of the form $S_j(r, s)$ or all of the form $T_j(r, s)$. Then w is a standard word.

Proof.

(i) Note that if $|w|_a = |w|_b$, then both numbers are equal to 1, and $w = ab$ or ba, which is a standard word. We thus may assume that $p = |w|_a > |w|_b = q$, exchanging the letters a and b if necessary. We may also assume that $q \ge 1$, since otherwise $p = 1, q = 0$, and $w = a$, a standard word. The words v_i cannot be all of the form $T_j(r, s)$, since the number of a's in the latter words is smaller than or equal to the number of b's. Hence, the words v_i are all of the form $S_j(r, s)$, and w contains both letters a and b.

Define the monoid homomorphism $\mu : \{a, b\}^* \to \mathbb{N}$, $\mu(a) = q, \mu(b) = -p$. Denote by $s(f)$ the slope of the word f, that is, $|f|_b/|f|_a$. For any word f, $\mu(f) = |f|_a|w|_b - |f|_b|w|_a$. Assuming that $|f|_a \ne 0$, we obtain

$$\mu(f) = |f|_a|w|_a(s(w) - s(f)) \tag{16.3}$$

(ii) We show that for any suffix v of w, one has $-p \le \mu(v) \le q$, with strict inequality except if $v = bv'$ and $\mu(v') = 0$ for the first inequality, and $v = av'$ and $\mu(v') = 0$

for the second one. We prove only the second inequality, the proof of the first being similar. For this, let $M(u)$ denote the maximum of the numbers $\mu(y)$ for the suffixes y of u. We have $\mu(w) = |w|_a\mu(a) + |w|_b\mu(b) = pq + q(-p) = 0$. Hence $M(w^2) = M(w)$, since each suffix of w is a suffix of w^2, and since each suffix of w^2 is either a suffix of w or of the form yw for some suffix y of w, and then $\mu(yw) = \mu(y)$.

Let v be a suffix of w such that $\mu(v) = M(w)$. Since $w = v_1 \cdots v_k$, we may write $v = hv_{i+1} \cdots v_k$, for some $i = 1, \ldots, k$, where h is a suffix of v_i. We claim that $\mu(v_i \cdots v_k) \leq 0$ and $\mu(v_{i+1} \cdots v_k) \leq 0$. Indeed, since $w^2 = v_1^2 \cdots v_k^2$, the word $hv_{i+1}^2 \cdots v_k^2$ is a suffix of w^2, and thus its image under μ is $\leq \mu(v)$ so that $\mu(h)+2\mu(v_{i+1} \cdots v_k) \leq \mu(h)+\mu(v_{i+1} \cdots v_k)$, which proves the second inequality. For the second one, one uses the fact that $hv_iv_{i+1}^2 \cdots v_k^2$ is a suffix of w^2, and the argument is similar.

We show that $\mu(v) \leq q$. We have $\mu(v) = \mu(h) + \mu(v_{i+1} \cdots v_k) \leq \mu(h)$. If h is the empty word, or $h = a$, or b, then $\mu(h) = 0, q$ or $-p$; hence $\mu(v) \leq q$, with strict inequality except if $h = a$ and $\mu(v_{i+1} \cdots v_k) = 0$; if $h = v_i$, then $\mu(h) = \mu(v_i \cdots v_k) \leq 0 < q$. Thus we may assume that $v_i = zcdy$, where $h = dy$ and $c, d \in \{a, b\}$, y non-empty. We have $\mu(v) = \mu(hv_{i+1} \cdots v_k) = \mu(d) + \mu(yv_{i+1} \cdots v_k) \geq \mu(yv_{i+1} \cdots v_k)$, since $yv_{i+1} \cdots v_k$ is a suffix of w and by maximality of $\mu(v)$; hence $\mu(d) \geq 0$, and $d = a$. Similarly, $\mu(v) \geq \mu(cv) = \mu(c) + \mu(v)$, hence $\mu(c) \leq 0$ and $c = b$. Thus $v_i = zbay$.

It follows from Exercise 16.3.4 that $s(y) > s(zba)$. Thus either $s(w) < s(y)$ or $s(w) > s(zba)$. In the first case, by (16.3) applied to $f = y$, we deduce that $\mu(y) < 0$ (if $|y|_a = 0$, this also holds); hence $\mu(v) = \mu(ayv_{i+1} \cdots v_k) = q + \mu(y) + \mu(v_{i+1} \cdots v_k) < q$. In the second case, using (16.3) applied to zba, we have $\mu(zba) > 0$; hence $\mu(v) = \mu(ayv_{i+1} \cdots v_k) = \mu(zbayv_{i+1} \cdots v_k) - \mu(zb) = \mu(v_i \cdots v_k) - \mu(zba) + \mu(a) < q$.

(iii) The set of words mapped to 0 under μ is a free submonoid of $\{a, b\}^*$; its basis B is the set of words z such that $\mu(z) = 0$ and that $\mu(u) \neq 0$ for all non-trivial proper prefixes (equivalently, suffixes) of z. We may factorize w in this basis: $w = z_1 \cdots z_m$, $z_i \in B$.

Note that since p and q are relatively prime, each non-empty word u with $\mu(u) = 0$ has the same slope as w and that its length is at least the length $n = p + q$ of w; in particular, it is at least 2. This applies to the words z_i.

Let v be a non-empty suffix of z_i. Then $vz_{i+1} \cdots z_m$ is a suffix of w. Since $\mu(z_j) = 0$ for all j, it follows from (ii) that

$$-p \leq \mu(v) \leq q \tag{16.4}$$

Moreover, the first (resp. the second) inequality is strict except if $vz_{i+1} \cdots z_m = bv'$ (resp. $= av'$) and $\mu(v') = 0$; note that $v \neq 1$; then $vz_{i+1} \cdots z_m = cv'$, $c \in A$, $\mu(v') = 0$ implies that $v = c$, $v' = z_{i+1} \cdots z_m$, since no word in B is a suffix of another word in B.

The word aa is not a suffix of z_i, by (16.4), since $\mu(aa) = 2q > q$; similarly bb is not a suffix of z_i. Hence z_i ends with ab or ba. Suppose first that it ends with ab. Let $A = \{a < b\}$. Since $|z_i| \geq n$, we may write $z_i = y\tilde{f}b$, where $y, f \in A^*$ with f of length $n-1$ and beginning with a. Then for any prefix u of f, $\tilde{u}b$ is a suffix of z_i so that, by (16.4), $-p \leq \mu(\tilde{u}b) < q$; because $\mu(b) = -p$, we deduce that $0 \leq \mu(\tilde{u}) = \mu(u) < n$. It follows from Exercise 15.1.3 that fb is the lower Christoffel word of slope q/p; hence $b\tilde{f}$ is the upper Christoffel word of slope q/p, $\tilde{f}b$ is a standard word with p letters a and q letters b; it follows in particular that $\mu(\tilde{f}b) = 0$, and since z_i is in the basis B, we must have $z_i = \tilde{f}b$.

Suppose now that z_i ends with ba. Then, by taking the ordered alphabet $A = \{b < a\}$ and replacing μ by $-\mu$, we argue similarly and find that z_i is a standard word with $|z_i|_a = p$, $|z_i|_b = q$.

(iv) We have proved that $w = z_1 \cdots z_m$, where the z_i are standard words having the same commutative image. It follows then, from the hypothesis that $|w|_a$ and $|w|_b$ are relatively prime, that $m = 1$. □

16.3.1 Exercises

Exercise 16.3.1 *Define the* Fibonacci words *by* $f_0 = b, f_1 = a$, *and* $f_{n+2} = f_{n+1}f_n$ *for any* $n \in \mathbb{N}$. *Define the endomorphism* $\varphi = (ab, a)$ *of* $\{a < b\}^*$. *Show that* $f_n = \varphi^n(f_0)$ *and that* $f_n = \varphi^{n-2}(ab)$ *if* $n \geq 2$. *Show that* $\varphi = GE = ED$. *Show that* $\varphi^n = (GD)^{n/2}$ *if n is even, and* $\varphi^n = ED(GD)^{(n-1)/2}$ *if n is odd. Deduce that* f_n *is a standard word. Show that*

$$f_7 = S_6S_1S_2S_4S_5S_3S_4S_1S_2, \quad f_7^2 = S_6^2S_1^2S_2^2S_4^2S_5^2S_3^2S_4^2S_1^2S_2^2$$

where we write S_i *for* $S_i(1,0)$ *(illustration of Theorem 16.3.1).*

Exercise 16.3.2 *Show that each word* $S_i(r,s)^2$ *and* $T_i(r,s)^2$ *has no non-trivial proper suffix that is a square.*

Exercise 16.3.3 *Let* $u = (ba)^r b$, $r \geq 1$, *and* $w = D(S_5(r,s)) = u^{s+1}ab$. *Prove the following three assertions:*

If r is even and s is odd, then

$$w^2 = ((T_4^2)^{r/2}T_1^2(T_2^2)^{r/2})^{(s-1)/2}T_6^2(T_5^2)^{(s-1)/2}(T_4^2)^{r/2-1}T_3^2(T_2^2)^{r/2}$$

If r is odd and s is even, then

$$w^2 = ((T_4^2)^{(r-1)/2}T_1^2(T_2^2)^{(r-1)/2})^{s/2}T_4^2T_5^2(T_5^2)^{s/2-1}(T_4^2)^{(r-1)/2}T_1^2(T_2^2)^{(r+1)/2}$$

If r and s are both odd, then

$$w^2 = ((T_4^2)^{(r-1)/2}T_3^2(T_2^2)^{(r-1)/2})^{(s-1)/2}T_6^2(T_5^2)^{(s-1)/2}(T_4^2)^{(r-1)/2}T_1^2(T_2^2)^{(r+1)/2}$$

Complete the case $i = 5$ *of the proof of Lemma 16.3.2 by proving that, in the three cases, w is equal to the expression obtained by removing the exponent 2 from the* T_j

in the three previous expressions. Hint: follow the calculations given in the proof of Lemma 16.3.3.

Exercise 16.3.4 *Denote by $s(w)$ the slope of the word $w \in \{a, b\}^*$, that is, $s(w) = |w|_b/|w|_a$. Suppose that v is a word of the form $S_i(r, s)$ defined in (16.1). Show that, assuming that the following word y is non-empty,*

(i) *if $v = zaby$, then $s(y) < s(zab)$;*
(ii) *if $v = zbay$, then $s(y) > s(zba)$.*

Hint: in the first case, take for example, $v = S_5 = (a^r b)^s a^{r+1} b$, and $zab = (a^r b)^i, y = (a^r b)^j a^{r+1} b$. Then the inverses of the slopes of zab and y are respectively r and $(jr + r + 1)/(j + 1) = r + 1/(j + 1)$. See Lemma 3.7 in [PS].

17
Lyndon words and Christoffel words

We give the basic properties of Lyndon words. Lower Christoffel words are particular Lyndon words. They are maximal for the lexicographical order among Lyndon words of a given slope (Borel and Laubie). The function which maps a lower Christoffel word onto its slope is increasing with respect to the lexicographical order (Borel and Laubie). Lower Christoffel words are the Lyndon words which are equilibrated (Melançon).

17.1 Slopes

We begin by comparing Christoffel words according to their slope and to the lexicographical order (defined in Section 5.6).

Theorem 17.1.1 *(Borel and Laubie [BL], p. 27) Let k and l be positive integers. Let u and v be two distinct lower Christoffel words of slope r and s, respectively. Then $r < s$ if and only if $u^k <_{lex} v^l$.*

We say that a lattice path p is *below* a lattice path q if, for each integral point (x, y) on p, either there is some integral point (x, y') of q with $y \leq y'$ or there is no integral point on q with first coordinate x.

Lemma 17.1.2 *Let u and v be two words in $\{a < b\}^*$ that encode, respectively, two lattice paths p, q having the same origin. If p is below q, then $u \leq_{lex} v$ or v is a proper prefix of u.*

Encoding of lattice paths is defined in Section 2.2.

Proof. We may assume that v is not a proper prefix of u. Therefore, if v is empty, then u must be empty too, and in that case we conclude. If v is non-empty, we conclude also if u is empty. So we may assume that both are non-empty. If $u = xu'$, $v = xv'$ where $x = a$ or b, then we conclude by induction that $u' \leq_{lex} v'$ (and then $u \leq_{lex} v$), since v' cannot be a proper prefix of u' and since the path associated with u' is below the path associated with v'. If $u = au'$ and $v = bv'$, we see that $u <_{lex} v$. The remaining case $u = bu'$, $v = av'$ is impossible because p is below q. □

Proof of Theorem 17.1.1. Since the orders are total and since $r \neq s$ (because $u \neq v$), it is enough to prove that if $r < s$, then $u^k <_{lex} v^l$. Note that u^k is the word coding the discretization from below of the segment $[(0, 0), k(|u|_a, |u|_b)]$ and that v^l discretizes $[(0, 0), l(|v|_a, |v|_b)]$. Hence, each prefix of u^k has a slope $\leq r$. Hence, v^l cannot be a

From Christoffel Words to Markoff Numbers. Second Edition. Christophe Reutenauer, Oxford University Press.
 DOI: 10.1093/9780197907535.003.0017

prefix of u^k, since its slope is s. The path discretizing the first segment is clearly below that discretizing the second. It follows from the lemma that $u^k \leq_{lex} v^l$, and they cannot be equal, having different slopes. □

17.2 Lyndon words

A *Lyndon word* on a totally ordered alphabet is a nonempty word w such that, for each non-trivial factorization $w = uv$, one has $w <_{lex} vu$. In particular, a Lyndon word is radical.

Theorem 17.2.1 *(Lyndon) The following conditions are equivalent for a nonempty word w:*

(i) w is a Lyndon word;
(ii) for each non-trivial factorization $w = uv$, one has $w <_{lex} v$.

The proof of the following property of the alphabetical order is left to the reader.

Lemma 17.2.2 *If $w <_{lex} v$ and w is not a prefix of v, then $wx <_{lex} vy$ for any words x and y.*

Proof of Theorem 17.2.1. If $w = uv$ is a Lyndon word with a non-trivial factorization, then assume by contradiction that v is a prefix of w. Then $w = vt$, and by definition we have $w <_{lex} tv$. Thus $uv <_{lex} tv$, and by Lemma 17.2.2, since u and t have the same length, $u <_{lex} t$, implying $vu <_{lex} vt = w$, a contradiction. Thus v is not a prefix of w.

Since $w <_{lex} vu$, $w <_{lex} v$ (otherwise, $v <_{lex} w$, and since v is not a prefix of w as we saw, we deduce by Lemma 17.2.2 that $vu <_{lex} w$, a contradiction).

Conversely, if $w <_{lex} v$ for any non-trivial factorization $w = vu$, then $w <_{lex} vu$, since $v <_{lex} vu$. □

The theorem implies that a Lyndon word of length at least 2 on the alphabet $\{a < b\}$ must begin with a and end with b.

The first part of the proof of the previous theorem also proves the following.

Corollary 17.2.3 *Each Lyndon word is unbordered.*

Corollary 17.2.4 *(Lyndon) If u and v are Lyndon words such that $u <_{lex} v$, then uv is a Lyndon word.*

Proof. Since $u <_{lex} v$, it follows from Lemma 17.2.2 that if u is not a prefix of v, then $uv <_{lex} v$. If, on the other hand, u is a prefix of v, then $v = uv'$, and then $v < v'$ by the previous theorem, and therefore $uv <_{lex} uv' = v$. Thus, in both cases, $uv <_{lex} v$.

Let s be a non-trivial proper suffix of uv, $s \neq v$. If s is a suffix of v, then by the theorem, $v <_{lex} s$; then, since $uv <_{lex} v$, $uv <_{lex} s$. Otherwise, $s = tv$, where t is a non-trivial proper suffix of u; then $u <_{lex} t$, and by Lemma 17.2.2, $uv <_{lex} tv = s$, since t is shorter than

u and hence not a prefix of u. In all cases, $uv <_{lex} s$, and uv is a Lyndon word by the theorem. □

Theorem 17.2.5 *(Lyndon) Let w be a Lyndon word of length at least 2, and let v be its longest proper suffix which is a Lyndon word. Then v is its smallest non-trivial proper suffix, $w = uv$, u is a Lyndon word, and $u <_{lex} v$.*

The factorization $w = uv$ given in the theorem is called the *right standard factorization* of w.

Proof. Let $u = u_1u_2$ be a non-trivial factorization. Since, by maximality of v, u_2v is not a Lyndon word, there is, by Theorem 17.2.1, a non-trivial factorization $u_2v = ps$, with $s <_{lex} u_2v$. We show that $s \leq_{lex} u_2$. Indeed, if $u_2 <_{lex} s$, then $u_2 <_{lex} s <_{lex} u_2v$, and by the properties of the lexicographical order, we must have $s = u_2s', s' <_{lex} v$; then $s' \neq 1$, and by the equality $u_2v = ps = pu_2s'$, we see that s' is shorter than v, whence s' is a suffix of v so that by Theorem 17.2.1, $v \leq_{lex} s'$, a contradiction. Thus $s \leq_{lex} u_2$, $u <_{lex} uv <_{lex} s$ (by Theorem 17.2.1, since s is a suffix of w) $\leq_{lex} u_2$, and by the same theorem, u is a Lyndon word.

Now, let v' be the smallest proper non-trivial suffix of w. Then v' is smaller than any of its proper non-trivial suffixes so that by Theorem 17.2.1, v' is a Lyndon word. Hence, by maximality of length, v is at least as long as v'. Thus $v = xv'$. If x is non-empty, then we must have, by the theorem again, $v <_{lex} v'$, contradicting the minimality of v'. Hence $v = v'$.

One has $u <_{lex} uv <_{lex} v$ by the same theorem. □

17.2.1 Exercises

Exercise 17.2.1 *(Lyndon's theorem) Show that each word w is uniquely a decreasing product of Lyndon words: $w = \ell_1 \cdots \ell_n, n \geq 0, \ell_1 \geq \cdots \geq \ell_n$. Hint: existence is proved using Corollary 17.2.4. For uniqueness, assume that $\ell_1 \cdots \ell_n = h_1 \cdots h_p$, and ℓ_1 longer than h_1; then $\ell_1 = h_1 \cdots h_j u, j \geq 1$, u is a non-empty prefix of h_{j+1} and deduce the contradiction $\ell_1 < \ell_1$.*

Exercise 17.2.2 *Show that if a Lyndon word is a circular factor of a Lyndon word, then it is a factor of this word. Hint: let x and w be Lyndon words, and x be a circular factor of w; suppose that $x = sp$, where s is a suffix of w and p a prefix of w; show that s or p is empty.*

17.3 Maximal Lyndon words

Lemma 17.3.1 *Lower Christoffel words are Lyndon words. Christoffel words are unbordered.*

Proof. The first assertion follows from Theorem 8.2.3. Hence, lower Christoffel words are unbordered by Corollary 17.2.3. Thus, upper Christoffel words are unbordered, since they are the reversals of the lower ones. □

Theorem 17.3.2 *(Borel and Laubie [BL], Théorème 1) Let r be a positive rational number. The lower Christoffel word of slope r in $\{a < b\}^*$ is the maximum, for the alphabetical order, of the Lyndon words of slope r.*

Proof. By Lemma 17.3.1, each lower Christoffel word is a Lyndon word.

Let w be the lower Christoffel word of slope $r = \frac{q}{p}$, and u be a Lyndon word of the same slope. Then the commutative image of w is (p, q) and that of u is $k(p, q)$ for some $k \in \mathbb{P}$. We may assume that $w \neq u$.

The word w determines a path p in the discrete plane from A to B, and u determines a path p' from A to some point C. Then $B = A + (p, q)$, $C = A + k(p, q)$. We also consider the path p_1 from B to C determined by w^{k-1}. Then we have two paths from A to C, namely pp_1 and p', respectively, determined by w^k and u. Suppose first that the path p' is below the path pp_1. This implies by Lemma 17.1.2 that $u <_{lex} w^k$ because w^k is not a prefix of u (indeed they have the same length and cannot be equal, since $w^k = u$ implies $k = 1$, u being unbordered, which implies $w = u$, a contradiction). Let i be the largest integer such that w^i is a prefix of u. Then $0 \leq i < k$, $u = w^i s <_{lex} w^k$, hence $s <_{lex} w^{k-i}$; actually, we have $s \leq_{lex} w$ (otherwise, $w <_{lex} s$, and since by maximality of i, w is not a prefix of s, Lemma 17.2.2 implies $w^{k-i} <_{lex} s$, a contradiction). Note that $s \neq 1$; otherwise, $u = w^i$, a contradiction. Indeed, we cannot have $i = 0$ since $u \neq 1$, nor $i = 1$ since $u \neq w$, nor $i \geq 2$ since u is radical. Since $s \neq 1$, we deduce by Theorem 17.2.1 that $u \leq_{lex} s \leq_{lex} w$, and finally, $u <_{lex} w$, since they are not equal.

Suppose now that the path p' is not below the path pp_1. Since pp_1 is the discretization from below of the segment from A to C, there is some point on p' located strictly above the segment from A to C. Let M be the last such point on p'; see Figure 17.1 for an example. Then, except for M and C, the points on the subpath q from M to C of the path p' are located strictly below the line AC and therefore below the segment from M to C. Let s code q, and let v^l be the word discretizing from below the segment from M to C, necessarily a power of some lower Christoffel v. Since q is below MC, it is also below the path coded by v^l, and therefore $s \leq_{lex} v^l$ by Lemma 17.1.2 (because they have the same length so that v^l is not a proper prefix of s). The slope of the segment MC is smaller than the slope r of the segment AC. Thus, by Theorem 17.1.1, $v^l <_{lex} w$. It follows that $s <_{lex} w$. Thus we cannot have $w \leq_{lex} u$: it would imply $s <_{lex} u$ and contradict Theorem 17.2.1, since s is a suffix of u (the path q being a suffix of the path p'). Hence $u <_{lex} w$. □

Note that the two cases in the proof may actually occur. For the first case, this is easy: for example, $w = ab$ and $u = aabb$ ($k = 2$). For the second case, $w = aabab$ and $u = a^4b^3a^3babab$ is an example; see Figure 17.1. Here, $k = 3$ and the path p' determined by u (in red) is not under the path pp_1 determined by $(aabab)^3$ (in blue).

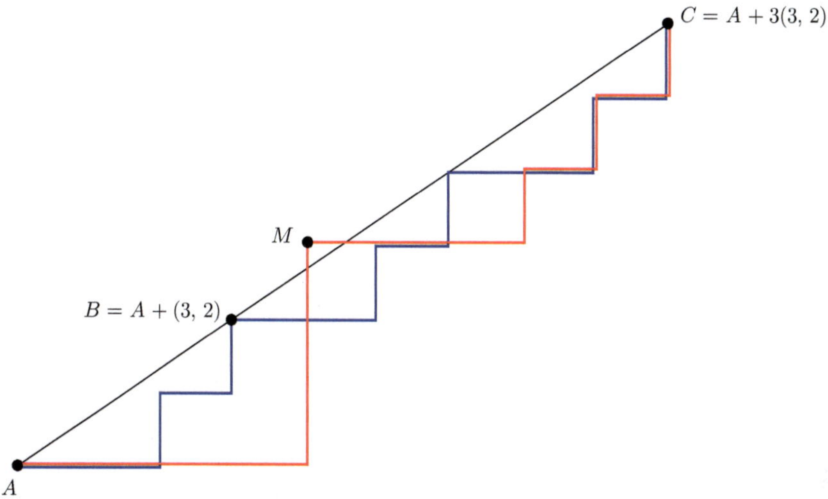

Figure 17.1 The lower Christoffel word $w = aabab$, its third power w^3, and the Lyndon word $a^4b^3a^3babab$

17.4 Equilibrated Lyndon words

We recursively define *equilibrated* Lyndon words as follows: letters are equilibrated, and a Lyndon word w of length at least 2 is equilibrated if, defining u (resp. v) to be the longest proper prefix (resp. suffix) of w that is a Lyndon word, u and v are equilibrated and one has $w = uv$. An example of a Lyndon word which is not equilibrated is $w = aabb$, since $u = aab$ and $v = abb$, and therefore $w \neq uv$.

Theorem 17.4.1 *(Melançon [Me1]) A word in $\{a < b\}^*$ is a lower Christoffel word if and only if it is an equilibrated Lyndon word.*

The next lemma is a particular case of Proposition 3.3 and Proposition 4.2 of Richomme [Ri1].

Lemma 17.4.2 *The endomorphisms $G = (a, ab)$ and $\tilde{D} = (ab, b)$ are increasing for the alphabetical order and preserve Lyndon words in $\{a < b\}^*$.*

Proof. Let $u <_{lex} v$. If u is a prefix of v, then clearly $G(u)$ is a prefix of $G(v)$. If u is not a prefix of v, then $u = xay$, $v = xbz$, and then $G(u) = G(x)aG(y)$, $G(v) = G(x)abG(z)$; if y is empty, then $G(u)$ is a proper prefix of $G(v)$; if y is non-empty, then $G(y) = at$ and then $G(u) = G(x)aat$. In all cases, $G(u) <_{lex} G(v)$. Thus G is increasing. The proof for $\tilde{D}$ is similar (and simpler).

Suppose now that m is a Lyndon word. We show that $w = G(m)$ is a Lyndon word. Let $w = uv$ be a non-trivial factorization. If v begins with a, then, using Lemma 14.1.4 (i), $m = xy$ is a non-trivial factorization, and $G(x) = u$, $G(y) = v$; by Theorem 17.2.1 we have $m < y$, hence $w <_{lex} v$, since G is increasing. If v begins by b, then $w <_{lex} v$ too, since w begins with a. Thus, by the same theorem, w is a Lyndon word. The proof for $\tilde{D}$ is similar. □

If w is an equilibrated Lyndon word of length ≥ 2, we use the notation $w = w'w''$, where, according to the definition, w' (resp. w'') denotes the longest proper prefix (resp. suffix) of w that is a Lyndon word. This factorization of the equilibrated Lyndon word w is called its *standard factorization.*

Lemma 17.4.3 *The endomorphisms* $G = (a, ab)$ *and* $\tilde{D} = (ab, b)$ *preserve equilibrated Lyndon words in* $\{a < b\}^*$ *by inverse image.*

Proof. Suppose that $w = f(m)$ is an equilibrated Lyndon word, where $f = G$ or $\tilde{D}$. Let $m = m_1m_2$ be a non-trivial factorization. Since f is injective, $w = f(m_1)f(m_2)$ is a non-trivial factorization, and therefore $f(m) = w <_{lex} f(m_2)f(m_1) = f(m_2m_1)$. Since, by Lemma 17.4.2, f is increasing and injective, and since the order is total, we see that $m <_{lex} m_2m_1$, which shows that m is a Lyndon word.

We may assume that m is of length at least 2. Then so is w, and $w = w'w''$. Suppose that $f = G$. If w'' begins with b, then $w'' = b$ (since it is a Lyndon word), and then w' must end with a (since w is in the image of G), and then $w' = a$ (since w' is a Lyndon word); then $w = ab$, $m = b$, which was excluded. Thus w'' begins with a, and we find by Lemma 14.1.4 (i), $m = uv$, $w' = G(u)$, $w'' = G(v)$. This works similarly for $\tilde{D}$, and then in both cases, we obtain $m = uv$, $w' = f(u)$, $w'' = f(v)$.

Since w' and w'' are shorter than w, we find by induction (since w' and w'' are equilibrated Lyndon words) that u and v are equilibrated Lyndon words. Suppose that m has a proper prefix longer than u which is a Lyndon word; then $w = f(m)$ has a prefix longer than w' which is a Lyndon word (since f preserves Lyndon words by Lemma 17.4.2), a contradiction. Thus the longest non-trivial proper prefix of m that is a Lyndon word is u. Arguing symmetrically for the suffixes of w, we conclude that m is an equilibrated Lyndon word. □

Proof of Theorem 17.4.1 (following [MR], Theorem 3). The proof is by induction. Let w be a lower Christoffel word. If it is a letter, it is equilibrated. Otherwise, it has length at least 2, and therefore a standard factorization $w = uv$. If a proper suffix of w is longer than v, then it has by Lemma 14.3.5 the non-trivial period $|v|$ and is, by Corollaries 14.3.2 and 17.2.3, not a Lyndon word. Hence the longest proper suffix which is a Lyndon word is v, since lower Christoffel words are Lyndon words by Lemma 17.3.1. Symmetrically, u is the longest proper prefix of w which is a Lyndon word. By induction, u and v are equilibrated, and therefore so is w.

Conversely, suppose that w is an equilibrated Lyndon word. We claim that w is of the form $G(m)$ or $\tilde{D}(m)$ for some equilibrated Lyndon word m; the claim is proved below. If w is a letter, it is a lower Christoffel word. Thus we may assume that w is not a letter, and we show that m is shorter than w. If $w = G(m)$, then $m \neq a$ (otherwise, $w = a$); hence m is not a power of a (since m is radical) so that m contains the letter b; therefore m is shorter than $w = G(m)$. If $w = \tilde{D}(m)$, similarly m is shorter than w. By induction, m is a lower Christoffel word so that w is too (by Corollary 2.6.2).

We now prove the claim. In view of the previous lemma, it is enough to prove that w is of the form $G(m)$ or $\tilde{D}(m)$. We may assume that w is not a letter. Write $w=w'w''$; then w' and w'' are equilibrated Lyndon words, and we may assume by induction that each of them is in the image of G or $\tilde{D}$. If both of them are images of G, or of $\tilde{D}$, we are done. In the two remaining cases, we show that one leads to the conclusion $w=G(m)$ or $\tilde{D}(m)$ and the other leads to a contradiction.

Case (i): $w' = G(u), w'' = \tilde{D}(v)$. Note that u and v are Lyndon words by Lemma 17.4.3.

Suppose that a appears in v. If $v = a$, then $w'' = \tilde{D}(a) = ab = G(b)$; so $w = G(ub)$ and we are done. If $v \neq a$, then v is of length at least 2 and begins by a, hence $w'' = \tilde{D}(v)$ has the proper prefix ab. If a appears in u, then either $u = a$ and $w' = G(a) = a$, or u begins with aa or ab and $w' = G(u)$ begins with aa. Either way, $w' < ab$, and therefore $w'ab$ is a Lyndon word (by Corollary 17.2.4). This is a proper prefix of w (because ab is a proper prefix of w''), that is longer than w', a contradiction. Hence a does not appear in u; thus $u = b$ and $w' = G(b) = ab = \tilde{D}(a)$, $w = \tilde{D}(av)$, and we are done.

Suppose that a does not appear in v. Then $v = b, w'' = b$. Thus $w = w'b$, and b is the longest suffix of w which is a Lyndon word. This implies by Corollary 17.2.4 that w' has no proper suffix which is a Lyndon word beginning with a. Now if w' is of length 1, then $w' = a$ (otherwise, $w' = b$, $w = bb$, a contradiction) and $w = ab = G(b)$. Alternatively, $(w')'' = b$. In this case, since $w' = (w')'(w')''$ is in the image of G, $(w')'$ must end with a; hence $(w')' = a$ and $w = abb = \tilde{D}(ab)$.

Case (ii): $w' = \tilde{D}(u), w'' = G(v)$. We know by the previous lemma that u and v are Lyndon words. Suppose, by contradiction, that a appears in v; hence v begins with a. Note that $w' = \tilde{D}(u)$ begins with ab or b. If v is not a letter, then $w'' = G(v)$ begins with aa, and therefore $w'' <_{lex} w' <_{lex} w$, contradicting Theorem 17.2.1. Hence v is a letter, $v = a$, and $w'' = a <_{lex} w' <_{lex} w$, again a contradiction.

Thus a does not appear in v. Then $v = b$ and $w'' = G(b) = ab$. Since w' is in the image of $\tilde{D}$, it is a product of ab's and b's. Since w is a Lyndon word of length ≥ 2, it does not begin with b. Thus w' begins with ab and w has ab as a proper prefix, but then $w'' <_{lex} w$, contradicting Theorem 17.2.1. □

The first part of the proof (the easiest part) implies, with the help of Theorem 17.2.5, the next result, which is useful among others for hand calculations.

Corollary 17.4.4 *Let $w = uv$ be a proper lower Christoffel word with its standard factorization. Then v is the smallest non-trivial proper suffix of w. Moreover, u (resp. v) is the longest proper prefix (resp. suffix) of w which is a Lyndon word.*

17.4.1 Exercises

Exercise 17.4.1 *Let $w = uv$ be a lower Christoffel word with its standard factorization. Show that if a lower Christoffel word x is a proper factor of w, then x is a factor of u or of v.*

Note: this result is extended to Lyndon words as follows. Let w be a Lyndon word, of length at least 2. Let u (resp. h) be its longest proper prefix (resp. suffix) which is a Lyndon word; let $w = uv$. Then v is a Lyndon word, hence a suffix of h, and if x is a proper factor of w, x is a factor of u, or of v, or is a suffix of h. See [Re5] (Lemma 18.3).

Exercise 17.4.2

(i) *Show that if the lower Christoffel word x is a circular factor of a lower Christoffel word $w = uv$ (standard factorization), then $x = w$ or x is a factor of u or of v. Hint: use Exercises 17.2.2 and 17.4.1. Deduce by induction that x is then an ancestor of w in the tree of Christoffel words. Deduce that all conjugates of x are circular factors of w. Hint: use Exercise 14.5.2.*

(ii) *Let x and w be lower Christoffel words and $n \geq 2$. Suppose that x^n is a circular factor of w. Show that x^n is a factor of w. Hint: use Exercise 17.2.2. Show that for some lower Christoffel word uv (standard factorization), uv is an ancestor of w in the tree of Christoffel words, and that either $x = u$ and $u^n v$ is an ancestor of w in the tree of Christoffel words (Case (i)), or $x = v$ and uv^n is an ancestor of w (Case (ii)). Hint: use Exercises 17.2.2, 17.4.1, and 2.5.3 and Corollary 2.4.3. Deduce that all conjugates of x^n are circular factors of w. Hint: use Exercise 14.5.2. Suppose now that n is maximum. Show that in Case (i) (resp. Case (ii)), uv is not a left child (resp. right child) in the tree of Christoffel word and that if uv is not the root, then its left parent (right parent) is u (resp. v).*

Exercise 17.4.3 *Show that if a factor set (see Exercise 15.3.6) contains the power of a Christoffel word, then it contains all the conjugates of this power. Hint: use Exercise 17.4.2.*

Exercise 17.4.4 *Show that if $w = uv$ is a Lyndon word, with u its longest proper prefix which is a Lyndon word, then v is also a Lyndon word and $u <_{lex} v$. Hint: use the results of Section 17.2 and Exercice 17.2.1; assume that v is not a Lyndon word; show that $v = \ell_1 \cdots \ell_n$, $n \geq 2$, $u \geq \ell_1$ by maximality of u, and then $\ell_n \leq u \leq w$, a contradiction with w Lyndon. Use Exercice 17.2.1 to show that $u <_{lex} v$.*

This factorization of w is called its left standard factorization *(see [V], pp. 15–16). For an equilibrated Lyndon word, its left and right standard factorizations coincide. Show that the words aabbab and aabaabb both satisfy this latter property but are not equilibrated. See [MR] for equilibrated Lyndon words on larger alphabets.*

18
Sturmian words

A Sturmian word is a factor of a Christoffel word. Note that in the literature it is often called a *finite* Sturmian word.

A word is Sturmian if and only if it is balanced (Dulucq and Gouyou-Beauchamps) if and only if it has a factor which is a Christoffel word of length n and has n as period (de Luca and De Luca). From the previous result, one may deduce a normal form for Sturmian words, which allows one to count them and to prove the formula of Lipatov and Mignosi. The unbordered Sturmian words are the Christoffel words (Chuan), and the Sturmian Lyndon words are the lower Christoffel words (Berstel and de Luca). A Lyndon word is a lower Christoffel word if and only if the word formed by the last letters of its conjugates, lexicographically ordered, is a weakly decreasing word (Mantaci, Restivo, Sciortino). A word m on the alphabet a, b is central if and only if it is a palindrome and if it is right special with respect to the set of Sturmian words (de Luca and De Luca) if and only if amb and bma are Sturmian (de Luca and Mignosi). A word is left special with respect to the set of Sturmian words if and only if it is prefix of some central word; in this case, its palindromic closure is a central word (de Luca).

18.1 Sturmian words and balanced words

Proposition 18.1.1 *For a word $w \in \{a, b\}^*$, it is equivalent to be a factor of either a Christoffel word or a lower one, or an upper one, or a standard word, or a central word, or to be a circular factor of some Christoffel word, or a lower one, or an upper one, or a standard word.*

A (finite) word is called *Sturmian* if it satisfies these equivalent conditions. Sturmian words are usually defined as finite factors of *Sturmian sequences*, which are the sequences obtained by discretizing half-lines in the plane. It is well known that the two notions coincide (see [dLM, Ch1]).

We obtain from Corollary 15.3.7 the following result.

Corollary 18.1.2 *The set of Sturmian words is closed under reversal.*

Lemma 18.1.3 *For each Christoffel word u and each natural number n, there is some lower Christoffel word w such that u^n is a factor of w.*

Proof. Suppose first that u is a lower Christoffel word. The result is clear if $u = b$, since ab^n is a lower Christoffel word. If $u \neq b$, there exists a Christoffel pair (u, v). Indeed, this is clear if $u = a$, and if $u \neq a, b$, we have the standard factorization $u = xv$; then (x, v) is a Christoffel pair and $(u, v) = (xv, v)$, too, by Corollary 2.4.3. We

From Christoffel Words to Markoff Numbers. Second Edition. Christophe Reutenauer, Oxford University Press.
 DOI: 10.1093/9780197907535.003.0018

can consider this pair as a morphism: it sends the lower Christoffel word $a^n b$ onto $u^n v$, which is, by Corollary 2.6.2, a lower Christoffel word because this morphism is a product of G and $\tilde{D}$ by Theorem 2.6.1.

If u is an upper Christoffel word, then u is conjugate to the lower Christoffel word $\tilde{u}$, and therefore u^n is a factor of $\tilde{u}^{n+1}$, and one is reduced to the previous case. □

We say that a word u is an *inner factor* of the word v if $v = xuy$ for some non-empty words x and y.

Proof of Proposition 18.1.1. Suppose that w is a circular factor of some Christoffel word, equivalently of a standard word (since such a word is conjugate to a Christoffel word), say u; then w is a factor of u^2, hence an inner factor of u^4; then, by Lemma 18.1.3, there is some lower Christoffel word amb such that w is an inner factor of amb; hence w is a factor of the central word m.

Conversely, if w is a factor of the central word m, then w is a factor, hence a circular factor, of the lower Christoffel word amb, of the upper Christoffel word bma, and of the standard word mab. □

Balanced words were defined in Section 15.2.

Theorem 18.1.4 *Let $m \in \{a, b\}^*$ be a non-empty word. The following properties are equivalent:*

(i) m is Sturmian;
(ii) m is balanced;
(iii) m has some factor x, which is a Christoffel word, and m has $|x|$ as a period.

In this case, $|x|$ is the least non-zero period of w.

In the sequel of this section, the order of the alphabet is $a < b$.

The equivalence of (i) and (ii) is due to Dulucq and Gouyou-Beauchamps, Théorème 3.1, in [DGB]. The equivalence of (i) and (iii) is due to de Luca and De Luca, Theorem 10, and its proof in [dLDL3]; they use a result of Mignosi and Zamboni on Duval extensions of Sturmian words [MZ]; the proof given below is direct, by extension of the proof of Dulucq and Gouyou-Beauchamps.

The last assertion of the theorem implies that the least period of a Sturmian word is equal to the maximal length of its unbordered factors; in general, the least period of a word may be larger than the maximal length of its unbordered factors (see Exercise 18.1.3). There are several interesting results on relations between unbordered factors and the least period of a word (see among others [ES, Du, HN2, Ho]).

Lemma 18.1.5 *Consider the monoid endomorphism of $\{a, b\}^*$ defined by $\phi(a) = ab^h, \phi(b) = ab^{h+1}, h \in \mathbb{N}$.*

(i) ϕ sends lower Christoffel words onto lower Christoffel words.
(ii) If x is any upper Christoffel word, then $\phi(x)$ begins with a, and $a^{-1}\phi(x)a$ is an upper Christoffel word.

Proof.

(i) One has $\phi = \tilde{D}^h G$ (Exercise 18.1.5), and we apply Corollary 2.6.2.
(ii) We consider automorphisms of the free group $F(a, b)$. Let γ be the conjugation by a in $F(a, b)$: $\gamma(g) = a^{-1}ga$. We leave to the reader the verifications of the following identities: $\gamma \circ \phi = (b^h a, b^{h+1} a) = E \circ \psi \circ E$, where $\psi = G^h \tilde{D}$ (Exercise 18.1.5). Now, E exchanges upper and lower Christoffel words, and ψ sends lower Christoffel words onto lower Christoffel words by Corollary 2.6.2. This implies (ii). □

Proof of Theorem 18.1.4. (i) implies (ii): the set of circular factors of a Christoffel word is balanced by Lemma 15.2.3. If m is a Sturmian word, it is a factor of some Christoffel word w. Then the set of factors of m is included in the set of circular factors of w and is therefore balanced.

(ii) implies (iii): suppose that m is a balanced word. We argue by induction on the length of m. Since balanced words and Christoffel words are closed under the automorphism E, and since E preserves the period, we may assume that m begins with a. We may also assume that m contains at least two a's, since otherwise $m = ab^i$, a lower Christoffel word.

Therefore, we may write

$$m = ab^{i_1}ab^{i_2}a \cdots b^{i_{n-1}}ab^{i_n}, \quad n \geq 2$$

Let h be the minimum of the exponents $i_j, j = 1, \ldots, n-1$ (thus excluding the last one). Then these exponents are equal to h or $h+1$; otherwise m contains the factors $ab^h a$ and b^{h+2} and hence is not balanced. Moreover, i_n must be $\leq h+1$ for the same reason. Now define $v = m$ if $i_n = h+1$, and $v = ab^{i_1}ab^{i_2}a \cdots b^{i_{n-1}}$ if $i_n \leq h$. Note that, in both cases, v is a factor of m. Consider the morphism $\phi = (ab^h, ab^{h+1})$ and define u such that $v = \phi(u)$. Observe that if $i_n = h+1$, then u ends with b.

We show that u is balanced. Indeed, if u is not balanced, it contains by Lemma 15.4.3 two factors of the form apa and bpb. Then v (hence m) contains the factors $\phi(apa) = ab^h\phi(p)ab^h$ and $\phi(bpb) = ab^{h+1}\phi(p)\, ab^{h+1}$, hence also the factor $b^{h+1}\phi(p)ab^{h+1}$ of length $|\phi(p)| + 2h + 3$. We claim that m also contains the factor $ab^h\phi(p)ab^h a$ of the same length. The claim will imply that m is not balanced, a contradiction.

In order to prove the claim, note that $u = u_1 apa u_2$, hence $v = \phi(u) = \phi(u_1)ab^h\phi(p)ab^h\phi(u_2)$. Suppose first that $i_n = h+1$; then u ends with b, hence u_2 is non-empty; thus $\phi(u_2)$ begins with a and $v = \phi(u_1)ab^h\phi(p)ab^h a \cdots$, which proves the claim in this case. Suppose now that $i_n \leq h$; then either apa is followed in u by some letter and then we argue as previously or apa is a suffix of u and then $\phi(apa)$ is followed in m by a, which proves the claim.

Thus u is balanced. In the case where $i_n = h+1$, u contains at least one b (as last letter); hence u is shorter than v. The only possibility of having $|u| = |v|$ is therefore: $i_n \leq h$, $h = 0$, and u contains no b; in this case, we have $u = a^{n-1}$, $v = a^{n-1}$, $m = a^n$, a case we may exclude since $m = a^n$ certainly satisfies (iii).

Thus we may assume that u is shorter than v, hence shorter than m. By induction, there is a factor x of u which is a Christoffel word and u is of period $|x|$; then u is a factor of x^N for some large enough N. Thus $\phi(u)$ is a factor of $\phi(x)^N$, hence $\phi(u)$ has the period $|\phi(x)|$.

Suppose that $i_n = h + 1$. Then $m = v = \phi(u)$ and u ends with b. If x is a lower Christoffel word, then $\phi(x)$ is a lower Christoffel word (Lemma 18.1.5 (i)), a factor of m, and therefore m has the period $|\phi(x)|$. If x is not a lower Christoffel word, then x is an upper Christoffel word, and we cannot have $x = b$ (because b is a lower Christoffel word); thus x ends with a. This implies that the factor x of u is not a suffix of u; hence the factor $\phi(x)$ of $\phi(u)$ is followed by a in $\phi(u)$, and the upper Christoffel word $a^{-1}\phi(x)a$ (Lemma 18.1.5 (ii)) is a factor of $m = \phi(u)$. In both cases, there is a Christoffel factor of m of length $|\phi(x)|$, and the latter is a period of m.

Suppose now that $i_n \leq h$. There exists a letter $c = a$ or b such that uc has the period $|x|$, and x is a factor of u, hence of uc. Then $\phi(uc)$ has the period $|\phi(x)|$ and $\phi(uc) = vab^h$ or vab^{h+1}, depending on $c = a$ or b. We also have $m = vab^{i_n}$, $0 \leq i_n \leq h$. Then m is a prefix of $\phi(uc)$; hence m has the period $|\phi(x)|$. Moreover, x is a factor of u; hence $\phi(x)$ is a factor of $\phi(u) = v$. If x is a lower Christoffel word, then $\phi(x)$ too, and $\phi(x)$ is a factor of m. Suppose that x is an upper Christoffel word; note that $\phi(x)$ begins by a, and that $u = u_1xu_2$, hence $m = \phi(u_1)\phi(x)\phi(u_2)ab^{i_n}$, and $\phi(x)$ is followed in m by an a; hence the upper Christoffel word $a^{-1}\phi(x)a$ (see Lemma 18.1.5 (ii)) is a factor of m. Thus (iii) is satisfied.

(iii) implies (i): m is then a factor of some power of x, hence of some Christoffel word (Lemma 18.1.3), hence is Sturmian.

The last assertion follows from the fact that x is unbordered by Lemma 17.3.1: if m had a non-zero period smaller than $|x|$, it would be a period of x, which would have a border. □

18.1.1 Exercises

Exercise 18.1.1

(i) *Show that aa and bb cannot simultaneously be factors of a Sturmian word.*

(ii) *Let $w = baaab$. Show that w is Sturmian but not conjugate to any Christoffel word. Show that the same holds for wa. Show that wb is not Sturmian.*

Exercise 18.1.2 *Using Exercise 14.5.4, show that a palindrome is Sturmian if and only if it is a central factor of some central word. See [dLDL1] (Theorem 3).*

Exercise 18.1.3

(i) *Show that if a word w has an unbordered factor x, whose length is equal to a period of w, then this period is the least period of w. See Duval [Du], Proposition 2.2.*

(ii) *Let $f = ba^{n+1}ba^nba^{n+2}ba^nba^{n+1}ba^n$. Show that the least period of f is $4n + 7$ and that the longest length of an unbordered factor of f is $3n + 6$. See Duval [Du], Remark 3.1.*

Exercise 18.1.4 *Let $m = b^iab^j$. Determine a Christoffel factor of m whose length is the least period of m. In which case, are there two such factors?*

Exercise 18.1.5 *Consider the automorphisms* ϕ, γ *of the free group* $F(a,b)$ *defined by* $\phi(a) = ab^h, \phi(b) = ab^{h+1}, \gamma(g) = a^{-1}ga$ *for any* $g \in F(a,b)$. *Show that* $\phi = \tilde{D}^h G$, $\gamma \circ \phi = (b^h a, b^{h+1} a) = E \circ \psi \circ E$, *where* $\psi = G^h \tilde{D}$.

Exercise 18.1.6 *Show that the smallest non-zero period of a Sturmian word may decrease after applying a Sturmian morphism by computing the smallest period of babaa and of its image under G. This illustrates why the proof of Theorem 18.1.4 is so technical.*

Exercise 18.1.7 *Show that if m is a Sturmian word with an unbordered factor f of length n, and no longer unbordered factor, then each factor of m of length n is a conjugate of f.*

Exercise 18.1.8 *Let C be a primitive conjugation class in* $\{a,b\}^*$. *Show that the following conditions are equivalent:*

(i) C is the set of conjugates of some Christoffel word;
(ii) for each word in C of the form aub (resp. bua), one has $u \geq_{lex} \tilde{u}$ *(resp.* $u \leq_{lex} \tilde{u}$*).*

Hint: for the "(i) implies (ii)" part, use Corollary 15.2.3 and Lemma 18.1.3 to show that C is contained in the set of factors of some balanced word, and then see Exercise 15.4.2. For the converse, see [Re7], Section 6.

Exercise 18.1.9 *Let w be a word, and consider its set F of factors. Let R be 1+ the maximum length of a word in F which is right special with respect to F. Let K be the minimum length of a word* $u \in F$ *such that* $uA \cap F = \emptyset$; *show that u is then the shortest suffix of w which is unioccurrent. Example: verify that for w = aabab (resp. w = ababb), one has* $R = 2, K = 3$ *(resp.* $R = 3, K = 2$*).*

Assume now that w is a Sturmian word. Denote by f(k) the number of factors of w of length k (f is called the complexity function *of w). Show that* $f(0) = f(|w|) = 1$. *Let* $m = \min(K,R), M = \max(K,R)$. *Show that f satisfies:* $\forall x, x+1 \in \{0,1,\ldots,m\}$ *(resp.* $\{m, m+1, \ldots, M\}$, *resp.* $\{M, M+1, \ldots, |w|\}$), $f(x+1) = f(x) + 1$ *(resp.* $f(x) = f(x+1)$, *resp.* $f(x+1) = f(x) - 1$*). Show that* $|w| = K + R$ *and that* $|F| = (K+1)(R+1)$. *See [dL2]. A word w satisfying* $|w| = K + R$ *is called* trapezoidal *(because the graph of f is then a trapezoid, which is isosceles); see [D'A] and [BDLF]. In Proposition 2.8 of the latter article are many equivalent conditions for a word to be trapezoidal (see also [LS], Property 3.2; [AC] pp. 128–129).*

Exercise 18.1.10 *Show that if w is a Christoffel word, with standard factorization w = uv, then* $K = |u|, R = |v|$, *with the notations of Exercise 18.1.9.*

Exercise 18.1.11 *Show that the set of Sturmian words is closed under Sturmian morphisms. Show that for u = bbaa, G(u) is Sturmian but not u. Hint: use Theorem 15.5.1, Corollary 2.6.2, and Corollary 2.6.3.*

Exercise 18.1.12 *Let $A = \{a, b\}^*$.*

(i) Let w be a Sturmian word. Deduce from Exercise 18.1.1 that either $w = a^{i_0}ba^{i_1}\cdots a^{i_{n-1}}ba^{i_n}$, or $w = b^{i_0}ab^{i_1}\cdots b^{i_{n-1}}ab^{i_n}$, for some $n \geq 0$ and exponents $i_0, i_n \geq 0, i_1, \ldots, i_{n-1} \geq 1$.

(ii) Let $w = a^{i_0}ba^{i_1}\cdots a^{i_{n-1}}ba^{i_n}$ be any non-empty word on A, with the previous conditions on the exponents. Show that w is Sturmian if and only if $a^{i_0-1}ba^{i_1-1}\cdots a^{i_{n-1}-1}ba^{i_n-1}$, where a^{i_0-1} (resp. a^{i_n-1}) is removed if $i_0 = 0$ (resp. $i_n = 0$) is a Sturmian word. Hint: use the morphism G and Exercises 2.6.2 and 18.1.11. See [BDJR].

(iii) Use the previous question to find an algorithm which decides if a given word on A is Sturmian. Other algorithms may be found in [BF]; [dLDL3], Section 4; [FKS], Theorem 3.1; and [Re5], Section 13.

Exercise 18.1.13 *Deduce from Exercise 15.5.1 that each Sturmian word is full. See [DJP] (Corollary 4).*

18.2 Unbordered Sturmian words

Theorem 18.2.1 *(Chuan [Ch2], Theorem 4.4; see also Harju and Nowotka [HN1], Theorem 10) Christoffel words are precisely the Sturmian words which are unbordered.*

Proof. We know that lower Christoffel words are Lyndon words and hence are unbordered by Corollary 17.2.3; since upper Christoffel words are the reversals of lower ones, they are unbordered too. They are also clearly Sturmian words.

Conversely, let m be an unbordered Sturmian word. By Theorem 18.1.4, m has some factor x which is a Christoffel word, and m has $|x|$ as a period. Since m is unbordered, its smallest non-trivial period is $|m|$ so that $|m| \leq |x|$; since x is a factor of m, $|x| \leq |m|$. Hence $m = x$ is a Christoffel word. □

Corollary 18.2.2 *(Berstel and de Luca [BdL], Theorem 3.2; see also [Me2], Corollary 4.2) The lower Christoffel words on the alphabet $\{a < b\}$ are the Sturmian Lyndon words.*

Proof. This follows from the theorem, since Lyndon words are unbordered by Corollary 17.2.3. □

Corollary 18.2.3 *A word is a conjugate of a Christoffel word if and only if it is radical and all its conjugates are Sturmian.*

Proof. A Christoffel word is radical, and all its conjugates are factors of w^2, hence are Sturmian by Lemma 18.1.3. Conversely, if w is radical and all its conjugates are Sturmian, then one of its conjugates is a Sturmian Lyndon word. By Corollary 18.2.2, it is a Christoffel word. □

18.2.1 Exercises

Exercise 18.2.1 *Show that the following conditions are equivalent:*

(i) *w is the conjugate of some power of some Christoffel word;*
(ii) *ww is Sturmian;*
(iii) *w and all its conjugates are Sturmian;*

Hint: use Corollary 18.2.3 and Lemma 18.1.3.

Exercise 18.2.2 *Let u be a word of length n. Let* $i, j = 1, \ldots, n$. *Define* $S(u, i, j) = \sum_{1 \le k \le i} |u_k|_b$, *where* $u_1, \ldots, u_n$ *are the n circular factors of length j of u, with multiplicities, ordered in such a way that* $|u_1|_b \le \cdots \le |u_n|_b$.

(i) *Show that* $S(u, n, j) = j|u|_b$.
Let w and v be two words of length n in the same commutation class of $\{a, b\}^*$. *Suppose that w is conjugate to a Christoffel word.*
(ii) *Show that* $S(v, i, j) \le S(w, i, j)$ *for any i and j.*
(iii) *Suppose that v is not conjugate to a Sturmian word. Show that* $S(v, i, j) < S(w, i, j)$ *for some i and j. See Jenkinson and Zamboni [JZ] (Theorem A).*

Exercise 18.2.3 *Let F be a factor set of length* $n \ge 2$ *(see Exercise 15.3.6). Suppose that F contains an unbordered word. Show that it contains the conjugation class C of a Christoffel word amb. Hint: Theorem 18.2.1 and Exercise 17.4.3. Show that it contains ama or bmb and that* $F = C \cup \{ama\}$ *or* $F = C \cup \{bmb\}$. *Hint: use Theorem 8.2.3 and Exercise 15.3.6 to prove that either amb is the smallest element of F or bma is the largest; then show that either* $a\ell(F) = amb$ *or* $b\ell(F) = bma$. *Suppose that F is the set of circular factors of the Christoffel word w of length > n ; show that ama (resp. bmb) is in F if and only if w is in the left (resp. right) subtree under amb. Hint: use Exercises 17.4.2 and 14.5.1. See [Re5] (Proposition 7.1).*

Exercise 18.2.4 *A factor set is of* type C *if it contains a Christoffel word; in this case, it is the set of conjugates of this word, together with another word (see Exercise 18.2.3). Let w be a Christoffel word and consider its set* $\mathcal{F}$ *of circular factors. Show that the factor sets of type C contained in* $\mathcal{F}$ *are associated with the ancestors of w in the Christoffel tree. Hint: use Exercise 14.5.2 and Exercise 17.4.2.*

Exercise 18.2.5 *Let* $\mathcal{F}$ *be the set of circular factors of the lower Christoffel word w.*

Show that if $\mathcal{F}$ *contains a conjugation class C, then it is the conjugation class of a power of some lower Christoffel word. Hint: use the definition of Lyndon words and Corollary 18.2.2.*

Let x be a lower Christoffel word and n an integer. Show that the conjugation class of x^n *is contained in* $\mathcal{F}$ *if and only if x satisfies the condition in Exercise 17.4.2.*

Exercise 18.2.6 *Let w be a radical word of length n on the alphabet* $\{a < b\}$. *Show that the following conditions are equivalent:*

(i) *w is a conjugate of a Christoffel word;*
(ii) *for any circular factors u and v of equal length of w, $u \leq_{lex} v$ implies $|u|_b \leq |v|_b$.*

For "(i) implies (ii)", use Theorem 15.2.1. For "(ii) implies (i)", let F be the set of factors of w^∞. Show that (iii) for any $u, v \in F$ of equal length, $u \leq_{lex} v$ implies $|u|_b \leq |v|_b$; by contradiction, assume that (i) does not hold; then F is not balanced by Theorem 18.1.4 and Corollary 18.2.3, hence $au_1a, bu_1b \in F$ for some word u_1 (Lemma 15.4.3); let u_2 be of maximal length such that $u_2au_1a, u_2bu_1b \in F$ (u_2 exists by Exercise 1.0.5); if $bu_2au_1a, au_2bu_1b \in F$, this contradicts (iii); hence $au_2au_1a \in F, au_2bu_1b \in F$, or $bu_2au_1a, bu_2bu_1b \in F$, or $au_2au_1a, bu_2bu_1b \in F$; the first two cases contradict the maximality of u_2 so that we must have the third case; iterating this, we find for any k, $au_ka \cdots au_1a, bu_kb \cdots bu_1b \in F$; since F is the set of factors of a periodic sequence, for any two words u and *v of equal length in F, $|u|_b - |v|_b$ must be bounded; thus we find a contradiction. See [JZ] (Theorem B).*

Exercise 18.2.7

(i) *Show that if $u, v \in A^*$ are conjugate words and if $f : A^* \to A^*$ is a monoid homomorphism, then $f(u), f(v)$ are conjugate. Show that f induces a mapping from the set of conjugation classes of A^* into itself. For a conjugation class C and $w \in C$, verify that $|w|$, $|w|_a$, and $|w|_b$ depend only on C; these numbers are denoted $|C|$, $|C|_a$, and $|C|_b$.*
(ii) *With $A = \{a < b\}$, show that $(a^{p+1}b, a^pb)$ is a Sturmian morphism.*
(iii) *Let $\mathcal{C}$ (resp. $\mathcal{C}_a$) be the set of conjugation classes of Christoffel words (resp. of conjugation classes C of Christoffel words such that $|C|_a \geq |C|_b$). Let $C \in \mathcal{C}_a$. Show that there exists a unique $C_1 \in \mathcal{C}$ such that $G(C_1) = C$. Show that there exists a unique minimal p and a unique $C_2 \in \mathcal{C}$ such that $(a^{p+1}b, a^pb)$ maps C_2 onto C. Show that $|C|_a = |C_1|$ and $|C|_b = |C_2|$.*
(iv) *Denote $L(C) = C_1$ and $R(C) = C_2$. For C in $\mathcal{C}$, define recursively a binary tree $\tau(C)$. If C is the class of a letter, $\tau(C)$ is a tree with a unique node. Otherwise, either $C \in \mathcal{C}_a$, and $\tau(C)$ is the tree with a root, left subtree $\tau(L(C))$, and right subtree $\tau(R(C))$; or $C \notin \mathcal{C}_a$ and then $E(C) \in \mathcal{C}_a$, and $\tau(C) = \tau(E(C))$. Show that $\tau(C) = \tau(C')$ if and only if $C = C'$ or $C = E(C')$. Hint: show by induction that if $C \in \mathcal{C}_a$, then $\tau(C)$ determines $|C|_a$ and $|C|_b$ uniquely; then show that the latter numbers determine C uniquely. See [CRS] (especially Proposition 14).*

Exercise 18.2.8 *Show that in the Lyndon factorization (Exercise 17.2.1) of a Sturmian word, the Lyndon factors are all lower Christoffel words.*

Related work: in [BLPR], Brlek, Lachaud, Provençal, and the author characterize digital convexity: *the factorization into Lyndon words of the words discretizing a convex curve (the so-called* convex words*) must all be lower Christoffel words. Further related properties are proved by Provençal [Pr]: he characterizes minimal non-convex words using the standard factorization of Christoffel words. In [Me2], Melançon gives the Lyndon factorization of characteristic Sturmian sequences; this factorization is an infinite decreasing product of Lyndon words, whose existence is* *guaranteed by a theorem of Siromoney, Matthew, Dare, and Subramanian [SMDS], Theorem 2.4, an extension of a theorem of Lyndon for finite words (see Exercise 17.2.1); the Lyndon words appearing in the factorization are then Christoffel words (by Theorem 18.2.2) and form an ω-division of the sequence (for each non-trivial permutation of the Lyndon factors, the sequence obtained is lexicographically smaller). See also [Ri2], where Richomme characterizes those factorizations in which each Lyndon word is the image of the preceding one by a fixed morphism.*

18.3 Normal forms and counting

Theorem 18.3.1 *([Re5]) Each Sturmian word m, which is not the power of a letter, has a unique factorization $m = sw^n p$, where w is a proper Christoffel word with palindromic factorization $w = v'u'$, $n \geq 1$, s is a proper suffix of u', and p is a proper prefix of w. A word of this form is necessarily Sturmian and has $|w|$ as smallest period.*

For example, $m = aba^2baba^2 = ab \cdot a^2bab \cdot a^2$, with $n = 1$, and $w = a^2bab = a^2 \cdot bab$ (palindromic factorization), ab a proper suffix of bab, a^2 a proper prefix of w.

Proof. Let m be a Sturmian word on the alphabet $\{a, b\}$, and suppose that m is neither a power of a nor of b. By a result of de Luca and De Luca (Theorem 18.1.4 (iii)), m admits a factor w which is a Christoffel word, whose length is equal to the smallest period of m; note that w is necessarily distinct from a and b, hence proper. We choose the leftmost occurrence of such a factor; since w is a periodic pattern of m, we may write $m = sw^n p$, where $n \geq 1$, s is a proper suffix of w and p is a proper prefix of w. Now, w has the palindromic factorization $w = v'u'$, and $u'v'$ is the conjugate Christoffel word $\tilde{w}$ (Theorem 14.2.1). This implies that s must be a proper suffix of u'; otherwise u' is a suffix of s, $s = s_1 u'$ and $\tilde{w} = u'v'$ is a factor of $m = sw^n p = s_1 u'(v'u')^n p$, which is to the left of w, a contradiction.

If $m = sw^n p$, with w Christoffel word, s suffix of w, and p a prefix of w, then m is a factor of w^∞, hence has the period $|w|$. It is necessarily the least period of m, since w is unbordered by Theorem 14.3.1.

For uniqueness, assume that $m = sw^n p$ with the indicated conditions; then we have $|w|$ = least period of m; hence it is enough to show that w is the leftmost occurrence in m of a Christoffel word of length $\pi = |w|$. The only Christoffel words conjugate to w are w and $\tilde{w}$. Note that in a product $\cdots v'u'v'u'v'u' \cdots$, the only occurrences of $w = v'u'$ and $\tilde{w} = u'v'$ are the apparent ones, w and $\tilde{w}$ being radical words. Thus, since s is a proper suffix of u', there is no occurrence of $\tilde{w}$ in $m = sw^n p$ to the left of the leftmost occurrence of w.

The last assertion follows from the fact that each power of a Christoffel word is a Sturmian word, as follows from Lemma 18.1.3. □

We call *left normal form* the normal form given in the previous theorem. The integer n is called the *exponent* of the normal form, and the word w its *periodic pattern*.

There is also a *right normal form* $sw^n p$, where this time s is a proper suffix of w and p is a proper prefix of v'.

We say that the left normal form $m = sw^n p$ is *short* if p is a proper prefix of v'; for such a word m, left and right normal forms coincide. Otherwise, v' is a prefix of p, and we say that the left normal form is *long*; then the right normal form is $(sv')(\tilde{w})^n p_1$, where $u'v' = \tilde{w}$ and where $p = v'p_1$ and p_1 is a proper prefix of u'.

For example, the word $m = aba^2baba^2$ has a long left normal form $ab \cdot a^2bab \cdot a^2$, since $w = a^2bab = a^2 \cdot bab$ (palindromic factorization), and m has the right normal form $aba^2{\cdot}baba^2{\cdot}1$. This example shows why one has to choose a left-right asymmetric normal form if one wants to have uniqueness of this form.

Normal forms of some classes of Sturmian words are determined in [Re5], Proposition 11.2; the normal forms of conjugates of Christoffel words are determined in Lapointe [La], Theorem 6.

We are on the way to counting the Sturmian words. To this end, denote by $\mathbb{Z}\langle\langle A\rangle\rangle$ the ring of *non-commutative formal power series* in the non-commuting variables a and b. Each such series is an infinite linear combination over $\mathbb{Z}$ of words in $\{a, b\}^*$, written $\sum_{w\in\{a,b\}^*} \alpha_w w$, where the coefficients α_w are integers. Sum and product in $\mathbb{Z}\langle\langle A\rangle\rangle$ of two series are defined by

$$\sum_{w\in\{a,b\}^*} \alpha_w w + \sum_{w\in\{a,b\}^*} \beta_w w = \sum_{w\in\{a,b\}^*} (\alpha_w + \beta_w) w$$

and

$$\Big(\sum_{w\in\{a,b\}^*} \alpha_w w\Big)\Big(\sum_{w\in\{a,b\}^*} \beta_w w\Big) = \sum_{w\in\{a,b\}^*} \Big(\sum_{w=uv} \alpha_u\beta_v\Big)\Big) w$$

In other words, the sum is performed coefficientwise, and the product extends the product in the free monoid $\{a, b\}^*$.

If L is a subset of $\{a, b\}^*$, its *characteristic series* is the sum $\sum_{w\in L} w \in \mathbb{Z}\langle\langle A\rangle\rangle$ of all words in L. We denote by w^+ the characteristic series of the set $\{w^n \mid n \geq 1\}$.

The fact that each Sturmian word, which is not a power of a nor of b, has a normal form, as indicated in Theorem 18.3.1, implies that the characteristic series of the set of Sturmian words is equal to the sum of all powers of the letters a and b, to which one adds the sum of all words of the form sw^kp, for all possible proper Christoffel words w, with palindromic factorization $v'u'$, with p a proper prefix of w and s a proper suffix of u'. By grouping together the Sturmian words having the same periodic pattern in their normal form, we obtain the following result.

Corollary 18.3.2 *In the algebra of non-commutative formal power series in a and b over the integers, the sum of all Sturmian words is equal to*

$$1 + a^+ + b^+ + \sum_w S(u')w^+P(w)$$

where the sum is over all proper Christoffel words w with palindromic factorization $w = v'u'$, with the following notations: $x^+ = \sum_{n>0} x^n$, $S(x)$ (resp. $P(x)$) is the sum of the proper suffixes (resp. prefixes) of the word x.

We recover the counting formula for Sturmian words (Lipatov [Lp] and independently Mignosi [Mi]). For that, we use the canonical ring homomorphism

$$\rho : \mathbb{Z}\langle\langle A\rangle\rangle \to \mathbb{Z}[[x]]$$

the ring of usual formal power series in one variable x over $\mathbb{Z}$: it sends a and b onto x, and in particular each word m of length n onto x^n. We use the notation $x^* = \sum_{n\geq 0} x^n$, also equal to $1 + x^+$ and to $(1-x)^{-1}$.

Corollary 18.3.3 *(Lipatov [Lp]; Mignosi [Mi]) Let s_n denote the number of Sturmian words of length n, $n \geq 0$. Then*

$$\sum_{n\geq 0} s_n x^n = x^* + (x^*)^2 \sum_{n\geq 1} \varphi(n)x^n$$

In other words, for any $n \geq 0$,

$$s_n = 1 + \sum_{i+j=n} (j+1)\varphi(i)$$

with the convention that $\varphi(0) = 0$.

We use the following pretty identities in the proof, expressing the generating series of the Euler function.

Lemma 18.3.4 *Denote by $s(M)$ the sum of all entries of a matrix M.*

(i) $\sum_{n\geq 2} \varphi(n)x^n = \sum_{M\in \mathrm{SL}_2(\mathbb{N})} x^{s(M)}$.

(ii) $\sum_{n\geq 2} \varphi(n)x^n = x^2 + \sum_M (x^{2a+b+2c+d} + x^{a+2b+c+2d})$.

(iii) $\sum_{n\geq 1} \varphi(n)x^n = x - x^2 + \sum_M (2x^{a+b+c+d} - x^{2a+b+2c+d} - x^{a+2b+c+2d})$,

where the summation of the second and third sums is the same as the first, with $M = \begin{pmatrix} a & b \\ c & d \end{pmatrix}$.

Proof.

(i) follows directly from Lemma 14.5.5.

(ii) By Corollary 14.5.4, $\mathrm{SL}_2(\mathbb{N})$ is a free monoid generated by the two matrices $\nu(a)$ and $\nu(b)$. It follows that each matrix in $\mathrm{SL}_2(\mathbb{N})$, except the identity matrix, is equal either to $M\nu(a)$ or $M\nu(b)$ for a unique matrix $M \in \mathrm{SL}_2(\mathbb{N})$, and these cases are mutually disjoint. Thus, by (i),

$$\sum_{n\geq 2} \varphi(n)x^n = x^2 + \sum_M x^{s(M\nu(a))} + \sum_M x^{s(M\nu(b))}$$

We have $M\nu(a) = \begin{pmatrix} a & b \\ c & d \end{pmatrix}\begin{pmatrix} 1 & 1 \\ 0 & 1 \end{pmatrix} = \begin{pmatrix} a & a+b \\ c & c+d \end{pmatrix}$ and $M\nu(b) = \begin{pmatrix} a & b \\ c & d \end{pmatrix}\begin{pmatrix} 1 & 0 \\ 1 & 1 \end{pmatrix} = \begin{pmatrix} a+b & b \\ c+d & d \end{pmatrix}$. Thus $s(M\nu(a)) = 2a + b + 2c + d$ and $s(M\nu(b)) = a + 2b + c + 2d$, which proves (ii).

(iii) is obtained by multiplying (i) by 2 and subtracting (ii), and adding x in order to take into account the case $n = 1$ in the summation. □

Proof of Corollary 18.3.3. We evaluate in two ways the image under ρ of the characteristic series of the set of Sturmian words. On one hand, it is equal to $\sum_n s_n x^n$.

For the second evaluation, we apply ρ to the series in Corollary 18.3.2. We obtain, with the notation of this corollary,

$$H = 1 + \rho(a^+) + \rho(b^+) + \sum_w \rho(S(u'))\rho(w^+)\rho(P(w))$$

Clearly, for a word m, $\rho(m^+) = \sum_{n\geq 1}(x^{|m|})^n$, $\rho(P(m)) = 1 + x + \cdots + x^{|m|-1|} = \rho(S(m))$. It follows that

$$H = 1 + 2\sum_{n\geq 1} x^n + \sum_w\left(\left(1 + x + \cdots + x^{|u'|-1|}\right)\left(\sum_{n\geq 1}\left(x^{|w|}\right)^n\right)\left(1 + x + \cdots + x^{|w|-1|}\right)\right)$$

Now, by Corollary 14.5.3, to each proper lower Christoffel word w, with standard factorization $w = uv$ and palindromic factorization $w = v'u'$, corresponds bijectively a matrix $M = \begin{pmatrix} a & b \\ c & d \end{pmatrix}$ in $SL_2(\mathbb{N})$, in such a way that $|w| = a + b + c + d$, $|u| = a + c$, $|v| = b + d$; since $|u| = |u'|$ (Theorem 14.2.1), we have also $|u'| = a + c$. Now, $\tilde{w} = u'v'$ is the corresponding upper Christoffel and $|v'| = b + d$.

Recall that $\sum_{n\geq 1} x^{kn} = \frac{x^k}{1-x^k}$. Recall also that the summation in H is on all proper Christoffel words, lower and upper, so that we have

$$H = x^* + xx^* + \sum_M (1 + x + \cdots + x^{a+c-1} + 1 + x + \cdots + x^{b+d-1})$$
$$\times \frac{x^{a+b+c+d}}{1 - x^{a+b+c+d}}(1 + x + \cdots + x^{a+b+c+d-1})$$

where the sum is over all matrices $M = \begin{pmatrix} a & b \\ c & d \end{pmatrix}$ in $SL_2(\mathbb{N})$. Since $(1 + x + \cdots + x^{k-1})/(1 - x^k) = 1/(1 - x)$, we obtain

$$H = x^* + xx^* + \sum_M (1 + x + \cdots + x^{a+c-1} + 1 + x + \cdots + x^{b+d-1})x^{a+b+c+d}x^*$$

It follows that

$$(1 - x)^2(H - x^*) = x(1 - x) + \sum_M (1 - x^{a+c} + 1 - x^{b+d})x^{a+b+c+d}$$

The right-hand side is equal to

$$x - x^2 + \sum_M (2x^{a+b+c+d} - x^{2a+b+2c+d} - x^{a+2b+c+2d})$$

By Lemma 18.3.4 (iii), we obtain $(1 - x)^2(H - x^*) = \sum_{n\geq 1} \varphi(n)x^n$ and finally $H = x^* + \frac{1}{(1-x)^2}\sum_{n\geq 1} \varphi(n)x^n$, which ends the proof of the first identity.

For the second, one uses the identity $(x^*)^2 = \sum_{n\geq 0}(n+1)x^n$. □

The previous proof is from [Re5]. Besides the first proofs of this result by Lipatov and Mignosi, there are several other proofs: geometrical proofs by Berstel and Pocchiola [BP] and by Bender, Patashnik, and Rumsey Jr. [BPR]); a combinatorial proof by de Luca and Mignosi [dLM] (see also Berstel and Séébold in [Lo], Chapter 2); and he work of Bédaride, Domenjoud, Jamet, and Rémy [BDJR] for a bivariate counting formula. There is a link with the Riemannn hypothesis, as noted by Mignosi [Mi] and also by [BPR]. Indeed, the previous function of n is asymptotically equal to $\frac{N^3}{\pi^2}$; moreover, the error term (actually the error term of a variant of this function) is $O(N^{\frac{3}{2}+\epsilon})$ for each positive ϵ if and only if the Riemann hypothesis holds ([BPR] pp. 315–316).

18.3.1 Exercises

Exercise 18.3.1 *Show that the left normal form of m is short if and only if there is a unique occurrence of a Christoffel factor in m; show that then necessarily the exponent of the normal form is 1. Show that the left normal form of aba has exponent 1 and is long. Show that if the left normal form is long, then there is a Christoffel word x such that x and $\tilde{x}$ are factors of m (use Theorem 14.2.1).*

Particular cases of Sturmian words are the so-called *triangular partitions*, or *corner cuts*, which are counted by area (see [CRST] and [BM]).

18.4 Conjugates lexicographically

A look at (15.1) shows that its last column is formed downwards by b's followed by a's. The following result characterizes Christoffel words; it implicitly uses the *Burrows–Wheeler transform* [BW].

Theorem 18.4.1 *(Mantaci, Restivo, Sciortino [MRS], Theorem 9) Let w be a Lyndon word on the alphabet $\{a < b\}$, with $|w|_a = p$, $|w|_b = q$, and p, q relatively prime. Let $w = w_1 <_{lex} w_2 <_{lex} \cdots <_{lex} w_{p+q}$ be its conjugates, lexicographically ordered. Let a_i be the last letter of w_i. Then w is a lower Christoffel word if and only if $a_1 \cdots a_{p+q} = b^q a^p$.*

We say that a word w on a totally ordered alphabet is *perfectly clustering* if, denoting by $w_1 <_{lex} w_2 <_{lex} \cdots <_{lex} w_n$ its conjugates, and by a_i the last letter of w_i, one has $a_1 \geq \cdots \geq a_n$. Equivalently, w is not perfectly clustering if it has two conjugates $u <_{lex} v$ with respective last letters a and b such that $a < b$.

Lemma 18.4.2 *[DLPPRT] A word w in A^* is not perfectly clustering if and only if, for some word m and some letters a, a', b, and b', with $a < b$ and $a' < b'$, the two words ama' and bmb' are circular factors of w.*

Proof. If w is not perfectly clustering, then it has two conjugates $u <_{lex} v$ whose respective last letters satisfy $a < b$. By definition of the lexicographical order, since u and v have the same length, one has factorizations $u = ma'u', v = mb'v'$, for some letters $a' < b'$ and some words m, u', and v'. Note that $u \neq v$ and that u' and v' have the same length. If u' and v' are non-empty, then $u = ma'u''a, v = mb'v''b$, and therefore ama' and bmb' are circular factors of w. Suppose that u' and v' are empty; then $u = ma$ and $v = mb$ are conjugates of w, and therefore, by counting letters, one must have $a = b$, hence $u = v$, and thus this case cannot occur.

Conversely, suppose that for some word m and some letters a, a', b, and b', with $a < b$ and $a' < b'$, the two words ama' and bmb' are circular factors of w. Then for some words x, y, x', *and* y', $xama'x'$ and $ybmb'y'$ are conjugates of w. Then $ma'x'xa <_{lex} mb'y'yb$ are two conjugates of w, with last letters $a < b$. Hence, w is not perfectly clustering. □

Proof of Theorem 18.4.1. We have to show that a Lyndon word $w \in \{a < b\}^*$ is perfectly clustering if and only if it is a lower Christoffel word. Suppose that w is a Lyndon word which is not a lower Christoffel word. Then, by Corollary 18.2.2, w is not Sturmian. Hence w is not balanced by Theorem 18.1.4. It follows by Lemma 15.4.3 that for some word m, ama, and bmb are factors of w. Hence, by Lemma 18.4.2, w is not perfectly clustering.

Conversely, suppose that w is a lower Christoffel word. Then the set of circular factors of w is balanced by Lemma 15.2.3. It follows that w cannot have two circular factors of the form xmx' and ymy', for some letters x, x', y, and y', with $x < y$ and $x' < y'$, and some word m. Indeed, in the contrary, since the alphabet has only the two letters $a < b$, we should have $x = a = x', y = b = y'$, and w would have the factors ama and bmb, a contradiction. □

18.4.1 Exercises

Exercise 18.4.1 *Let w be a Lyndon word with two factorizations $w = uv = xy$. Show that $vu <_{lex} yx$ if and only if $v <_{lex} y$. See [HR] (Proposition 1).*

Exercise 18.4.2 *The* Burrows–Wheeler matrix *(or tableau) of a radical word w of length n is the matrix $M = (m_{ij})$ such that the words $m_{i,1}m_{i,2} \cdots m_{i,n}$, $i = 1, \ldots, n$, are the conjugates of w in lexicographical order. Suppose that w is a Christoffel word w on the alphabet $\{a < b\}$.*

(i) *In M, choose a column j and a row i such that $m_{i,j} = a, m_{i+1,j} = b$, where $i + 1$ is taken modulo n. Consider the two submatrices of M corresponding to columns 1 to j and rows 1 to i for the first matrix, and rows $i + 1$ to n for the second. Show that these two matrices have a central symmetry.*

(ii) *Consider now the matrix K of circular factors of length $k < n$ of w, ordered lexicographically. In the last column, there is a letter a followed by a letter b immediately under it. Deduce from (i) that the two submatrices, obtained by cutting K between the two corresponding rows, have a central symmetry.*

(iii) *Suppose that* $|w|_a = p, |w|_b = q$. *Show that the word, obtained by reading downwards the first column of M, is* $a^p b^q$, *and for* $i = 2, \dots, n$, *the i-th column is obtained from the* $i-1$*-th column by shifting cyclically the latter downwards by p steps (use Exercise 15.1.2). Deduce that* $\Gamma^p M = M\Gamma$, *where* Γ *is the matrix of the circular permutation* $(1, 2, \dots, p+q)$.

18.5 Characterizations of central words

Theorem 18.5.1 *(de Luca and Mignosi [dLM], Lemma 6 and Corollary 1) Let m be a word on the alphabet* $\{a < b\}$. *The following conditions are equivalent:*

(i) *m is central;*
(ii) *ama, bmb, amb, and bma are Sturmian;*
(iii) *amb and bma are Sturmian.*

Lemma 18.5.2 *If amb and bma are balanced, then m is a palindrome.*

Proof. Let p be the longest common prefix of m and $\tilde{m}$. If m is not a palindrome, then $m = pxu = vy\tilde{p}$, for some x, y, such that $\{x, y\} = \{a, b\}$. Then $xmy = xpxuy = xvy\tilde{p}y$ has the factors xpx and $y\tilde{p}y$ and is not balanced. □

Proof of Theorem 18.5.1. (i) implies (ii): since m is central, amb and bma are Christoffel words so that they are Sturmian. Moreover, for any Christoffel word u different from b, there is a Christoffel pair (u, v) (see the proof of Lemma 18.1.3). Taking $u = amb$, we deduce by Theorem 14.2.1 that bmb is the second factor of the palindromic factorization of some Christoffel word; hence it is Sturmian. Similarly, ama is Sturmian.

That (ii) implies (iii) is evident.

(iii) implies (i): suppose that amb and bma are Sturmian. By the previous lemma, m is a palindrome. By hypothesis, amb is a factor, hence a circular factor of some Christoffel word $w = apb$. Hence so is $\widetilde{amb} = bma$ by Lemma 15.3.7. Thus m is a right-special circular factor of w. Hence it is a suffix of p by Corollary 15.3.5. Thus it is a central word by Corollary 14.1.7. □

Theorem 18.5.3. *([dL1, dLDL1]) A word m on the alphabet* $\{a, b\}$ *is central if and only if it is a palindrome and if m is right special with respect to the set of Sturmian words.*

Lemma 18.5.4 *For each Sturmian word u, there exist letters x* and *y such that xuy is Sturmian.*

Proof. Indeed, u is a factor of some lower Christoffel word w, hence an inner factor of w^3, hence, by Lemma 18.1.3, an inner factor of some lower Christoffel word. The result follows. □

Lemma 18.5.5 *If ma* and *mb are balanced, then so are cma* and *cmb for some letter* $c = a$ *or b.*

Proof, following [Lo], proof of Lemma 2.2.35. Assume, by contradiction, that bma and amb are not balanced. Then, by Theorem 18.1.4 and Lemma 18.5.4, ama and bmb are balanced. By Lemma 15.4.3, there exists a palindrome u such that aua and bub are factors of bma. Then bub is a factor of bm. Since bm is balanced, aua is not a factor of bm; hence aua is a suffix of bma. Similarly, bub is a prefix of bma. Thus we have $bma = bubsa = bpaua$ and $m = ubs = pau$. Considering amb, we find symmetrically that $m = u'as' = p'bu'$. Then $ubs = u'as'$ and $pau = p'bu'$, and we may assume that $|u| < |u'|$. Thus $u' = ubx$ and $u' = yau$. Finally, $yauas' = u'as' = m = p'bu' = p'bubx$ is not balanced, since it has the factors aua and bub, a contradiction.

We conclude that, for example, amb is balanced. By Lemma 18.5.4 and Theorem 18.1.4, there exists a letter c in $\{a, b\}$ such that cma is balanced. If $c = a$, we are done. Otherwise $c = b$; then amb and bma are balanced, and we conclude using Theorem 18.5.1 and Theorem 18.1.4. □

Proof of Theorem 18.5.3. The "only if" part follows from Theorem 18.5.1. For the converse, we apply the previous lemma, and we may assume that amb is balanced and hence is Sturmian by Theorem 18.1.4. Then amb is a circular factor of some Christoffel word, and we conclude as in the "(iii) implies (i)" part of the proof of Theorem 18.5.1. □

Corollary 18.5.6 *(de Luca [dL1], Theorem 3) A word w is left special with respect to the set of Sturmian words if and only if w is a prefix of some central word. In this case, $w^{(+)}$ is a central word, and w is left special with respect to the set of factors of some Christoffel word.*

Note that the last assertion is not quite evident: indeed, if w is left special with respect to the set of Sturmian words, then aw and bw are Sturmian, which means that each of them is a factor of some Christoffel word, but these two Christoffel words may be distinct. The last assertion implies that they are factors of the same Christoffel word.

Proof.

(i) If w is prefix of some central word m, then $\tilde{w}$ is a suffix of m, and $\tilde{w}$ is right special with respect to the set of circular factors of amb (Corollary 15.3.5). Since these words are Sturmian, and since the set of Sturmian words is closed under reversal (Corollary 18.1.2), w is left special with respect to this set.

(ii) Now let w be some Sturmian word that is left special with respect to the set of Sturmian words. Write $w = mw'$, where m is the longest palindromic prefix of w. Then m is left-special with respect to the set of Sturmian words. Hence m is central by Theorem 18.5.3 and Corollary 18.1.2. If $w' = 1$, we are done. So let $w' = cu$, for some letter c. Since $m = \text{Pal}(f)$ for some word f (Theorem 14.1.2), mc is a prefix of the central word $\text{Pal}(fc) = \text{Pal}(f)c\cdots = mcv$. The word v cannot be a prefix of u; otherwise mcv is a prefix of $mcu = w$, contradicting the maximality of m. If u is a prefix of v, $w = mcu$ is a prefix of mcv, and we are done. Otherwise, no one of the words u and v is a prefix of the other one; hence we may write $u = gdu'$, $v = gev'$ with $\{d, e\} = \{a, b\}$. Thus $w = mcgdu'$, $mcv = mcgev'$ are left special (the first by hypothesis, the second since it is central). Thus $mcgd$ and $mcge$ are left special so that $xmcgy$ is Sturmian for any $x, y \in \{a, b\}$. Hence mcg is central by Theorem 18.5.1, hence a palindrome. This contradicts the maximality of the palindrome m as a prefix of $w = mcgdu'$.

(iii) Suppose that w is a prefix of a central word m. By Proposition 14.1.6, with its notations, for some $i = 0, \ldots, n-1$, we have the prefix order inequalities $Pal(a_1 \cdots a_i)a_{i+1} \leq w \leq \mathrm{Pal}(a_1 \cdots a_{i+1})$. It follows that the right palindromic closure of w is $\mathrm{Pal}(a_1 \cdots a_{i+1})$; indeed, this closure could not be shorter, since the right palindromic closure of $\mathrm{Pal}(a_1 \cdots a_i)a_{i+1}$ is $\mathrm{Pal}(a_1 \cdots a_{i+1})$.

Now, since w is a prefix of m, aw is prefix of the lower Christoffel word amb, and bw is a factor of the upper Christoffel word bma. Since amb and bma are conjugate (Theorem 8.2.3), they are circular factors of amb; thus aw and bw too. Hence, they are factors of some Christoffel word (Exercise 18.5.1). □

18.5.1 Exercises

Exercise 18.5.1 *Show that if finitely many words are circular factors of some Christoffel word, then they are factors of some Christoffel word. Hint: use Lemma 18.1.3.*

Exercise 18.5.2 *Let ls_n be the number of words of length n which are left special with respect to the set of Sturmian words. Let sbs_n be the number of these words which are* strictly bi-special, *that is, the words w such that awa, awb, bwa, and bwb are Sturmian. Show that a word is strictly bi-special if and only if it is a central word. Show that $ls_0 = 1$, $ls_n = ls_{n-1} + sbs_{n-1}$. Hint: using Lemma 18.5.5, show that the mapping $w \mapsto w^-$ (the latter word is w without its last letter) is a surjection from the set of non-empty left-special words onto the set of left-special words, such that the preimage of each word has one or two elements, the second case occurring exactly when the image is a bi-special word. Using Theorem 18.5.1 and Exercise 14.1.1, show that $sbs_n = \varphi(n+2)$. Deduce that $ls_n = \varphi(1) + \varphi(2) + \cdots + \varphi(n+1)$. These results are due to de Luca and Mignosi; see [dLM], proof of Theorem 7 (see also the proof of Theorem 2.2.36 in [Lo]).*

Exercise 18.5.3 *Let $n \geq 1$. Consider the mapping from the set L of left-special words, with respect to the set of Sturmian words, of length $n-1$, into the set of lower Christoffel words, which maps h onto $w = ah^{(+)}b$. Show that this mapping is well defined. Hint: use Corollary 18.5.6. Show that if $w = uv$, the standard factorization of w, then $|w| \geq n+1$ and $|u|, |v| \leq n$. Hint: if $|u| > n$, then $u = amb$, m a central word; h is a prefix of m, hence $h^{(+)}$ is a prefix of m, and w is not longer than u, a contradiction. Show that if $g, h \in L$, and $g <_{lex} h$, then $ag^{(+)}b <_{lex} ah^{(+)}b$ so that the mapping is injective. Show that it is surjective by using Exercise 18.5.2 and Exercise 14.5.5. Hence the mapping is a bijection from L onto the set of lower Christoffel words w, of length $\geq n+1$, whose standard factorization $w = uv$ satisfies $|u|, |v| \leq n$. See [Re5] (Proposition 5.1).*

Exercise 18.5.4 *Let $n \geq 1$. Consider the mapping $\mathcal{F}$ from the set L of left-special words of length $n-1$, with respect to the set of Sturmian words, into the set of factors sets of length n (see Exercise 15.3.6), defined by $h \mapsto \mathcal{F}(h)$, the set of circular words of length n of the Christoffel word $ah^{(+)}b$.*

(i) *Show that ah is the smallest element of $\mathcal{F}(h)$ in alphabetical order. Hint: use Theorem 8.2.3. Show that if $h <_{lex} h'$, then $\min(\mathcal{F}(h)) <_{lex} \min(\mathcal{F}(h'))$ (minimum for alphabetical order), and deduce that $\mathcal{F}$ is injective.*

(ii) *Let F be a factor set of length n, equal to the set of circular factors of some Christoffel word amb of length $> n$. With the notations of Exercise 15.3.6, show that $\ell(F)^{(+)}$ is a prefix of m. Hint: use Proposition 14.1.6. Deduce that $a\ell(F)^{(+)}b$ is an ancestor of amb in the tree of Christoffel word. Hint: use Corollary 14.1.7. Deduce that $F = \mathcal{F}(\ell(F))$ and that $\mathcal{F}$ is surjective. Hint: use Exercise 15.2.4. See [Re5] (Proposition 5.1).*

(iii) *Show that the mapping ℓ of Exercise 15.3.6 is a bijection from the set of factor sets of length n onto the set L. Hint: show that the inverse mapping is $\mathcal{F}$. Deduce that the number of factor sets of length n is $\varphi(1) + \cdots + \varphi(n)$. Hint: use Exercise 18.5.2. The bijection ℓ and the counting formula for factor sets are due to Richomme and Séébold [RS], Theorem 4.2 and Proposition 4.4.*

(iv) *Show that the factors sets contained in the set of circular factors of a given Christoffel word amb are the sets of the form $\mathcal{F}(p)$, where p runs over the set of prefixes of the central word m. Hint: use Theorem 15.3.3 and Corollary 15.3.5.*

Exercise 18.5.5 *Show that a Sturmian word in $\{a, b\}^*$ is left and right special with respect to the set of Sturmian words if and only if it is the maximal inner factor of some power of some Christoffel word. In other words, it is of the form $(uyx)^n u$ for some central word u, some exponent $n \geq 0$, with $\{x, y\} = \{a, b\}$. Deduce that the number of these words which have length n is $2(n+1) - \varphi(n+2)$. These results are due to Fici; see [Fi2], Theorem 3.11 and Corollary 4.2.*

Exercise 18.5.6 *Say that a word is a* Markoff word *if it is of the form m, as in the first sentence of Section 4.2, for some bi-infinite sequence s. Show that m is a Markoff word if and only if it is a central word (Glen, Lauve, and Saliola [GLS], Theorem 1).*

19
Stern–Brocot tree

Christoffel words have a strong connection with continued fractions of rational numbers, as was recognized by Christoffel himself in [C2]. They are in some sense equivalent to them, being a way to encode the expansion of these numbers into continued fractions.

The Stern–Brocot tree (see Graham, Knuth, and Patashnik [GKP]) is another way to compute continued fractions; it is a numerical specialization of the tree of Christoffel words by their slopes. Among other things in this chapter, we show the equivalence between semi-convergents of rational numbers, nodes in the Stern–Brocot tree, palindromic prefixes of central words, and points on the frontier of the convex hull of some subset of $\mathbb{Z}^2$ associated with a rational number. The Raney tree (or Calkin–Wilf tree) is another specialization of the tree of Christoffel pairs by the lengths of the two words in the pair; it is dual to the Stern–Brocot tree. This duality is extended to the Christoffel words (Berthé, de Luca, and the author).

Many of these concepts may be extended to irrational numbers, but we have chosen to stay in the world of finite words.

19.1 Stern–Brocot tree and continued fractions

The *Stern–Brocot tree* is the tree obtained from the tree of Christoffel words (see Section 14.5) by replacing each node w with its slope $\frac{|w|_b}{|w|_a}$. Since each proper lower Christoffel word appears exactly once on the tree of Christoffel words and since these words are in bijection with positive rational numbers, via their slope, we see that each positive rational number appears exactly once on the Stern–Brocot tree. Use Proposition 14.5.2 and, in particular, the homomorphism ν: if $w \in \{a, b\}^*$ encodes the path from the root to some node, with as usual a meaning "left" and b meaning "right", then the corresponding number of the Stern–Brocot tree is

$$\frac{|w|_b}{|w|_a} = \frac{k+l}{i+j}, \qquad \nu(w) = \begin{pmatrix} i & j \\ k & l \end{pmatrix} \tag{19.1}$$

The direct construction of the tree of Christoffel words given in Theorem 14.5.1 implies the direct construction of the Stern–Brocot tree, starting with the root $\frac{1}{1}$: consider the path from some node $\frac{q}{p}$ to the root, and let $\frac{q'}{p'}$ (resp. $\frac{q''}{p''}$) be the node immediately after the first north-west (resp. north-east) step in this path; if no such step exists, then take $\frac{q'}{p'} = \frac{0}{1}$ (resp. $\frac{q''}{p''} = \frac{1}{0}$). Then $\frac{q}{p}$ is the *mediant* of $\frac{q'}{p'}$ and $\frac{q''}{p''}$, that is, $p = p' + p''$ and $q = q' + q''$. For example, $\frac{4}{11}$ is the mediant of $\frac{1}{3}$ and $\frac{3}{8}$, and $\frac{1}{3}$ is the mediant of $\frac{0}{1}$ and $\frac{1}{2}$.

From Christoffel Words to Markoff Numbers. Second Edition. Christophe Reutenauer, Oxford University Press.
 DOI: 10.1093/9780197907535.003.0019

There are two algorithms that allow us to compute the location of a given positive rational number ξ in the Stern–Brocot tree. Let $w(\xi)$ denote the word coding the path from the root to ξ, thereby defining a bijection $\xi \mapsto w(\xi)$ from the set of positive rational numbers onto the free monoid $\{a, b\}^*$.

Theorem 19.1.1 *(Hurwitz [H2], p. 424, Graham, Knuth, Patashnik [GKP], pp. 121–122) Let $\xi = \frac{q}{p}$. Then the word $w = w(\xi)$ coding the location of ξ in the Stern–Brocot tree is obtained by either of the two following algorithms:*

(i) If $\xi = 1$, $w = 1$; if $q > p$, then $w(\frac{q}{p}) = bw(\frac{q-p}{p})$; if $q < p$, then $w(\frac{q}{p}) = aw(\frac{q}{p-q})$.

(ii) Define the matrix function $f\left(\begin{pmatrix} i & j \\ k & l \end{pmatrix}\right) = \frac{i+j}{k+l}$; initialize $w = 1$ (the empty word) and $M =$ the identity matrix of order 2; while $\xi \neq f(M)$, replace w with wa and M with $M\nu(a)$ if $\xi < f(M)$, and otherwise replace w with wb and M with $M\nu(b)$.

Lemma 19.1.2 *If r is a node in the Stern–Brocot tree, then each node in its left (resp. right) subtree is smaller (resp. larger) than r.*

For subsequent use, let $u \in \{a, b\}^*$, $\nu(u) = \begin{pmatrix} i & j \\ k & l \end{pmatrix}$, and recall that $i, l > 0$. Then

$$\nu(a)\nu(u) = \begin{pmatrix} i+k & j+l \\ k & l \end{pmatrix}, \quad \nu(b)\nu(u) = \begin{pmatrix} i & j \\ i+k & j+l \end{pmatrix} \tag{19.2}$$

Proof. Let x be an ancestor of y in the tree of Christoffel words, and $w \in \{a, b\}^*$ be the path from x to y, where a means "left" and b means "right". Denote by $s(f)$ the slope of the word f.

Suppose that y is in the left subtree of x; then w begins by a: $w = au$. Thus $\nu(w) = \nu(a)\,\nu(u)$, and by (19.2), $\nu(w)\cdot 1 > 1$, with the notations of (6.7).

Denote by $w_1 \in \{a, b\}^*$ the path from the root ab to x. It follows from (19.1) that $s(x)^{-1} = \nu(w_1)\cdot 1$ and $s(y)^{-1} = \nu(w_1 w)\cdot 1$. Note also that the real function $t \mapsto \nu(w_1)\cdot t$ is strictly increasing, since the determinant of $\nu(w_1)$ is equal to 1. We have $s(y)^{-1} = \nu(w_1 w)\cdot 1 = \nu(w_1)\cdot(\nu(w)\cdot 1) > \nu(w_1)\cdot 1$ (because the function is strictly increasing) $= s(x)^{-1}$.

Hence the slope of y is smaller than the slope of x, and it follows that in the Stern–Brocot tree, each node in the left subtree of some node is smaller than the latter. The proof for right subtrees is similar. □

Proof of Theorem 19.1.1.

(i) If $w = au$ (resp. $w = bu$), then by (19.1) and (19.2), $\xi = \frac{q}{p} = \frac{k+l}{i+j+k+l} < 1$ (resp. $\xi = \frac{q}{p} = \frac{i+j+k+l}{i+j} > 1$). Define η by $w(\eta) = u$. Then $\eta = \frac{k+l}{i+j} = \frac{k+l}{(i+j+k+l)-(k+l)} = \frac{q}{p-q}$ (resp. $\eta = \frac{k+l}{i+j} = \frac{(i+j+k+l)-(i+j)}{i+j} = \frac{q-p}{p}$).

(ii) The second algorithm follows from (19.1) and the lemma. □

In order to relate the representation of positive rational numbers by the Stern–Brocot tree to their expansion into continued fractions, recall that each *positive* rational number ξ has two expansions into continued fractions: $\xi = [n_0, n_1, \ldots, n_k+1] = [n_0, n_1, \ldots, n_k, 1]$, where $k \geq 0$, $n_0 \in \mathbb{N}$, and $n_i \in \mathbb{P}$ for $i = 1, \ldots, k$ (for $k = 0$, this means $\xi = [n_0 + 1] = [n_0, 1]$). A *convergent* of ξ is any number in the sequence

$$[n_0], [n_0, n_1], \ldots, [n_0, n_1, \ldots, n_{k-1}], [n_0, n_1, \ldots, n_k + 1]$$

Thus $[n_0, n_1, \ldots, n_k]$ is not considered a convergent.

Write $\xi = [m_0, m_1, \ldots, m_k]$ so that $m_i = n_i$ for $i = 0, \ldots, k-1$ and $m_k = n_k + 1$. A *semi-convergent* of ξ is any number of the form $[m_0, \ldots, m_{i-1}, m]$, where $i = 0, \ldots, k-1$ and $m = 1, 2, \ldots, m_i$.

Thus each non-zero convergent is a semi-convergent; 0 is not a semi-convergent, and it is a convergent if and only if $n_0 = 0$, equivalently if $0 < \xi < 1$. Moreover, $[n_0, n_1, \ldots, n_k]$ is a semi-convergent.

For example, the convergents of $\xi = [0, 2, 1, 1, 2]$ are $[0] = 0, [0, 2] = \frac{1}{2}, [0, 2, 1] = \frac{1}{3}, [0, 2, 1, 1] = \frac{2}{5}$, and $\xi = \frac{5}{13}$ itself. The other semi-convergents are $[0, 1] = 1$ and $[0, 2, 1, 1, 1] = \frac{3}{8}$.

Theorem 19.1.3 *([GKP], Section 6.7) Let ξ be some positive rational number and $p \in \{a, b\}^*$ the word coding the path from the root of the Stern–Brocot tree to ξ. Then ξ has the continued fraction expansion $\xi = [n_0, n_1, \ldots, n_k + 1] = [n_0, n_1, \ldots, n_k, 1]$ if and only if $p = b^{n_0} a^{n_1} b^{n_2} \cdots (a \text{ or } b)^{n_k}$.*

Proof. Consider the word

$$p = b^{n_0} a^{n_1} b^{n_2} \cdots b^{n_k} \text{ or } p = b^{n_0} a^{n_1} b^{n_2} \cdots a^{n_k}$$

according to the parity of k, with $k \geq 0$, with the exponents n_i in $\mathbb{N}$ and $n_1, \ldots, n_k > 0$. Let w be the lower Christoffel word corresponding to the path from the root to w, coded by p, in the tree of Christoffel words. Similarly, let ξ be the node of the Stern–Brocot tree corresponding to the path from the root to ξ, coded by p. By definition of the Stern–Brocot tree, the slope of w is ξ. We show that $\xi = [n_0, n_1, \ldots, n_k, 1]$. This will complete the proof by uniqueness of continued fractions and uniqueness of the exponents in the expression above for p.

By Theorem 2.6.1, $w = \rho(p)(ab)$, where ρ sends a onto G and b onto $\tilde{D}$. Define endomorphisms of the free monoid $\{a < b\}^*$ by $\pi_n = (a^n b, a)$ and $\tilde{\pi}_n = (ba^n, a)$. They both have the matrix $P(n)$ as abelianization (see Chapter 1 for this notion and its properties, which we use below). Moreover, $G^n = \pi_n E$ and $\tilde{D}^n = E\tilde{\pi}_n$.

Suppose that k is even so that p has the first form above. One has

$$\rho(p) = \tilde{D}^{n_0} G^{n_1} \tilde{D}^{n_2} \cdots \tilde{D}^{n_k} = E\tilde{\pi}_{n_0} \pi_{n_1} E E \tilde{\pi}_{n_2} \cdots E\tilde{\pi}_{n_k}$$

Since $\pi_1(a) = ab$ and $E^2 = id$, $Ew = E\rho(p)(ab) = \tilde{\pi}_{n_0} \pi_{n_1} \tilde{\pi}_{n_2} \cdots \tilde{\pi}_{n_k} \pi_1(a)$. The abelianization of a is ${}^t(1, 0)$ and that of Ew is ${}^t(|w|_b, |w|_a)$. Therefore ${}^t(|w|_b, |w|_a)$ is equal to $P(n_0)P(n_1)P(n_2) \cdots P(n_k)P(1)\, {}^t(1, 0)$. It follows from (5.5) and (5.7) that the slope $|w|_b / |w|_a$ of w is equal to $[n_0, n_1, n_2, \ldots, n_k, 1]$.

When k is odd and p has the second form above, a similar calculation shows that $Ew = \tilde{\pi}_{n_0} \pi_{n_1} \tilde{\pi}_{n_2} \cdots \pi_{n_k} E(ab)$, and since $E(ab) = ba = \tilde{\pi}_1(a)$, one obtains that the slope of w is as before. □

The theorem has the following consequence:

Corollary 19.1.4 *Let ξ be some positive rational number. The semi-convergents of ξ are the nodes on the path from the root to ξ in the Stern–Brocot tree. The convergents of ξ, distinct from $0, \xi$, are the nodes on this path that precede those nodes where the path has a turn to the left or the right.*

In other words, let p encode the path from the root to ξ in the Stern–Brocot tree. The convergents of ξ, distinct from $0, \xi$, are the nodes of the Stern–Brocot tree corresponding to the words u such that $p = uab\cdots$ or $p = uba\cdots$.

Continuing the example before Theorem 19.1.3, we see on Figure 19.1 that the semi-convergents of $\xi = \frac{5}{13}$ are $\frac{1}{1}$, $\frac{1}{2}$, $\frac{1}{3}$, $\frac{2}{5}$, and $\frac{3}{8}$, and ξ, whereas the non-zero convergents are $\frac{1}{2}$, $\frac{1}{3}$, and $\frac{2}{5}$, and ξ.

Proof. The semi-convergents are the numbers $\alpha = [m_0, \ldots, m_{i-1}, m]$, where $i = 0, \ldots, k-1$ and $m = 1, 2, \ldots, m_i$, with the notations before the previous theorem. By the theorem, this number is the node on the tree which corresponds to the path coded by $p_{i,m} = b^{m_0} a^{m_1} b^{m_2} \cdots (b \text{ or } a)^{m_{i-1}} (a \text{ or } b)^{m-1}$ (note that, by the theorem, this is also true if $m = 1$). The $p_{i,m}$ are exactly all prefixes of the word $p = p_{k,m_k}$, which corresponds to ξ.

Now α is a convergent, distinct from 0 and ξ, if and only if $m = m_i$ and $i < k$. Then the prefix $p_{i,m}$ is followed by ab or ba in the word p. Thus the corollary follows. □

Corollary 19.1.5 *Let $w = amb$ be a proper lower Christoffel word of slope ξ.*

(i) The semi-convergents of ξ are the slopes of the words on the path from the root ab to w in the Christoffel tree.

(ii) The semi-convergents of ξ are the numbers $\frac{|v|_b+1}{|v|_a+1}$ for the central words v on the path from the root 1 to m in the tree of central words (equivalently, the palindromic prefixes v of m).

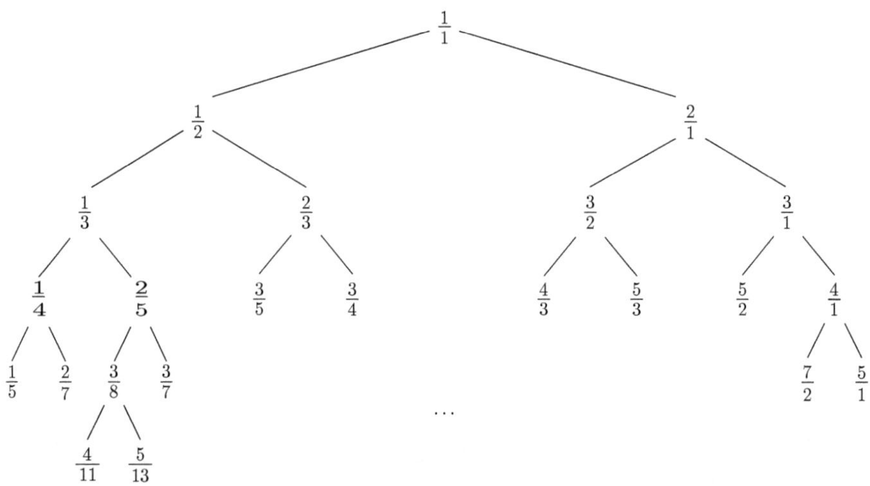

Figure 19.1 The Stern–Brocot tree

Proof. (ii) follows from Corollary 14.1.7. □

Corollary 19.1.6 *Let $w = uv$ be a Christoffel word of slope ξ, with its standard factorization. Let r (resp. s) be the slope of the shortest (resp. longest) of the words u and v. Then r is the penultimate convergent of ξ and s is the penultimate semi-convergent of ξ.*

Take, for example, $w = a^3ba^3ba^2ba^3ba^2b$, of slope $\xi = \frac{5}{13}$, with standard factorization $w = uv$, $u = a^3ba^3ba^2b$, $v = a^3ba^2b$. The slopes of these words are $s = \frac{3}{8}$ and $r = \frac{2}{5}$ in accordance with the corollary; see the previous example.

Proof. We may assume that w is a lower Christoffel word. Then the result follows from Corollary 19.1.4, taking into account the direct construction of the tree of Christoffel words; see Theorem 14.5.1. □

Convergents and semi-convergents may be defined for any real number $\xi = [n_0, n_1, n_2, \ldots]$ by the same formulas as above. We know that each convergent $\frac{p}{q}$ (written in reduced form) satisfies $|\xi - \frac{p}{q}| < \frac{1}{q^2}$ and that if a fraction $\frac{p}{q}$ satisfies $|\xi - \frac{p}{q}| < \frac{1}{2q^2}$, then it is a convergent. It is shown by Hurwitz ([H2], Satz 5, p. 424) that if $\frac{p}{q}$ is a positive rational number such that $|\xi - \frac{p}{q}| < \frac{1}{q^2}$, then $\frac{p}{q}$ is a semi-convergent of ξ; the converse does not hold, however, as shown by $\xi = n \geq 2$, a natural integer, which has $\frac{1}{1}$ as semi-convergent.

19.1.1 Exercises

Exercise 19.1.1 *Prove the* mediant inequality: *if p, q, p',* and *q' are positive numbers such that $\frac{q}{p} < \frac{q'}{p'}$, then $\frac{q}{p} < \frac{q+q'}{p+p'} < \frac{q'}{p'}$.*

Exercise 19.1.2 *Let $\xi = \frac{q}{p}$ (irreducible fraction) be a positive rational number. Assume that $p > q$, and let $\eta = \frac{q}{p-q}$. Show that the mapping $\frac{s}{r} \mapsto \frac{s}{r-s}$ is a bijection from the set of semi-convergents of ξ, different from 1, onto the set of semi-convergents of η. Hint: use Theorem 19.1.1 (i), noting that $w(\xi) = aw(\eta)$.*

Exercise 19.1.3 *Let $0 < \xi < 1$. Show that 1 is a convergent of ξ if and only if $\xi > \frac{1}{2}$. Hint: $\xi = [0, a_1, a_2, \ldots]$ and 1 is a convergent if and only if $a_1 = 1$, or use Corollary 19.1.4.*

Exercise 19.1.4 *Let w be a lower Christoffel word, and $n_1, \ldots, n_k$ the lengths of the successive left and right subpaths from the root to w in the tree of Christoffel words. Show that the length of w is $p(1, n_1, \ldots, n_k, 1)$, where p is the continuant polynomial. Hint: it is the sum of the numerator and of the denominator of the rational number located at the same place in the Stern–Brocot tree.*

Exercise 19.1.5 *Let x, w, $w \neq a, b$, $x \neq w$, be lower Christoffel words such that x is a circular factor of w. Then x is a factor of w, and an ancestor of w in the tree of Christoffel words (Exercise 17.4.2). Let h be the largest exponent such that x^h is a*

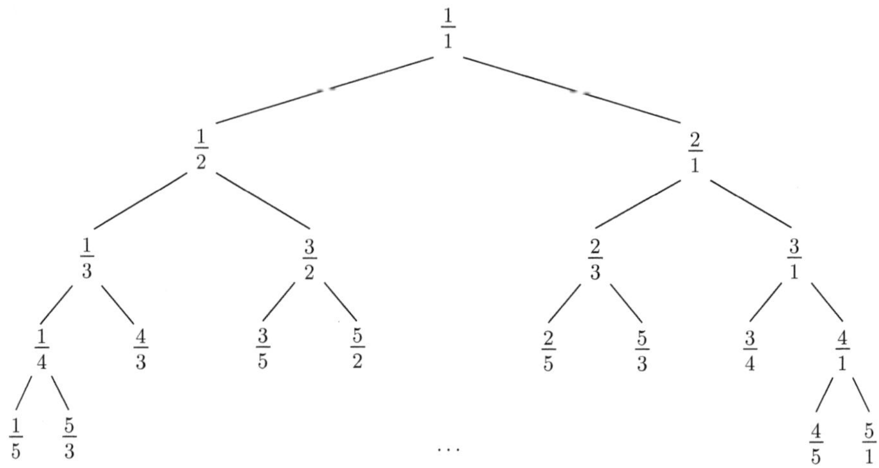

Figure 19.2 The Raney tree

circular factor of w. Let ξ be the slope of w, with continued fraction $\xi = [n_0, \ldots, n_k, 1]$. Show that if $x = b$, then $h = n_0 + 1$. Otherwise, show that either the slope of x is a convergent $[n_0, \ldots, n_i]$ of ξ and $h = n_{i+1} + 1$, with $i < k$, or the slope of x is a semi-convergent, but not a convergent, of ξ and $h = 1$. Hint: use Exercise 17.4.2.

Note: exponents of words in Sturmian sets have been more generally studied in [DL].

Exercise 19.1.6 *Let F be the set of circular factors of the proper Christoffel word w of slope $\xi = [n_0, n_1, \ldots, n_k + 1] = [n_0, n_1, \ldots, n_k, 1]$. Show that F contains a conjugation class of words of length ℓ if and only if $\ell = 0$ or 1, or if ℓ is equal to $r + s$ for some semi-convergent $\frac{s}{r}$ of ξ, or if ℓ is equal to $j(r + s)$, for some convergent $\frac{s}{r} = [n_0, \ldots, n_i]$ of ξ, where $i = 0, \ldots, k - 1, j = 2, \ldots, n_{i+1} + 1$. Use Exercise 19.1.5. These results are a finitary version of results of Fici and De Luca, who studied the* Lie complexity *of Sturmian sequences [FiDL].*

19.2 The Raney tree and dual words

The *Raney tree* (also called the *Calkin–Wilf tree*[1]) is obtained from the tree of Christoffel pairs by replacing each node (u, v) with the rational number $\frac{|u|}{|v|}$. Note that $|u|$ and $|v|$ are relatively prime by Corollary 2.5.1. It may be constructed directly by applying the following rule: the root is $\frac{1}{1}$; if $\frac{q}{p}$ is a node, then its left child is $\frac{q}{p+q}$ and its right child is $\frac{p+q}{p}$. This follows directly from the rule for constructing the tree of Christoffel pairs (see Figure 2.9).

The Stern–Brocot and Raney trees are closely related; we use the involution on positive rational numbers introduced in the paragraph before Corollary 14.5.6.

[1] The first terminology is from [BdL], while the second is motivated by the article [CW].

Corollary 19.2.1 *([BdL]) Let $\frac{j}{i}$ be some node on the Stern–Brocot tree. Then the corresponding node on the Raney tree is $\frac{j^*}{i^*}$.*

Proof. We have to prove that if $w = uv$ is a proper lower Christoffel word, with its standard factorization, and $\frac{|w|_b}{|w|_a} = \frac{j}{i}$, then $\frac{|u|}{|v|} = \frac{j^*}{i^*}$. By Corollary 14.5.3, the matrix $\begin{pmatrix} |u|_a & |v|_a \\ |u|_b & |v|_b \end{pmatrix}$ is in $SL_2(\mathbb{N})$. Now $j = |u|_b + |v|_b$, $i = |u|_a + |v|_a$. By definition, i^* is the inverse of i modulo $i+j = |u|_a+|v|_a+|u|_b+|v|_b$; thus, by Lemma 14.5.7, $i^* = |v|_a+|v|_b = |v|$. Similarly, $j^* = |u|_a + |u|_b = |u|$. □

Corollary 19.2.2 *Each positive rational number appears exactly once on the Raney tree.*

Corollary 19.2.3 *If a central word m of length $p+q-2$ has the relatively prime periods p and q, and if it is not the power of a single letter, then the lengths of the factors of the standard factorization of the lower Christoffel word amb are p and q.*

Proof. Note that $p, q \neq 1$; otherwise, m is the power of a single letter. Since each positive rational number appears exactly once on the Raney tree, each pair (i, j) of relatively prime positive integers is of the form $(|u|, |v|)$ for exactly one Christoffel pair (u, v). Thus there exist two Christoffel pairs (u, v) and (u', v'), such that $(|u|, |v|) = (p, q)$ and $(|u'|, |v'|) = (q, p)$; these two pairs are distinct, since $(p, q) \neq (q, p)$ (otherwise, $p = q = 1$). The two corresponding central words, call them m_1 and m_2, which are distinct, have the periods p and q by Corollary 14.3.8. Since $p, q \neq 1$, these Christoffel pairs are not on the extreme branches of the tree of Christoffel pairs; therefore, the two central words m_1 and m_2 are not a power of a single letter. Thus, by the uniqueness statement in Corollary 14.3.7, m must be equal to m_1 or m_2. Therefore amb is equal to $am_1b = uv$ or $am_2b = u'v'$, and the corollary follows. □

Given a proper lower Christoffel word w of slope $\frac{j}{i}$, we denote by w^* the proper lower Christoffel word of slope $\frac{j^*}{i^*}$ and call it the *dual Christoffel word*.

Proposition 19.2.4 *([BdLR], Proposition 3.1) Let w be a proper lower Christoffel word on the alphabet $\{a < b\}$, and let u be the unique word on this alphabet such that $w = a\mathrm{Pal}(u)b$. Then $w^* = a\mathrm{Pal}(\tilde{u})b$. Equivalently, if $u \in \{a, b\}^*$ encodes the node w on the tree of Christoffel words, then w^* is coded by $\tilde{u}$.*

Proof. By Theorem 2.6.1 and Lemma 14.1.5, $w = a\mathrm{Pal}(u)b$ is the node coded by u on the tree of Christoffel words.

Let $\nu(u) = \begin{pmatrix} p & q \\ r & s \end{pmatrix}$. Then $\nu(\tilde{u}) = \begin{pmatrix} s & q \\ r & p \end{pmatrix}$, since the mapping of $SL_2(\mathbb{Z})$ onto itself, obtained by transposing and conjugating by the matrix $\begin{pmatrix} 0 & 1 \\ 1 & 0 \end{pmatrix}$, maps the first matrix onto the second one and is an anti-automorphism fixing the matrices $\nu(a)$ and $\nu(b)$. By Proposition 14.5.2, w is the lower Christoffel word of slope $\frac{r+s}{p+q}$, and $a\mathrm{Pal}(\tilde{u})b$ is the lower Christoffel word of slope $\frac{p+r}{q+s}$. Now, by Lemma 14.5.7, w^* has slope $\frac{p+r}{q+s}$. Hence $w^* = a\mathrm{Pal}(\tilde{u})b$. □

Corollary 19.2.5 *Let p and q be relatively prime positive integers and w the lower Christoffel word w of slope $\frac{q}{p}$.*

(i) *If p or $q = 1$, then $w^* = w$.*
(ii) *If $p, q \neq 1$, then w^* is completely defined by the conditions $w^* = amb$, m is a word having both letters a and b, m has the periods p and q, and m begins with a if and only if $q^* < p^*$.*

Proof.

(i) is left to the reader.
(ii) Let $w^* = amb$; its slope is $\frac{q^*}{p^*}$. Note that since $p, q \neq 1$, we have also $p^*, q^* \neq 1$. Hence w^* is not of the form $a^n b$ or ab^n, and m contains both letters a and b. Thus m is, by Corollary 14.3.7, completely defined by its periods up to the exchange of the letters.

Since amb is the lower Christoffel word of slope $\frac{q^*}{p^*}$ and m is not empty (otherwise, p and q are both equal to 1), m begins with a if and only if the slope of w^* is < 1 (Exercise 2.2.3), that is, if $q^* < p^*$. We conclude, since we know that m has the periods p and q by Corollary 14.3.8 and Corollary 14.5.6. □

Note that, generalizing Corollary 19.2.5 (i), Carpi and de Luca [CdL] study Christoffel words which are equal to their dual, which they call *harmonic*.

Corollary 19.2.6 *If $w = amb$ is a Christoffel word of slope $\frac{q}{p}$, then the central word m has the periods p^* and q^*.*

Proof. This follows from Corollary 14.3.8 and Corollary 14.5.6. □

19.2.1 Exercises

Exercise 19.2.1 *Show that the denominator of a node in the Raney tree and the numerator of the one to its right are equal and that the the denominator of the last node in a row is equal to the numerator of the first one in the next row (and equal to 1). Deduce that each positive rational number is of the form $b(n-1)/b(n)$ for some unique $n \geq 1$, where $b(n)$ is the Stern diatomic sequence (Stern's theorem). Use Exercise 2.5.4, with $b(-1) = 0, b(0) = 1$. Deduce from this exercise that $b(2n+1) = b(n)$ and $b(2n) = b(n-1) + b(n)$ for any $n \geq 0$ and, moreover, that $b(n)$ is equal to the number of hyperbinary representations of n (these results are due to Calkin and Wilf [CW]).*

Exercise 19.2.2 *Show that a Christoffel word and its dual have the same length. Deduce that* $\mathrm{Pal}(v)$ *and* $\mathrm{Pal}(\tilde{v})$ *have the same length. Note that this latter equality is true for any alphabet (see [J] Theorem 2.3).*

19.3 Convex hulls

Say that a convex polygon in the plane is *integral* if its vertices are integral points. Recall that an *extreme point* of a convex set C is a point P of C such that $C \setminus P$ is convex. Thus, the extreme points of a convex polygon are its vertices. Since a convex polygon is the convex hull of its extreme points (Krein–Milman's theorem), a convex polygon is integral if and only if it is the convex hull of the integral points that it contains.

Let p and q be non-zero relatively prime natural numbers, and consider the segment from $(0,0)$ to (p,q). Let $\xi = q/p$; call *rectangle defining* ξ the rectangle whose four vertices are $(0,0)$, $(p,0)$, (p,q), and $(0,q)$. This segment divides the rectangle into two subsets. The *upper set* (resp. *lower set*) *of* ξ is the set of integral points in the rectangle which lie strictly above (resp. under) the segment, together with the point (p,q), which is therefore in both sets; see Figure 19.3, where the upper set is represented by red points and the lower set by blue points, except $(p,q) = (13,5)$, which is represented in black.

The convex hull of the upper set (resp. lower set) of ξ is an integral convex polygon; we call it the *upper* (resp. *lower*) *polygon of* ξ. In the upper and lower polygons, we consider the part of the frontier which is in the *interior* of the rectangle, called for convenience the *upper frontier* and *lower frontier*. In Figure 19.3, the upper polygon is a pentagon, and the lower polygon is a quadrilateral; the upper frontier is in red and the lower frontier in blue.

Theorem 19.3.1 *(Felix Klein [K1], pp. 17–20; [K2], pp. 328–329)*[2] *Let* $\xi = \frac{q}{p}$ *be a positive rational number. Identify a rational number* $\frac{j}{i}$ *(in reduced form) with the point* (i,j)*. The convergents of* ξ*, distinct from* 0 *and* ξ*, are the extreme points of the upper and lower polygons located on the upper and lower frontiers of* ξ*, whereas the semi-convergents* $\neq \xi$ *are all integral points on them.*

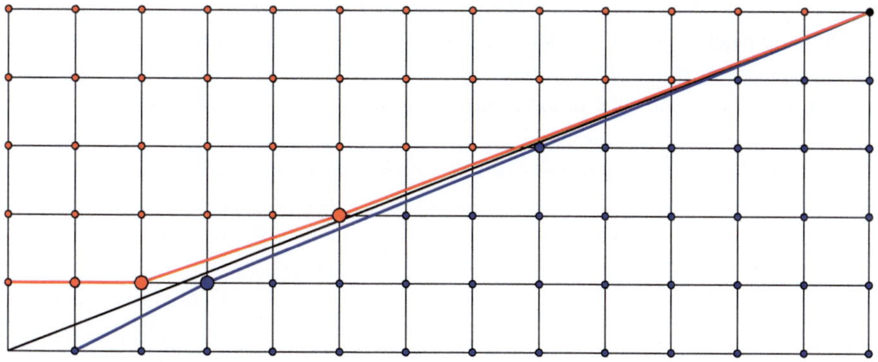

Figure 19.3 Convex hull of $\xi = \frac{5}{13}$

[2] This result is sometimes attributed to Smith [S] (see, for example, Davenport [Da], p. 114 or [Fo], p. 321), but a look at p. 14 of Smith's article shows that he states rather the optimality result on continued fractions (attributed to Lagrange) given in [S], Section 5.5, which also has a geometrical interpretation.

See Figure 19.3, where the convergents $\neq 0, \xi$ correspond to very fat points, and the other semi-convergents, $\neq \xi$, to fat points. Compare to the example after Corollary 19.1.4 and to the path from the root to the node 5/13 in the Stern–Brocot tree, Figure 19.1, in light of this corollary.

Lemma 19.3.2 *Let P be an integral convex polygon, D a line passing through two distinct integral points of the frontier of P, H an open half-plane determined by D, $\bar{H}$ the corresponding closed half-plane, and $P' = P \cap \bar{H}$. Then:*

(i) P' is an integral convex polygon;
(ii) the intersection with H of the frontier of P is equal to the intersection with H of the frontier of P'.

The verification of this lemma is left to the reader.

Proof of Theorem 19.3.1. We prove the result by induction on $p + q$. The lowest value of this number is 2. If $p + q = 2$, then $p = q = 1$ and the result is clear: there is no extreme point in the interior of the rectangle and no semi-convergent except ξ.

Suppose that $p + q > 2$. By symmetry of the result, we may assume that $p > q$, that is, the slope ξ of the diagonal is < 1. Then the word $w(\xi)$ coding ξ on the Stern–Brocot tree begins with a: $w(\xi) = aw(\eta)$, and $\eta = \frac{q}{p-q}$, by Theorem 19.1.1 (i).

The rectangle defining η has the four vertices $(0,0), (p-q,0), (p-q,q)$, and $(0,q)$. Now we change the basis, taking $(1,0)$ and $(1,1)$ as the new basis for the latter rectangle; since

$$r(1,0) + s(1,1) = (r+s, s)$$

these four points, expressed in the new basis, are $(0,0)$, $(p-q,0)$, (p,q), and (q,q). These are the four vertices of the parallelogram, represented in Figure 19.4; this parallelogram is, in the new basis, the rectangle defining η.

The construction of the upper and lower frontiers above perfectly makes sense for a parallelogram instead of a rectangle; we call them the upper and lower frontiers of η. By induction, the integral points of the upper and lower frontiers of η are

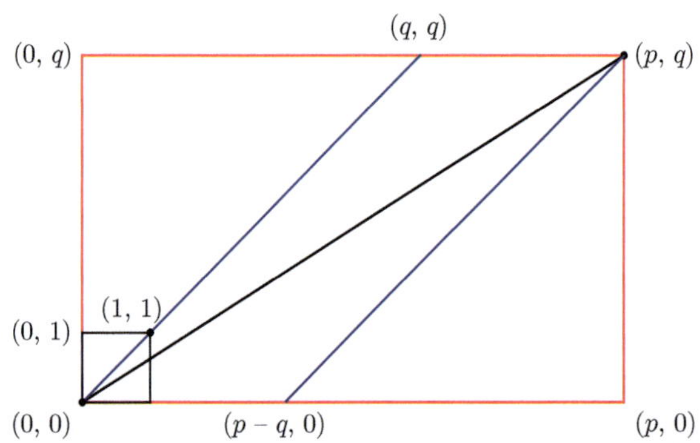

Figure 19.4 Recursive construction of the frontier

the semi-convergents $\neq \eta$ of η. We claim that all integral points on the upper and lower frontiers of ξ are in the parallelogram. The claim is proved below. Using the lemma, applied to the upper polygon of ξ and the line from $(0,0)$ to (q,q), and to the lower polygon of ξ and the line from $(p-q,0)$ to (p,q), it follows that the integral points of the upper and lower frontiers of η are exactly the integral points of the upper and lower frontiers of ξ, except that $(1,1)$ is such a point for ξ and not for η.

The statement on the semi-convergents follows by induction on $p+q$, using Corollary 19.1.4: indeed, $1/1$ is a semi-convergent of ξ, and the mapping $\frac{s}{r} \mapsto \frac{s}{r-s}$ is a bijection from the set of semi-convergents of ξ different from $1/1$ onto the set of semi-convergents of η (Exercise 19.1.2).

Concerning convergents, we only have to verify that $(1,1)$ is an extreme point of the upper polygon of ξ if and only if $\frac{1}{1}$ is a convergent of ξ. By Exercise 19.1.3, 1 is a convergent of ξ if and only if $\xi > \frac{1}{2}$. This is equivalent to $(2,1)$ being below the diagonal of the rectangle (see Figure 19.4). But this holds if and only if $(1,1)$ is an extreme point. Note that the case where the slope is $< \frac{1}{2}$ is illustrated in Figure 19.3, showing that $(1,1)$ is not an extreme point in this case.

We now prove the claim. Let M be any integral point in the interior of the rectangle and outside the parallelogram. Suppose first that M is above the diagonal of the rectangle. The convex hull of the upper set of ξ clearly contains the quadrilateral with vertices $(0,1)$, $(1,1)$, (q,q), and $(0,q)$, and M is in the interior of it (see Figure 19.4). Hence a small disc around M is completely in this quadrilateral, hence in the upper polygon, and therefore M is not on the upper frontier of ξ (since, for a point on the frontier of a convex polygon, each disc around it intersects the inside and the outside of the polygon). If M is under the diagonal of the rectangle, then M is in the interior of the triangle with vertices $(p-q,0)$, $(p,0)$, and (p,q), and one argues similarly to show that M is not on the lower frontier of ξ. Thus the claim follows. □

We now show that there is a linear relation among the integral points of the upper and lower frontiers, that is, according to Theorem 19.3.1, among the semi-convergents. Let us illustrate this with an example; see Figure 19.3. These points are totally ordered with respect to the natural partial order of $\mathbb{N}^2$. We list them according to this order: $(1,1), (2,1), (3,1), (5,2), (8,3), (13,5)$. Each vector is the sum of the previous one and of some vector: $(2,1) = (1,1) + (1,0)$, $(3,1) = (2,1) + (1,0)$, $(5,2) = (3,1) + (2,1)$, $(8,3) = (5,2) + (3,1)$, $(13,5) = (8,3) + (5,2)$. The adding vector is the last point before a crossing of the diagonal, with $(1,0)$ as initial value.

Corollary 19.3.3 *Suppose that $\xi = \frac{q}{p} \neq 1$. Let $M_1 = (1,1), M_2, \ldots, M_n = (p,q)$ be the integral points on the upper and lower frontiers, in increasing order for the natural order of $\mathbb{N}^2$. Let $M_0 = (1,0)$ if $\xi < 1$ and $M_0 = (0,1)$ if $\xi > 1$. For $i = 2, \ldots, n$, let j, $1 \leq j \leq i-2$, be maximal, such that M_j is not on the same side of the diagonal as M_{j+1} if it exists, and $j = 0$ otherwise. Then $M_i = M_{i-1} + M_j$.*

Proof. The points M_j and M_{j+1} are on different sides of the diagonal if and only if M_j is an extreme point; this follows from the previous proof, by induction, noting the initial case $M_1 = (1, 1)$. Thus these points M_j are the convergents, that is, the nodes on the path from the root to ξ immediately preceding a node where there is a left or right turn by Corollary 19.1.4. Thus the corollary follows from the direct construction of the Stern–Brocot tree given at the beginning of Section 19.1. □

A byproduct of the previous result and proof is that the branches of the path from the root to ξ in the Stern–Brocot tree are seen on the plane: the branches correspond to the segments of the polygonal line $M_1, \ldots, M_n$. For example, for $\xi = 5/13$, there are four segments: the first includes the points $(1, 1)$, $(2, 1)$, and $(3, 1)$; the second, the two points $(3, 1)$ and $(5, 2)$; then $(5, 2)$ and $(8, 3)$; and finally $(8, 3)$ and $(13, 5)$. They correspond to the four branches seen in Figure 19.1.

For another approach to the geometry of continued fractions, see [Bn]. Not only do convergents have a geometric interpretation in Euclidean geometry, as is shown above, but also in hyperbolic geometry; the convergents appear naturally in the hyperbolic plane, using the *Farey tessellation* (see [H, Se1, GK]).

20
Free group on two generators

This chapter is motivated by the fact that Sturmian morphisms are exactly the positive automorphisms of the free group on two generators a and b and that conjugates of Christoffel words are exactly the positive primitive elements of this group. In order to establish these results, one needs several results on the automorphisms of this group, which go back to Nielsen and which are completely proved here: generators of this group and equality of the subgroup of inner automorphisms and of the kernel of the abelianization. We use a result of Weinbaum which has, as a consequence, a set of representatives of the automorphism group modulo this subgroup. A byproduct is the proof of Nielsen's criterion, stating that two elements u and v form a basis of this group if and only if the commutator $uvu^{-1}v^{-1}$ is conjugate to $aba^{-1}b^{-1}$ or its inverse (we follow Shpilrain's proof by Fox derivations for one direction of this characterization). We also present special bases called Christoffel bases, which are representatives of all bases up to conjugation; this goes back to Osborne and Zieschang. A very efficient algorithm (from Kassel and the author, based on Séébold's theory of conjugate Sturmian morphisms) that checks if two given elements form a basis is given. Generators of the monoid of Sturmian morphisms are given (a result due to Berthé, de Luca, and the author, a variant of a result of Mignosi and Séébold). As a consequence, we obtain the result of Wen and Wen showing that positive automorphisms have the same set of generators.

20.1 Bases and automorphisms

Recall that $\mathrm{F}(a, b)$ denotes the free group with basis a, b. It is well known that each element in $\mathrm{F}(a, b)$ is represented by words in the free monoid $\{a, b, a^{-1}, b^{-1}\}^*$; this monoid has an involutory anti-automorphism called *inversion*, written $w \mapsto w^{-1}$, which exchanges x and x^{-1} for $x = a, b$. Each element in $\mathrm{F}(a, b)$ has in $\{a, b, a^{-1}, b^{-1}\}^*$ a unique representative which is *reduced*, that is, without a factor of the form xx^{-1} with $x = a, b, a^{-1}, b^{-1}$. The *length* of an element in $\mathrm{F}(a, b)$ is the length of its unique reduced representative.

It is a common abuse of notation to use the same notation for a word in the free monoid $\{a, b, a^{-1}, b^{-1}\}^*$ and for the element in the free group that it represents. This may sometimes be misleading, but we shall use sentences as "in $\mathrm{F}(a, b)$" or "in $\{a, b, a^{-1}, b^{-1}\}^*$" to avoid confusion. For example, one says, "Let g be a reduced element in the free group", which means that g is the reduced representative in $\{a, b, a^{-1}, b^{-1}\}^*$ and also the represented element in $\mathrm{F}(a, b)$. Also, we say that the "product gh of two reduced elements is reduced" if g and h are reduced and if the product gh in the monoid is reduced, that is, if the last letter of g is not the inverse of the first letter of h; equivalently, we say that "there is no simplification in the product".

From Christoffel Words to Markoff Numbers. Second Edition. Christophe Reutenauer, Oxford University Press.
 DOI: 10.1093/9780197907535.003.0020

A word in $\{a, b, a^{-1}, b^{-1}\}^*$ is *cyclically reduced* if it is reduced and, if moreover, its first letter is not the inverse of the last; equivalently, the word and its square in the monoid are reduced.

Lemma 20.1.1 *Each reduced element w of the free group $\mathrm{F}(a, b)$ has in $\{a, b, a^{-1}, b^{-1}\}^*$ a unique reduced expression $w = xyx^{-1}$, where x is reduced, y is cyclically reduced, and where the products xy and yx^{-1} are without simplification. Then, for any $k \neq 0$, xy^kx^{-1} is reduced.*

Proof. If $w = 1$, we take $x = y = 1$. Otherwise, if w is cyclically reduced, take $x = 1, w = y$. In the other cases, choose in $\{a, b, a^{-1}, b^{-1}\}^*$ a prefix x of maximum length of w such that x^{-1} is a suffix; then $w = xyx^{-1}$, as desired.

By construction, since w is reduced, the products xy and yx^{-1} are without simplification. Since y is cyclically reduced, this implies the last assertion. □

A *basis* is a pair (u, v) such that the mapping sending a onto u and b onto v extends to an automorphism of $\mathrm{F}(a, b)$. It is useful to also have the notion of an *unordered basis*: this is a two-element set $\{u, v\}$ such that the couple (u, v) (or equivalently the couple (v, u)) is a basis. We say simply basis for both notions. The *length* of $\{u, v\}$ is the sum of the lengths of its elements.

Theorem 20.1.2 *(Nielsen [N], pp. 387–389) Let u and v be reduced elements which generate $\mathrm{F}(a, b)$.*

(i) The pair $\{u, v\}$ is conjugate in $\mathrm{F}(a, b)$ to a pair such that both elements of the pair are cyclically reduced and which is not longer than $\{u, v\}$.

(ii) If both elements in the pair $\{u, v\}$ are cyclically reduced, then either $\{u, v\}$ is one of the four pairs $\{a, b\}, \{a^{-1}, b\}, \{a, b^{-1}\}$, and $\{a^{-1}, b^{-1}\}$, or $\{u, v\}$ is conjugate in $\mathrm{F}(a, b)$ to a pair which has the same length and such that one of its elements, or its inverse, is, as an element of $\{a, b, a^{-1}, b^{-1}\}^*$, a prefix or a suffix of the other one.

If $g_1 \cdots g_n$ is a product of reduced elements in the free group $\mathrm{F}(a, b)$, we say that g_i $(1 \leq i \leq n)$ is *absorbed by its neighbours* if one has the reduced products $g_{i-1} = xu^{-1}$, $g_i = uv$, $g_{i+1} = v^{-1}y$ (with the convention $g_0 = 1 = g_{n+1}$). For example, in the product of three elements $(ab)(b^{-1}ab^{-1})(ba^{-1}b^{-1})$, the middle factor is absorbed by its two neighbours.

Lemma 20.1.3 *Let $g_1, \ldots, g_n$, $n \geq 2$, be reduced elements in the free group F_2 such that $g_1 \cdots g_n = 1$. Then for some i, $1 \leq i \leq n - 1$, g_i is absorbed by its neighbours.*

Note that in the conclusion, $i < n$. There is also a symmetric statement with $i > 1$.

Proof. If all $g_i = 1$, this is clear. Otherwise, there must be some j, $1 \leq j \leq n - 1$, some elements h_j, h_{j+1}, and some $c \in \{a, b, a^{-1}, b^{-1}\}$ such that one has the reduced products $g_j = h_jc$, $g_{j+1} = c^{-1}h_{j+1}$. Define $h_k = g_k$ for $k \neq j, j + 1$. Then the elements h_i

are all reduced, and their product is 1. By induction on the total length, there exists i, $1 \le i \le n-1$, such that h_i is absorbed by its neighbours h_{i-1} and h_{i+1}. We have, therefore, $h_{i-1} = xu^{-1}$, $h_i = uv$, $h_{i+1} = v^{-1}y$, and all these products are reduced. We show below that g_i is absorbed by its neighbours. There are six cases.

If $i < j-1$, then $i-1, i, i+1 < j$ so that $g_{i-1} = h_{i-1} = xu^{-1}$, $g_i = h_i = uv$, $g_{i+1} = h_{i+1} = v^{-1}y$.

If $i = j-1$, then $i-1, i < j$ so that $g_{i-1} = h_{i-1} = xu^{-1}$, $g_i = h_i = uv$, $g_{i+1} = g_j = h_jc = h_{i+1}c = v^{-1}yc$, and all these products are reduced.

If $i = j$, then $g_{i-1} = g_{j-1} = h_{j-1} = h_{i-1} = xu^{-1}$, $g_i = g_j = h_jc = h_ic = uvc$, $g_{i+1} = g_{j+1} = c^{-1}h_{j+1} = c^{-1}h_{i+1} = c^{-1}v^{-1}y$, and all these products are reduced.

The three other cases $i = j+1$, $i = j+2$, and $i > j+2$ are similar. □

Proof of Theorem 20.1.2, following Nielsen [N]. Note that $F(a, b)$ cannot be generated by one element; this follows by abelianization from the same property for $\mathbb{Z}^2$.

By hypothesis, a and b may both be written as a product of the form $(*)$ $u^{n_1}v^{p_1} \cdots u^{n_k}v^{p_k}$ for some $k \ge 1$ and for some integers n_i and p_i, which are non-zero except perhaps n_1 and p_k.

(i) If u is not cyclically reduced, then we have the reduced product $u = cwc^{-1}$, with $c \in \{a, b, a^{-1}, b^{-1}\}$. Note that at least one of the two products $(*)$ must involve u and v simultaneously: indeed, otherwise, either a or b are both powers of v, which is not possible, or one of the two products involves only u and then cannot be equal to a or b. Indeed, by Lemma 20.1.1, we have $u = xyx^{-1}$ with $x, y \ne 1$, and therefore u^k is either 1 (if $k = 0$) or of length at least 3 by the lemma.

Write also $v = x'y'x'^{-1}$ according to Lemma 20.1.1. Then the given product is a product of factors which alternate between powers $u^k = xy^kx^{-1}$ and $v^{k'} = x'y'^{k'}x'^{-1}$, $k, k', \in \mathbb{Z} \setminus 0$, and these elements are reduced, both appear, and u^k is of length at least 3, and $v^{k'}$ of length at least 1. The product is equal to a or b. Hence, there must be some simplification. This implies that v or v^{-1} must begin with c or end with c^{-1}. Hence v begins with c or ends with c^{-1}: $v = cv'$ or $v = v'c^{-1}$. Thus the pair (u, v) is conjugate in $F(a, b)$ to the shorter pair $(w, v'c)$ or $(w, c^{-1}v')$, and we conclude by induction.

(ii) Suppose that the two elements of the pair $\{u, v\}$ are cyclically reduced. Then a and b are equal to products of the form $(*)$; these two products may therefore be written in $\{a, b, a^{-1}, b^{-1}\}^*$ under the form hak and $h'b'k$, respectively, with $h = k = h' = k' = 1$ in $F(a, b)$.

Assume that $h \ne 1$ in $\{a, b, a^{-1}, b^{-1}\}^*$. Then the first factor in $(*)$, which is either u^{n_1} (if $n_1 \ne 0$) or v^{p_1} (if $n_1 = 0$), must be a proper prefix of h in $\{a, b, a^{-1}, b^{-1}\}^*$; otherwise, h is a prefix of this factor and one could not have $h = 1$ in $F(a, b)$, since u and v are cyclically reduced, and since a prefix in the monoid of a reduced word is reduced. Assume that $n_1 \ne 0$ (the case $n_1 = 0$ is quite similar). Then, with $\epsilon = \pm 1$, either $h = u^\epsilon \cdots u^\epsilon v^{\pm 1} \cdots p$ where p is a prefix of $u^{\pm 1}$ or $v^{\pm 1}$, or $h = u^\epsilon \cdots u^\epsilon p$ where p is a prefix of $v^{\pm 1}$. Using Lemma 20.1.3, we see that some factor u or v, or the inverse of one of them, must be absorbed by its neighbours, which is also one of these four elements (in the lemma, the condition $i \leq n - 1$ ensures that it is not the last factor p that is absorbed, but a preceding factor; moreover, if the absorbed element is absorbed on the right by p, then *a fortiori* by the word of which p is a prefix). Since u and v are cyclically reduced, there is no simplification in the square product of any of these four elements, and using Lemma 20.1.3, we see that either u or u^{-1} is a prefix or a suffix of v or v^{-1}, or conversely, which proves (ii) in this case; or there are, in $\{a, b, a^{-1}, b^{-1}\}^*$, non-trivial factorizations $v^{\pm 1} = xm^{-1}$, $u^{\pm 1} = mp$, $v^{\pm 1} = p^{-1}y$, or the three equalities obtained by exchanging u and v in the three previous ones.

We omit these last three cases, since they are quite similar to the previous one, and in this latter case, we may assume that the exponent of u is 1 (taking inverses if necessary): $v^{\epsilon} = xm^{-1}$, $u = mp$, $v^{\eta} = p^{-1}y$, with $\epsilon, \eta = \pm 1$. If $\epsilon = 1, \eta = -1$, then $v = xm^{-1} = y^{-1}p$, so that the last letter in p is equal to the last one in m^{-1}, that is, the inverse of the first one in m, and u is not cyclically reduced. Similarly, if $\epsilon = -1, \eta = 1$. So we must have $\epsilon = \eta$. Suppose that they are both equal to 1. Then $v = xm^{-1}$, $u = mp$, $v = p^{-1}y$. This holds in $\{a, b, a^{-1}, b^{-1}\}^*$. Since $xm^{-1} = p^{-1}y$, p^{-1} is a prefix of x or conversely. The pair $\{u, v\} = \{mp, xm^{-1}\}$ is conjugate in $F(a, b)$ to $\{u', v'\} = \{pm, m^{-1}x\}$, which is cyclically reduced and of the same length since u and v are cyclically reduced. If $u = mp$ is longer than $v = xm^{-1}$, then p is longer than x and hence x is a prefix of p^{-1}; therefore, x^{-1} is a suffix of p and $v'^{-1} = x^{-1}m$ is a suffix of $u' = pm$ (and (ii) holds). If u is shorter than v, then p is shorter than x and hence p^{-1} is a prefix of x; therefore, $u'^{-1} = m^{-1}p^{-1}$ is a prefix of $v' = m^{-1}x$ (and (ii) holds). If u and v have the same length, then $x = p^{-1}$ and therefore $u = mp = mx^{-1} = v^{-1}$, which is impossible. The case where $\epsilon = \eta = -1$ is similar.

The case $h' \neq 1$ is similar to the case $h \neq 1$. Likewise for the cases $k \neq 1$ and $k' \neq 1$ by taking the symmetric statement of Lemma 20.1.3.

If all h, k, h' and k' are all equal to 1 in $\{a, b, a^{-1}, b^{-1}\}^*$, that is, if the two products $(*)$ are of length 1 in $\{a, b, a^{-1}, b^{-1}\}^*$, then there is no simplification at all, and these products reduce to a single u, v, or one of their inverses; then the pair $\{u, v\}$ is equal to one of the four pairs $\{a, b\}$, $\{a^{-1}, b\}$, $\{a, b^{-1}\}$, and $\{a^{-1}, b^{-1}\}$. □

The first part of the proof also shows the following useful fact.

Corollary 20.1.4 *(Nielsen) If the reduced elements u and v generate* F(a, b) *and if u is not cyclically reduced, then for some $c \in \{a, b, a^{-1}, b^{-1}\}$ one has the reduced products $u = cwc^{-1}$ and either $v = cv'$ or $v = v'c^{-1}$. In particular, the pair (u, v) is conjugate to the shorter pair $(w, v'c)$ or $(w, c^{-1}v')$, whose second element is cyclically reduced if v is cyclically reduced.*

Each endomorphism of the free monoid $\{a, b\}^*$ extends uniquely to an endomorphism of the free group F(a, b) and is denoted by (u, v), where this pair is the image of the pair (a, b). Among them are the endomorphisms $G = (a, ab)$, $D = (ba, b)$, $\tilde{G} = (a, ba)$ and $\tilde{D} = (ab, b)$, $E = (b, a)$, whose extensions are automorphisms of F(a, b), and which we denote by the same letter. Consider also the automorphism $S = (a^{-1}, b)$ of F(a, b). Recall that an *inner automorphism* of F(a, b) is an automorphism of the form $g \mapsto fgf^{-1}$ for some fixed $f \in$ F(a, b).

Corollary 20.1.5 *The group* Aut(F(a, b)) *is generated by the three automorphisms G, E, and S.*

The proof of the following result is left to the reader.

Lemma 20.1.6 *The subgroup $\mathcal{B}$ of* Aut$(F(a, b))$ *generated by E and S is the set of the eight automorphisms*

$$(x^{\epsilon}, y^{\eta}), \{x, y\} = \{a, b\}, \epsilon, \eta = \pm 1 \tag{20.1}$$

Proof of Corollary 20.1.5. We first show that Aut(F(a, b)) is generated by the inner automorphisms, together with the *elementary automorphisms*, which are the eight automorphisms in (20.1) and the eight automorphisms of the form

$$(a, ab), (a, ba), (ab, b), (ba, b), (a, a^{-1}b), (a, ba^{-1}), (b^{-1}a, b), (ab^{-1}, b) \tag{20.2}$$

It is enough to show that, given any ordered basis (u, v) of F(a, b), there exists a product of inner and elementary automorphisms that sends the basis (a, b) onto (u, v). We do this by induction on the length of (u, v). If this length is 2, then this basis is the image of (a, b) under an elementary automorphism of the form (20.1). Suppose now that the length is ≥ 3. If u or v is not cyclically reduced, then, by Corollary 20.1.4, (u, v) is the image under some inner automorphism of a basis which is shorter, and we conclude by induction in this case. If both are cyclically reduced, then by Theorem 20.1.2 (ii), (u, v) is the image under some inner automorphism of an equally long basis such that one of its elements or its inverse is a prefix or a suffix of the other element; in other words, (u, v) is equal to one of the eight couples

$$(x, xy), (x, yx), (xy, y), (yx, y), (x, x^{-1}y), (x, yx^{-1}), (y^{-1}x, y), (xy^{-1}, y)$$

where x and y are reduced words and the products are without simplification; then in each case (x, y) is a basis which is shorter than (u, v) so that, by induction, $(x, y) = (f(a), f(b))$ for some automorphism f which is a product of elementary and inner automorphisms; then (u, v) is the image of (a, b) under the automorphism obtained by multiplying f on the right by one of the eight elementary automorphisms in (20.2); this ends the induction step.

We have $D = EGE$, $T := (a, b^{-1}) = ESE$, $I := (a^{-1}, b^{-1}) = ST$, $\tilde{G} = IGI$, $\tilde{D} = E\tilde{G}E$. This proves that the four first elementary automorphisms in (20.2) (which are G, $\tilde{G}$, $\tilde{D}$, and D) are in the subgroup generated by G, E, and S; the four last ones are their inverses. Finally, one has $G^{-1}\tilde{G} = (a, a^{-1}ba)$ and $D^{-1}\tilde{D} = (b^{-1}ab, b)$, which generate the subgroup of inner automorphisms.

Thus, by Lemma 20.1.6, the corollary follows. □

The proof also shows that the following result holds, since we used only the fact that u and v generate F(a, b).

Corollary 20.1.7 *If u and v generate* F(a, b), *then they form a basis of* F(a, b). *Schmutz Schaller [Sc3] gives other results on bases of the free group on two generators, related to the theory of Christoffel words.*

20.2 Inner automorphisms

Denote by $\mathrm{Inn}(\mathrm{F}(a,b))$ the subgroup of inner automorphisms of $\mathrm{F}(a,b)$. As is well known, it is a normal subgroup of $\mathrm{Aut}(\mathrm{F}(a,b))$.

Theorem 20.2.1 *(Weinbaum [W], Theorem 2*[1]*) Let $\mathcal{B}$ denote the subgroup of* $\mathrm{Aut}(\mathrm{F}(a,b))$ *generated by E and S, and let $\mathcal{E}$ (resp. $\mathcal{E}^+$) be the submonoid (resp. the subsemigroup) generated by G and $\tilde{D}$. Then* $\mathrm{Aut}(\mathrm{F}(a,b)) = \mathrm{Inn}(\mathrm{F}(a,b))(\mathcal{B} \cup \{1,S\}\mathcal{E}^+\mathcal{B})$.

Note that the subgroup $\mathcal{B}$ is the set of eight automorphisms in (20.1).

Proof. Let $\mathcal{G} = \mathrm{Inn}(\mathrm{F}(a,b))(\mathcal{B}\cup\{1,S\}\mathcal{E}^+\mathcal{B})$. It is a subset of $\mathrm{Aut}(\mathrm{F}(a,b))$. The identity belongs to $\mathcal{G}$. Thus, by Corollary 20.1.5, it is enough to show that $\mathcal{G}$ is closed under the four operations $\varphi \mapsto \varphi G$, $\varphi \mapsto \varphi G^{-1}$, $\varphi \mapsto \varphi E$, and $\varphi \mapsto \varphi S$ (since $E^{-1} = E, S^{-1} = S$). This is clear for the two last operations because $E, S \in \mathcal{B}$.

One has $EGE = D$ and $SGS = G^{-1}$. Therefore, $EG = DE$, $EG^{-1} = D^{-1}E$, $ED = GE$, $ED^{-1} = G^{-1}E$, $SG = G^{-1}S$, $SG^{-1} = GS$. Moreover, $SDS = \tilde{D}^{-1}$. Since $D^{-1}\tilde{D} = (b^{-1}ab, b) \in \mathrm{Inn}(\mathrm{F}(a,b))$, $\tilde{D} \equiv D$, where $\equiv$ denotes equivalence modulo the normal subgroup $\mathrm{Inn}(\mathrm{F}(a,b))$, and we have $SD = \tilde{D}^{-1}S$, $SD^{-1} = \tilde{D}S$, and therefore $SD \equiv D^{-1}S$ and $SD^{-1} \equiv DS$.

For subsets A, B of $\mathrm{Aut}(\mathrm{F}(a,b))$, we write $A \sqsubseteq B$ if $\mathrm{Inn}(\mathrm{F}(a,b))A \subset \mathrm{Inn}(\mathrm{F}(a,b))B$. Note that $A \subset B$ implies $A \sqsubseteq B$. The previous equalities then imply that $X\{G,D,G^{-1},D^{-1}\} \sqsubseteq \{G,D,G^{-1},D^{-1}\}X$ for $X = E$ and $X = S$. Since S, E are involutions, they generate $\mathcal{B}$ as a monoid, and therefore

$$\mathcal{B}\{G,D,G^{-1},D^{-1}\} \sqsubseteq \{G,D,G^{-1},D^{-1}\}\mathcal{B}$$

as follows inductively, using $A\mathrm{Inn}(\mathrm{F}(a,b)) = \mathrm{Inn}(\mathrm{F}(a,b))A$, since $\mathrm{Inn}(\mathrm{F}(a,b))$ is a normal subgroup.

Now, we have the equalities $GD \equiv G\tilde{D}$ (since $\tilde{D} \equiv D$), $GD^{-1} = (b^{-1}, ab) = \tilde{D}ES$, $\tilde{D}D \equiv \tilde{D}^2$, $\tilde{D}D^{-1} \equiv \tilde{D}\tilde{D}^{-1} = 1$, $\tilde{D}G^{-1} = (ab, b^{-1}a^{-1}b) \equiv (ba, a^{-1}) \equiv (ab, a^{-1}) = GSE$.

The previous equalities imply that $\{G,\tilde{D}\}\{G,D,G^{-1},D^{-1}\} \sqsubseteq \mathcal{E}\mathcal{B}$; hence, since $\mathcal{E}^+ = \mathcal{E}\{G,\tilde{D}\}$, we have

$$\mathcal{E}^+\{G,D,G^{-1},D^{-1}\} \sqsubseteq \mathcal{E}\mathcal{B}$$

Next we have

$$\{G,D,G^{-1},D^{-1}\} \sqsubseteq \{1,S\}\mathcal{E}^+\mathcal{B}$$

since $\tilde{D} \equiv D$, $G^{-1} = SGS$ and $D^{-1} = S\tilde{D}S$. We have therefore

$(\mathcal{B}\cup\{1,S\}\mathcal{E}^+\mathcal{B})\{G,G^{-1}\} \subset (\mathcal{B}\cup\{1,S\}\mathcal{E}^+\mathcal{B})\{G,D,G^{-1},D^{-1}\}$
$= \mathcal{B}\{G,D,G^{-1},D^{-1}\} \cup \{1,S\}\mathcal{E}^+\mathcal{B}\{G,D,G^{-1},D^{-1}\}$
$\sqsubseteq \{G,D,G^{-1},D^{-1}\}\mathcal{B} \cup \{1,S\}\mathcal{E}^+\{G,D,G^{-1},D^{-1}\}\mathcal{B}$
$\sqsubseteq \{1,S\}\mathcal{E}^+\mathcal{B}\mathcal{B} \cup \{1,S\}\mathcal{E}\mathcal{B}\mathcal{B} \subset \{1,S\}\mathcal{E}^+\mathcal{B} \cup \{1,S\}\mathcal{B}$ (since $\mathcal{B}\mathcal{B} = \mathcal{B}$ and $\mathcal{E} = \{1\}\cup\mathcal{E}^+$) $\subset \mathcal{B}\cup\{1,S\}\mathcal{E}^+\mathcal{B}$ (since $\{1,S\}\mathcal{B} = \mathcal{B}$). Thus $\mathcal{G}$ is closed under the product on the right by G and G^{-1}, which ends the proof. □

[1] Weinbaum states that $\mathrm{F}(a,b) = \mathrm{Inn}(\mathrm{F}(a,b))\mathcal{B}\mathcal{E}\mathcal{B}$, but his proof may be adapted so that the present, stronger statement holds.

Corollary 20.2.2 *The set* $\mathcal{B} \cup \{1, S\}\mathcal{E}^+\mathcal{B}$ *is a set of representatives of* Aut(F(a, b)) *modulo* Inn(F(a, b)).

We need the following lemma.

Lemma 20.2.3 *Let P be in* $\mathrm{GL}_2(\mathbb{Z})$. *Then either P is one of the eight matrices*

$$\begin{pmatrix}1 & 0\\ 0 & 1\end{pmatrix}, \begin{pmatrix}-1 & 0\\ 0 & 1\end{pmatrix}, \begin{pmatrix}1 & 0\\ 0 & -1\end{pmatrix}, \begin{pmatrix}-1 & 0\\ 0 & -1\end{pmatrix}$$
$$\begin{pmatrix}0 & 1\\ 1 & 0\end{pmatrix}, \begin{pmatrix}0 & 1\\ -1 & 0\end{pmatrix}, \begin{pmatrix}0 & -1\\ 1 & 0\end{pmatrix}, \begin{pmatrix}0 & -1\\ -1 & 0\end{pmatrix} \tag{20.3}$$

which constitute the subgroup $\mathcal{Y}$ *generated by the two matrices* $\begin{pmatrix}-1 & 0\\ 0 & 1\end{pmatrix}$ *and* $\begin{pmatrix}0 & 1\\ 1 & 0\end{pmatrix}$. *Or (and the two cases are mutually exclusive) P has a unique expression of the form $P = XMY$, where X is in the subgroup* $\mathcal{X}$ *generated by* $\begin{pmatrix}-1 & 0\\ 0 & 1\end{pmatrix}$, *$M$ is in the subsemigroup generated by* $\begin{pmatrix}1 & 1\\ 0 & 1\end{pmatrix}$ *and* $\begin{pmatrix}1 & 0\\ 1 & 1\end{pmatrix}$, *and* $Y \in \mathcal{Y}$.

Recall that the abelianization map is the canonical mapping Ab from Aut(F(a, b)) onto $\mathrm{GL}_2(\mathbb{Z})$. It is surjective; see Chapter 1. Note that the abelianizations of S, G, $\widetilde{D}$, and E are respectively $\begin{pmatrix}-1 & 0\\ 0 & 1\end{pmatrix}$, $\begin{pmatrix}1 & 1\\ 0 & 1\end{pmatrix}$, $\begin{pmatrix}1 & 0\\ 1 & 1\end{pmatrix}$, and $\begin{pmatrix}0 & 1\\ 1 & 0\end{pmatrix}$. Note also that the abelianization of an inner automorphism is the identity matrix.

Proof. It follows by abelianization from Theorem 20.2.1 that $P \in \mathcal{Y}$ or $P = XMY$. The two cases are mutually exclusive, since the first case corresponds to the fact that P has exactly two non-zero entries and the second to the fact that P has at least three non-zero entries.

Concerning uniqueness, we have to show that, with evident notations, $XMY = X'M'Y'$ implies that $X = X', M = M'$, and $Y = Y'$. Actually, by group properties, it is enough to show that $XMY = M'$ implies that $X = Y = 1$ and $M = M'$. Let $M = \begin{pmatrix}p & q\\ r & s\end{pmatrix}$. The subgroup $\mathcal{Y}$ is described in the statement, and $\mathcal{X}$ has two elements, which are the first two in the list of (20.3). Thus XMY is one of the following 16 matrices (of which the first eight are obtained by taking $X = 1$ and Y equal, respectively, to the eight matrices in the previous list):

$$\begin{pmatrix}p & q\\ r & s\end{pmatrix}, \begin{pmatrix}-p & q\\ -r & s\end{pmatrix}, \begin{pmatrix}p & -q\\ r & -s\end{pmatrix}, \begin{pmatrix}-p & -q\\ -r & -s\end{pmatrix}$$
$$\begin{pmatrix}q & p\\ s & r\end{pmatrix}, \begin{pmatrix}-q & p\\ -s & r\end{pmatrix}, \begin{pmatrix}q & -p\\ s & -r\end{pmatrix}, \begin{pmatrix}-q & -p\\ -s & -r\end{pmatrix}$$
$$\begin{pmatrix}-p & -q\\ r & s\end{pmatrix}, \begin{pmatrix}p & -q\\ -r & s\end{pmatrix}, \begin{pmatrix}-p & q\\ r & -s\end{pmatrix}, \begin{pmatrix}p & q\\ -r & -s\end{pmatrix}$$
$$\begin{pmatrix}-q & -p\\ s & r\end{pmatrix}, \begin{pmatrix}q & -p\\ -s & r\end{pmatrix}, \begin{pmatrix}-q & p\\ s & -r\end{pmatrix}, \begin{pmatrix}q & p\\ -s & -r\end{pmatrix} \tag{20.4}$$

If such a matrix has only non-negative entries, the entries with a minus sign must vanish. Moreover, M' has at least three non-zero entries. Hence, since $XMY = M'$, M' must be the first or the fifth in the list. But the latter has determinant -1. Thus M' is the first matrix and therefore $X = Y = 1$ and $M = M'$. □

Proof of Corollary 20.2.2. By Theorem 20.2.1, it is enough to show that the elements of $\mathcal{B} \cup \{1, S\}\mathcal{E}^+\mathcal{B}$ are distinct modulo $\mathrm{Inn}(\mathrm{F}(a, b))$. This follows from Lemma 20.2.3, since $\mathrm{Inn}(\mathrm{F}(a, b))$ is mapped onto 1 by abelianization. □

Corollary 20.2.4 *(Nielsen [N] 9, p. 393) The kernel of the abelianization map from* $\mathrm{Aut}(\mathrm{F}(a, b))$ *onto* $\mathrm{GL}_2(\mathbb{Z})$ *is the subgroup of inner automorphisms of* $\mathrm{F}(a, b)$.

Proof. Clearly, $\mathrm{Inn}(\mathrm{F}(a, b))$ is contained in the kernel. Conversely, let φ be some automorphism in the kernel of the abelianization map. Then, by Corollary 20.2.2, multiplying φ by some inner automorphism, we may assume that either $\varphi \in \mathcal{B}$, or $\varphi = \alpha\beta\gamma$, where $\alpha = 1$ or S, $\beta \in \mathcal{E}^+$, and $\gamma \in \mathcal{B}$.

In the second case, the abelianization of φ, which is by hypothesis equal to 1, is also equal to XMY, with the notations of Lemma 20.2.3. This lemma then implies that $X = M = Y = 1$. Note that the restriction of the abelianization map to $\mathcal{B}$ is injective, as follows from (20.1) and (20.3). Likewise, the restriction to $\mathcal{E}$ (hence to $\mathcal{E}^+$) is injective; indeed, the submonoid generated by $\begin{pmatrix} 1 & 1 \\ 0 & 1 \end{pmatrix}$ and $\begin{pmatrix} 1 & 0 \\ 1 & 1 \end{pmatrix}$ is free (Corollary 14.5.4). We deduce that $\alpha = \beta = \gamma = 1$, and hence that $\varphi = 1$. In the first case, we see by injectivity that $\varphi = 1$. □

20.3 Christoffel bases

We define, for all relatively prime natural numbers i and j, the elements $g(\pm i, \pm j)$ of $\mathrm{F}(a, b)$ as follows: $g(i, j)$ is the lower Christoffel word of slope j/i; moreover,

$$g(-i, -j) = g(i, j)^{-1}, g(-i, j) = Sg(i, j), g(i, -j) = Sg(i, j)^{-1} \tag{20.5}$$

For example,

$$g(1, 1) = ab, g(-1, -1) = b^{-1}a^{-1}, g(-1, 1) = a^{-1}b, g(1, -1) = b^{-1}a$$

Note that when $i = 0, j = 1$, and when $i = 1, j = 0$, the double definition of $g(\pm i, \pm j)$ is not contradictory, as is easily verified.

The proof of the following result is left to the reader.

Lemma 20.3.1

(i) *The commutative image of* $g(i, j)$ *is* (i, j), *and the mapping* $g(i, j) \mapsto (i, j)$ *is injective.*

(ii) *An element* g *in* $\mathrm{F}(a, b)$ *is equal to some* $g(i, j)$, $i, j \in \mathbb{Z}$, *if and only if* $g = w$, $g = w^{-1}$, $g = Sw$, *or* $g = Sw^{-1}$ *for some lower Christoffel word* w.

Let $M = \begin{pmatrix} p & q \\ r & s \end{pmatrix}$ be in $\mathrm{GL}_2(\mathbb{Z})$; the pair $B_M = (g(p,r), g(q,s))$ is called a *Christoffel basis*. We see below that it is indeed a basis of $\mathrm{F}(a,b)$). We say that two pairs (u,v) and (x,y) in $\mathrm{F}(a,b)^2$ are *conjugate* if, for some $g \in \mathrm{F}(a,b)$, one has $u = gxg^{-1}$ and $v = gyg^{-1}$; in this case, if one pair is a basis of $\mathrm{F}(a,b)$, then the other is too.

Theorem 20.3.2 *(Cohn [Co2], Theorem 2.1; [KR], Theorem 4.2 and Corollary 4.3 (a)). Given a matrix* $M = \begin{pmatrix} p & q \\ r & s \end{pmatrix}$ *in* $\mathrm{GL}_2(\mathbb{Z})$, B_M *is a basis of* $\mathrm{F}(a,b)$. *Any basis of* $\mathrm{F}(a,b)$ *is conjugate to a unique Christoffel basis.*

Proof. We prove that any basis of $F(a,b)$ is conjugate to a unique pair B_M; then we show that, for all $M \in \mathrm{GL}_2(\mathbb{Z})$, some basis is conjugate to B_M; from this it will follow that each B_M is a basis of $F(a,b)$.

Note that two automorphisms in $\mathrm{Aut}(\mathrm{F}(a,b))$ are in the same class modulo the subgroup of inner automorphisms if and only if the two bases, which are the images of (a,b) under these automorphisms, are conjugate.

Thus it follows from Corollary 20.2.2, and by Lemma 20.1.6, that each basis of $\mathrm{F}(a,b)$ is conjugate to a unique basis of the form given in (20.1) or of the form $\alpha\beta\gamma(a,b)$, where $\alpha = 1$ or S, $\beta \in \mathcal{E}^+$, and $\gamma \in \mathcal{B}$.

In the first case, it is a Christoffel basis, corresponding to one of the eight matrices in (20.3).

In the second case, identifying as usual automorphisms and bases, we know by Theorem 2.6.1 that $\beta = (u,v)$, where (u,v) is a Christoffel pair, different from (a,b) (since $\beta \neq id$). Hence $(u,v) = (g(p,r), g(q,s))$ for some $M = \begin{pmatrix} p & q \\ r & s \end{pmatrix}$ in $\mathrm{SL}_2(\mathbb{N})$, $M \neq 1$, by Corollary 14.5.3. By Lemma 20.1.6, γ is of the form (x^ϵ, y^η) with $\{x,y\} = \{a,b\}$ and $\epsilon, \eta = \pm 1$. Hence $\alpha\beta\gamma(a,b) = (m^\epsilon, p^\eta)$ or (Sm^ϵ, Sp^η), with $\{m,p\} = \{u,v\}$. Thus, by Lemma 20.3.1, and since the abelianization of $\alpha\beta\gamma$ is in $\mathrm{GL}_2(\mathbb{Z})$, $\alpha\beta\gamma(a,b)$ is a Christoffel basis. Moreover, the abelianization of $\alpha\beta\gamma$ is one of the 16 matrices in (20.4). According to Lemma 20.2.3 and its proof, we obtain all of these matrices, since the automorphism $\alpha\beta\gamma$ is arbitrary. Hence each B_M is a basis of $\mathrm{F}(a,b)$. □

An element g in $\mathrm{F}(a,b)$ is called *primitive* if there exists an element h such that g and h form a basis of $\mathrm{F}(a,b)$.

Corollary 20.3.3 *(Osborne-Zieschang [OZ]) The elements* $g(i,j)$ $(i,j \in Z$, *relatively prime) constitute a set of representatives of the conjugation classes of primitive elements in* $\mathrm{F}(a,b)$.

Proof. By the theorem, each primitive element is conjugate to some $g(i,j)$. Uniqueness follows from the fact that the commutative image of $g(i,j)$, which is (i,j), completely determines it (Lemma 20.3.1 (i)), and from the fact that if two elements are conjugate, then they have the same commutative image. □

Corollary 20.3.4 *(Nielsen) Suppose that the two elements* x *and* y *of a basis of* $\mathrm{F}(a,b)$ *are cyclically reduced. Then either* x *and* y, *or* x *and* y^{-1}, *both lie in* M, *where* M

is one of the four submonoids $\{a,b\}^*$, $\{a^{-1},b\}^*$, $\{a,b^{-1}\}^*$, *or* $\{a^{-1},b^{-1}\}^*$ *of* $\mathrm{F}(a,b)$. *Moreover, each cyclically reduced primitive element of* $\mathrm{F}(a,b)$ *lies in one of these submonoids.*

Proof.

(i) Assume that (x,y) is conjugate to a pair lying in $\{a,b\}^*$. It follows from Exercise 20.3.1 that $x,y \in \{a,b\}^*$.

(ii) If $\begin{pmatrix} p & q \\ r & s \end{pmatrix} \in \mathrm{GL}_2(\mathbb{Z})$, then the vectors (p,r) and (q,s) must lie in the same quadrant of $\mathbb{Z}^2$ or in opposite ones: indeed, if not, and if, for example, (p,r) lies in the first and (q,s) in the second, then $p,r,s>0$ and $q<0$ so that $ps-qr \geq 2$, a contradiction. Note that $g(i,j)$, by its definition, lies in one of the four submonoids, which is determined by the quadrant where (i,j) lies. Thus, if (u,v) is a Christoffel basis, then either u and v lie in one of the four submonoids of the statement or u and v^{-1} do.

(iii) By Theorem 20.3.2, (x,y) is conjugate to a Christoffel basis (u,v). By 2, either u and v lie in one of the four submonoids of the statement or u and v^{-1} do. Thus, replacing, if necessary, y by y^{-1}, we may assume that u and v lie in one of the four submonoids of the statement and that (x,y) is conjugate to (u,v). We have seen in (i) that the result is true if (u,v) lies in the submonoid $\{a,b\}^*$. For the three other submonoids, the proof is quite similar, since all these monoids are isomorphic.

(iv) Now, let y be some cyclically reduced primitive element. Then (x,y) is a basis for some element x. If x is cyclically reduced, we conclude, by what has just been proved, that y is in one of these four submonoids. Otherwise, by Corollary 20.1.4, we have the reduced products $y=cy'$ or $y=y'c^{-1}$ ($c=a,b,a^{-1}$ or b^{-1}), and (x,y) is conjugate to a shorter pair $(z,y'c)$ or $(z,c^{-1}y)$; by induction, $y'c$ or $y'c^{-1}$ is in one of these four submonoids, and therefore so is y. □

We say that an element of the free group $\mathrm{F}(a,b)$ is *positive* if it is an element of $\{a,b\}^*$; likewise, we say that a basis (resp. an endomorphism) of $\mathrm{F}(a,b)$ is *positive* if the two elements (resp. if the images of a,b) are in $\{a,b\}^*$.

Recall that two words u and u in $\{a,b\}^*$ are conjugate if $u=xy, u=yx$ for some words $x,y \in \{a,b\}^*$. Observe that conjugation is the equivalence relation of $\{a,b\}^*$ generated by the relation linking u and u' if, for some letter c and some word u, one has $u=cu'', u'=u''c$.

We use this observation to extend *conjugation* to pairs of words; it is the equivalence relation of $\{a,b\}^* \times \{a,b\}^*$ generated by the relation $\to$, which is defined as follows:

$$(u,v) \to (u',v') \tag{20.6}$$

if, for some letter $c=a$ or b, one has $u=cu'', v=cv'', u'=u''c, v'=v''c$; in equivalent notations, this is written as $(u,v)=c(u'',v''), (u',v')=(u'',v'')c$. In words, this means that u and v must have the same first letter and that u',v' are obtained from u and v

by removing this letter and putting it at the end (and therefore u' and v' have the same last letter).

We leave to the reader the proof that if two words in $\{a, b\}^*$ are conjugate in $F(a, b)$, then they are conjugate in $\{a, b\}^*$; likewise, if two pairs of elements of $\{a, b\}^*$ are conjugate in $F(a, b)$, then they are conjugate in $\{a, b\}^*$. This follows from Exercise 20.3.1.

Corollary 20.3.5 *If $\{x, y\}$ is a positive basis of $F(a, b)$, then it is conjugate in $\{a, b\}^*$ to a basis of the form $\{u, v\}$ for some Christoffel word with standard factorization $w = uv$.*

Lemma 20.3.6 *Let $\begin{pmatrix} p & q \\ r & s \end{pmatrix} \in SL_2(\mathbb{N})$. Then the standard factorization of the Christoffel word $g(p + q, r + s)$ is $(g(p, r), g(q, s))$.*

Note that $(p + q)s - q(r + s) = 1$ so that $p + q, r + s$ are relatively prime, and $g(p + q, r + s)$ is well defined. Similarly for $g(p, r)$ and $g(q, s)$.

Proof. In the bijection of Corollary 14.5.3, let w be the lower Christoffel word corresponding to the matrix in the statement, and let uv be its standard factorization. Since a lower Christoffel word is completely determined by its commutative image, we have $w = g(p + q, r + s)$, $u = g(p, r)$, and $v = g(q, s)$, which proves the corollary. □

Proof of Corollary 20.3.5. Since $\{x, y\}$ has its commutative image in the first quadrant, the Christoffel basis to which it is conjugate must be positive. Conjugation in $F(a, b)$ implies conjugation in $\{a, b\}^*$, and the Christoffel basis is of the required form by Lemma 20.3.6. □

Corollary 20.3.7 *A word in $\{a, b\}^*$ is a conjugate of a Christoffel word if and only if it is a positive primitive word of the free group $F(a, b)$.*

Proof. A lower Christoffel word is primitive by Theorem 2.6.1, and primitivity is preserved by conjugation. Conversely, suppose that v is a positive primitive element of the free group $F(a, b)$. For some element u, (u, v) is a basis of $F(a, b)$. It is conjugate to some Christoffel basis (x, y) by Theorem 20.3.2. Then v and y have the same commutative image, so that the the commutative image of y lies in the first quadrant; therefore y is a lower Christoffel word by Lemma 20.3.1(ii). Since v is conjugate to y in $F(a, b)$, it is conjugate to y in $\{a, b\}^*$. □

20.3.1 Exercises

Exercise 20.3.1

(i) *Suppose that $x, y \in F(a, b)$ are cyclically reduced and that, for some $u, v \in \{a, b\}^*$ and $g \in F(a, b)$, one has $x = gug^{-1}, y = gvg^{-1}$. Show that $x, y \in \{a, b\}^*$ and that (x, y) and (u, v) are conjugate in $\{a, b\}^*$. Hint: induction on the length of g.*

(ii) *Show that for* $h = (x, y) \in \{a, b\}^* \times \{a, b\}^*$, *there is a unique maximal finite or bi-infinite sequence* (h_n) *of elements of* $\{a, b\}^* \times \{a, b\}^*$ *such that* h *is equal to one of the* h_n *and that* $h_n \to h_{n+1}$ *for any* n. *Show that it is infinite if and only if* x and y *are powers of the same word. Show that the pairs conjugate in* $\{a, b\}^*$ *of* h *are the* h_n.

20.4 Nielsen's criterion

Theorem 20.4.1 *(Nielsen [N], p. 393) An endomorphism of the free group* $\mathrm{F}(a, b)$ *is an automorphism if and only if it sends* $aba^{-1}b^{-1}$ *onto a conjugate of itself or of its inverse.*

Equivalently, u and v is a basis of $\mathrm{F}(a, b)$ if and only if $uvu^{-1}v^{-1}$ is conjugate to $aba^{-1}b^{-1}$ or its inverse. This result of Nielsen has a striking resemblance to a theorem of Dicks stating that an endomorphism of $K\langle a, b\rangle$ (the algebra of non-commutative polynomials in the variables a and b over the field K) is an automorphism if and only if it sends the commutator $ab - ba$ onto a scalar multiple of itself (see [Di], [Ch], Theorem 9.4). Also, it is similar to the *jacobian conjecture* for automorphisms of commutative polynomial algebras and to a characterization of automorphisms of free Lie algebras which uses a non-commutative jacobian matrix ([Re1], [Sh1], [U]).

For the proof of the "if" part of Theorem 20.4.1, we follow Shpilrain [Sh2], Proposition 2.4. This requires some preparation.

Lemma 20.4.2 *Let* F *be a set,* R *an equivalence relation on* F, $\mathbb{Z}F$ *the free* $\mathbb{Z}$*-module with basis* F, M *the submodule spanned by the elements* $u - v$, *for all* $u, v \in F$ *such that* uRv. *If* $x, y \in F$, *then* $x - y \in M$ *if and only if* xRy.

Proof. The "if" part is clear. We prove the converse. Choose for each equivalence class C of R an element $u_C \in C$. Then the set of all $u - u_C$, $u \in C \setminus u_C$, together with the set of all u_C, for all equivalence classes C, is a basis of $\mathbb{Z}F$. Note that M is spanned by the elements $u - u_C$, for all $u \in C$, since $u, v \in C$ implies $u - v = u - u_C - (v - u_C)$.

Suppose first that C_1 and C_2 are two classes such that $u_{C_1} - u_{C_2} \in M$. Then this element is a linear combination of elements $u - u_C$; this contradicts the basis property above, except if $C_1 = C_2$. Now let $u \in C_1$, $v \in C_2$, and suppose that $u - v \in M$. Then $u_{C_1} - u_{C_2} = (u_{C_1} - u) + (u - v) + (v - u_{C_2}) \in M$, hence $C_1 = C_2$ by what we have just shown; thus uRv. □

Corollary 20.4.3 *Let* F *be a group and* S *be a subset of* F. *Let* I *be the left ideal of the group algebra* $\mathbb{Z}\mathrm{F}$ *over* $\mathbb{Z}$ *generated by the elements* $s - 1$, $s \in S$. *If* $g \in \mathrm{F}$ *satisfies* $g - 1 \in I$, *then* g *is in the subgroup of* F *generated by* S.

Proof. Let R be the equivalence relation on F which is compatible with multiplication on the left and which is associated with the subgroup H generated by S; in other words, uRv if and only if $v \in uH$. Let M be the submodule of $\mathbb{Z}\mathrm{F}$ spanned by the elements $u - v$, $u, v \in \mathrm{F}$, such that uRv. By the previous lemma, if $g \in \mathrm{F}$ satisfies $g - 1 \in M$, then $gR1$, that is, $g \in H$. It is therefore enough to show that $M = I$. Clearly,

$s-1 \in M$ for any $s \in S$, and M is a left ideal of $\mathbb{Z}\mathrm{F}$ since R is left compatible. It follows that $I \subset M$.

Conversely, we show first that if $g \in H$, then $g-1 \in I$. We may write $g = s_1 \cdots s_n$ with $s_i \in S \cup S^{-1}$, and we argue by induction on n. If $n = 0$ it is clear. Otherwise, $g = hs$ or $g = hs^{-1}$ with $s \in S$, $h \in H$, and $h-1 \in I$ by induction. Then $g-1 = h(s-1)+h-1$ in the first case, and $g-1 = h(s^{-1}-1)+h-1$ in the second case. Since $s-1 \in I$ and $s^{-1}-1 = -s^{-1}(s-1) \in I$, we see that $g-1 \in I$ in both cases.

Now, if uRv, then $v^{-1}uR1$, hence $v^{-1}u \in H$. Therefore $v^{-1}u-1 \in I$, by what we have shown above, and thus $u-v = v(v^{-1}u-1) \in I$. It follows that $M \subset I$. □

Let F be a free group, freely generated by X, and ϵ the *augmentation map* of the group algebra $\mathbb{Z}\mathrm{F}$, that is, the homomorphism of this ring into $\mathbb{Z}$ that sends each $g \in \mathrm{F}$ onto 1 (in other words, ϵ sends $\sum_{g\in \mathrm{F}} a_g g$ onto $\sum_{g \in \mathrm{F}} a_g$). Given any mapping $D : X \to \mathbb{Z}\mathrm{F}$, it extends uniquely to a $\mathbb{Z}$-linear mapping from $\mathbb{Z}\mathrm{F}$ into itself, satisfying the rule

$$\forall u, v \in \mathbb{Z}\mathrm{F}, D(uv) = D(u)v + \epsilon(u)D(v) \tag{20.7}$$

This follows from the fact that, for each $x \in X$, the matrix $M(x) = \begin{pmatrix} x & 0 \\ D(x) & \epsilon(x) \end{pmatrix}$ is invertible in the ring $(\mathbb{Z}\mathrm{F})^{2\times 2}$ so that the mapping $x \mapsto M(x)$ extends uniquely to a group homomorphism $\mathrm{F} \to \mathrm{GL}_2(\mathbb{Z}\mathrm{F})$; the latter then extends uniquely by $\mathbb{Z}$-linearity to a $\mathbb{Z}$-algebra homomorphism $\mathbb{Z}\mathrm{F} \to \mathbb{Z}^{2\times 2}$ that we still denote M. For $u \in \mathbb{Z}\mathrm{F}$, write $M(u) = \begin{pmatrix} u & 0 \\ D(u) & \epsilon(u) \end{pmatrix}$. Thus (20.7) holds.

A mapping satisfying this rule is called an *ϵ-derivation* of $\mathbb{Z}\mathrm{F}$. It follows from the rule that $D(1) = 0$ (since $D(1) = D(1 \cdot 1) = D(1) \cdot 1 + \epsilon(1)D(1)$ and $\epsilon(1) = 1$), $D(u^{-1}) = -D(u)u^{-1}$ for $u \in \mathrm{F}$ (since $0 = D(1) = D(uu^{-1}) = D(u)u^{-1} + \epsilon(u)D(u^{-1})$ and $\epsilon(u) = 1$), and that $D(u_1 \cdots u_n) = \sum_{1\le i \le n} D(u_i)u_{i+1} \cdots u_n$ for any $u_1, \ldots, u_n \in \mathrm{F}$.

The *right Fox derivations* are the ϵ-derivations D_x, $x \in X$, defined by $D_x(y) = \delta_{xy}$ for any $y \in X$. One has $D_x(y^{-1}) = -\delta_{xy}x^{-1}$.

Proof of Theorem 20.4.1.

(i) For the free group $\mathrm{F}(a, b)$, one has

$$D_a(aba^{-1}b^{-1}) = a^{-1}b^{-1}(baba^{-1}b^{-1} - 1), D_b(aba^{-1}b^{-1}) = b^{-1}(ba^{-1}b^{-1} - 1).$$

Indeed, by the previous formula for the product, $D_a(aba^{-1}b^{-1}) = ba^{-1}b^{-1} - a^{-1}b^{-1}$ and $D_b(aba^{-1}b^{-1}) = a^{-1}b^{-1} - b^{-1}$.

Next, if $u, v \in \mathrm{F}(a, b)$, then $uvu^{-1}v^{-1} - 1 = uv((u^{-1}-1)(v^{-1}-1) - (v^{-1}-1)(u^{-1}-1))$. It follows that we have $uvu^{-1}v^{-1} - 1 = p(u^{-1}-1) + q(v^{-1}-1)$ for some $p, q \in \mathrm{Ker}(\epsilon)$. Applying to this equality an ϵ-derivation D, we obtain $D(uvu^{-1}v^{-1} - 1) = D(p)(u^{-1}-1) + D(q)(v^{-1}-1) \in I$, where I is the left ideal generated by the elements $h-1$, for $h \in H$, the subgroup generated by u, v.

(ii) Suppose that φ is an endomorphism of $\mathrm{F}(a, b)$ such that $\varphi(aba^{-1}b^{-1}) = aba^{-1}b^{-1}$. Let $u = \varphi(a), v = \varphi(v)$ so that $aba^{-1}b^{-1} = uvu^{-1}v^{-1}$. Then

$a^{-1}b^{-1}(baba^{-1}b^{-1} - 1) = D_a(aba^{-1}b^{-1} - 1) = D_a(uvu^{-1}v^{-1} - 1) \in I$ and therefore $baba^{-1}b^{-1} - 1 \in I$. Thus, by Corollary 20.4.3, $baba^{-1}b^{-1} \in H$. Similarly, $ba^{-1}b^{-1} \in H$. It follows that $ba \in H$ and then $b \in H$ and finally $a \in H$. Thus $H = \mathrm{F}(a, b)$, which shows that u and v generate $\mathrm{F}(a, b)$ so that φ is an automorphism by Corollary 20.1.7.

Suppose that φ is an endomorphism of $\mathrm{F}(a, b)$ such that $\varphi(aba^{-1}b^{-1}) = g^{-1}(aba^{-1}b^{-1})^{\pm 1}g$. Let $\psi = (gag^{-1}, gbg^{-1}) \circ \varphi$. Then $\psi(aba^{-1}b^{-1}) = (aba^{-1}b^{-1})^{\eta}$, $\eta = \pm 1$. If $\eta = 1$, ψ is an automorphism by what has been proved above, and φ too. If $\eta = -1$, then $\psi(aba^{-1}b^{-1}) = bab^{-1}a^{-1}$ and $E\psi(aba^{-1}b^{-1}) = aba^{-1}b^{-1}$, and we conclude also that φ is an automorphism.

(iii) We now prove the "only if" part. By induction and Corollary 20.1.5, it suffices to show that G, E, S send $aba^{-1}b^{-1}$ onto a conjugate of itself or its inverse. One has $G(aba^{-1}b^{-1}) = a(ab)a^{-1}(ab)^{-1} = aaba^{-1}b^{-1}a^{-1}$, clearly a conjugate of $aba^{-1}b^{-1}$. Next, $E(aba^{-1}b^{-1}) = bab^{-1}a^{-1} = (aba^{-1}b^{-1})^{-1}$. Finally, $S(aba^{-1}b^{-1}) = a^{-1}bab^{-1} = a^{-1}bab^{-1}a^{-1}a = a^{-1}(aba^{-1}b^{-1})^{-1}a$. □

20.4.1 Exercises

Exercise 20.4.1 *Let $\sigma : \{a, b\}^* \to \mathrm{Aut}(\mathrm{F}(a, b))$ send a onto G and b onto D. Let $K = aba^{-1}b^{-1}$. Show that for any word $v \in \{a, b\}^*$, one has $\sigma(v)(K) = Pal(v)KPal(v)^{-1}$. Hint: use Justins's formula, Theorem 14.1.3.*

20.5 An algorithm for the bases

Recall the relation $\to$ defined in (20.6). We consider it as a *rewriting system* on the set $\{a, b\}^* \times \{a, b\}^*$.

By Exercise 20.3.1, any pair $(u, v) \in \{a, b\}^* \times \{a, b\}^*$ is contained in a unique maximal chain of this rewriting system, whose length (the number of arrows) may be finite or infinite. For example, if u and v are powers of the same word, then this chain is infinite (that is, the rewriting never stops). As another example, (aab, ab) is contained in the maximal chain $(aba, ab) \to (baa, ba) \to (aab, ab) \to (aba, ba)$ of length 3.

The previous notion of chain is essentially equivalent to the notion of *conjugate morphisms*, due to Séébold [Sb2], p. 96, and the next result is also essentially due to him [Sb2], Theorem 4 (see also [Lo], Section 2.3.4).

Theorem 20.5.1 *Let $(u, v) \in \{a, b\}^* \times \{a, b\}^*$. The pair (u, v) is a basis of* $\mathrm{F}(a, b)$ *if and only if the maximal chain containing it is of length $|u| + |v| - 2$.*

Lemma 20.5.2 *Let $w = uv$ be a proper lower Christoffel word with its standard factorization. Then the maximal chain containing (u, v) is of length $|u| + |v| - 2$. In the case where u and v are proper Christoffel words, the first element in the chain is (hba, lab), where $u = ahb, v = alb$.*

Proof.

(i) Write $w = amb$. Let p (resp. s) be the prefix of length $|v| - 1$ (resp. the suffix of length $|u| - 1$) of w. Then, by Lemma 14.3.5, mbp has the period $|v|$ and symmetrically sam has the period $|u|$. For $i = 0, \ldots, |u| + |v| - 2$, let f_i be the factor of length $|w|$ of swp such that $swp = g_i f_i h_i$ and g_i of length i. This is well defined, since $|u| + |v| - 2 + |w|$ is equal to the length of swp. Now let $f_i = u_i v_i$, with $|u_i| = |u|$ and therefore $|v_i| = |v|$.

We show that $(u_0, v_0) \to (u_1, v_1) \to \cdots \to (u_{|u|+|v|-2}, v_{|u|+|v|-2})$ is the maximal chain, of length $|u| + |v| - 2$, containing (u, v). It has the required length and it contains $(u, v) = (u_{|s|}, v_{|s|})$, since for $i = |s|$, g_i has length $|s|$ so that $g_i = s$, $f_i = uv$, and hence $u = u_i, v = v_i$.

Next, let $i = 0, \ldots, |u| + |v| - 3$. Let x be the first letter of v_i; then the length of $g_i u_i x$ is $i + |u| + 1 \leq |u| + |v| - 3 + |u| + 1 = |u| - 1 + |w| - 1 = |sam|$, so that $g_i u_i x$, being a prefix of $swp = sambp$, is a prefix of sam; therefore, it has the period $|u|$, and we conclude that the first letter of u_i is equal to x. By symmetry, the last letter of v_{i+1} is equal to the last letter of u_{i+1}; that is, the letter following v_i in swp is equal to the letter following u_i in swp, which is x. Thus we have $u_i = xu_i'$, $v_i = xv_i'$, $u_{i+1} = u_i'x$, and $v_{i+1} = v_i'x$, which shows that $(u_i, v_i) \to (u_{i+1}, v_{i+1})$.

Now, u_0 is the prefix of length $|u|$ of swp, and $|s| = |u| - 1$ so that the last letter of u_0 is a; moreover, by Theorem 14.2.1, the palindromic factorization of w is $v'u'$, with v' of length $|v|$ and u' begins by b; hence, since $u_0 v_0$ is the prefix of length $|uv|$ of $swp = sv'u'p$, the last letter of v_0 is this b. Thus the chain cannot be extended to the left, and neither to the right, by symmetry. It is a maximal chain.

(ii) Suppose that u, v are proper Christoffel words. We may write $u = ahb, v = alb$. Then, By Theorem 14.2.1, the palindromic factorization is $w = (ala)(bhb)$. Hence $p = al$ and $s = hb$. It follows that $swp = hbalabhbal$, and therefore $f_0 = hbalab$ and $u_0 = hba, v_0 = lab$. □

Proof of Theorem 20.5.1.

(i) It follows from Lemma 20.5.2 and Corollary 20.3.5 that the maximal chain containing a positive basis is of the required length. Indeed, by Exercise 20.3.1, the elements of a chain containing $(u, v) \in \{a, b\}^* \times \{a, b\}^*$ are exactly the pairs conjugate in $\{a, b\}^*$ to (u, v).

(ii) Conversely, suppose that the maximal chain containing (u, v) is of length $|u| + |v| - 2$. We may assume that (u, v) is the leftmost element of the chain. This implies that the two infinite words u^∞ and v^∞ have a common prefix m of length $|u| + |v| - 2$.

Suppose first that u and v are both of length ≥ 2. Since $|u| + |v| - 2 \geq |u|, |v|$, the words u and v are both prefixes of m. The word m has the two periods $|u|$ and $|v|$. If these two numbers were not relatively prime, then m would be of length $\geq |u| + |v| -$

$\gcd(|u|, |v|)$ and therefore m would have the period $\gcd(|u|, |v|)$, by Corollary 14.3.9; thus u and v would also have this period, and they would be powers of a common word of length $\gcd(|u|, |v|)$; then the chain would be infinite, a contradiction.

Thus $|u|$ and $|v|$ are relatively prime. It follows that m is a central word by Theorem 14.3.3. Then $w = amb$ is a lower Christoffel word. It follows from Corollary 19.2.3 that the lengths of the factors of the standard factorization of w are $|u|$ and $|v|$ (since m is not the power of a single letter; otherwise u and v too are powers of this letter, and the chain is infinite). By Theorem 14.2.1, the standard factorization is $w = (ahb)(alb)$, and its palindromic factorization is $w = (ala)(bhb)$. Hence $m = hbal = labh$. We have $\{|u|, |v|\} = \{|ahb|, |alb|\}$, and since u and v are prefixes of m, we obtain that $\{u, v\} = \{hba, lab\}$. By Lemma 20.5.2, the first pair in the maximal chain containing (ahb, alb) is (hba, lab). Hence the latter pair is a basis of $F(a, b)$, being conjugate to a Christoffel pair, and therefore, (u, v) is a basis of $F(a, b)$.

If u is of length 1, $u = a$, say, we see directly that v must be of the form $v = a^n b$, so that (u, v) is a basis. The case where v is of length 1 is similar. □

The following result was obtained by Kassel and the author [KR], Corollary 5.2.

Theorem 20.5.3 *A pair (u, v) of reduced elements in $F(a, b)^2$ is a basis of $F(a, b)$ if and only if the following algorithm stops positively in Step (iv):*

(i) *If u, v are not both cyclically reduced, find $c \in \{a, b, a^{-1}, b^{-1}\}$ which is a common prefix or a common suffix of u and v. Then conjugate by c. Continue until u and v are both cyclically reduced.*
(ii) *If u and v are not both in one of the four submonoids $\{a, b\}^*$, $\{a^{-1}, b\}^*$, $\{a, b^{-1}\}^*$, or $\{a^{-1}, b^{-1}\}^*$ of $F(a, b)$, replace v by v^{-1} and check if the two new elements are in one of these four submonoids.*
(iii) *If u and v are not in $\{a, b\}^*$, apply simultaneously to u, v the operations: inversion, the automorphism S, or both, so that the new elements are in this submonoid.*
(iv) *Verify that the length of the maximal chain containing the pair is of length $|u| + |v| - 2$.*

Proof. If u and v form a basis, then step (i) is valid by Corollary 20.1.4 and step (ii) by Corollary 20.3.4. Step (iii) clearly applies. Then step (iv) follows from Theorem 20.5.1. □

20.5.1 Exercises

Exercise 20.5.1 *With the notations at the beginning of this section, label the arrow in the rewriting system as follows $(u, v) \xrightarrow{c} (u, v)$. Consider the maximal chain containing a standard pair (u, v). Show that the successive labels of the arrows form the central word associated with the Christoffel word uv. Hint: see the second part of the proof of Theorem 20.5.1.*

Exercise 20.5.2 *Show that a word in* $\{a, b\}^*$ *is a conjugate of a Christoffel word if and only if it is the image of some letter by some Sturmian morphism. See Séébold [Sb2] and Chuan [Ch3]).*

20.6 Positive automorphisms

Corollary 20.6.1 *Sturmian morphisms coincide with positive automorphisms of* $\mathrm{F}(a, b)$.

Proof. Let φ be a positive automorphism of $\mathrm{F}(a, b)$. It sends positive primitive words onto positive primitive words so that each conjugate of a Christoffel word is sent by Corollary 20.3.7 onto the conjugate of a Christoffel word. Thus φ is a Sturmian morphism. Conversely, each Sturmian morphism is a positive automorphism of $\mathrm{F}(a, b)$ by Theorem 15.5.1. □

Corollary 20.6.2 *(Wen and Wen [WW]) The submonoid of positive automorphisms of* $\mathrm{Aut}(\mathrm{F}(a, b))$ *is generated by* $E, G, \tilde{G}$ *or* $E, D, \tilde{D}$.

Note that Shpilrain [Sh3] has shown that if an endomorphism of $\mathrm{F}(a, b)$ sends primitive elements onto primitive elements, then it is an automorphism. This result has been extended to arbitrary free groups by Lee [Le].

21
The commutator subgroup $SL_2(\mathbb{Z})'$

The commutator subgroup $SL_2(\mathbb{Z})'$ of $SL_2(\mathbb{Z})$ is a free group on two generators. This gives a natural free group on two generators. A 3-matrix is a matrix in $SL_2(\mathbb{Z})$ whose trace is three times its upper right coefficient. A pair of 3-matrices is a 3-pair if their product is also a 3-matrix. These definitions are due to Schmutz-Schaller and are motivated by earlier work of Cohn [Co1] and Aigner (the "Cohn triples" in [A]).

If a matrix in $SL_2(\mathbb{Z})$ is symmetric, or a 3-matrix, then it is in $SL_2(\mathbb{Z})'$ if and only if its trace is positive (Schmutz-Schaller). If A and B are 3-matrices, then (A, B) is a 3-pair if and only if $AB^{-1}A^{-1}B = -L^6$; and if (A, B) is a 3-pair, then (A, AB), (AB, B), $(A, A^{-1}B)$, (AB^{-1}, B) are 3-pairs (Schmutz-Schaller). For each basis (A, B) of the commutator subgroup $SL_2(\mathbb{Z})'$, the third of the triple $(\mathrm{tr}(A), \mathrm{tr}(AB), \mathrm{tr}(B))$ is a Markoff triple (Perrine and Schmutz-Schaller for the positivity).

21.1 $SL_2(\mathbb{Z})'$ is a free group on two generators

Recall that the *commutator subgroup*, or *derived subgroup*, of a group G is the subgroup of G generated by the elements $ghg^{-1}h^{-1}$, $g, h \in G$. It is denoted G'. The commutator subgroup is a normal subgroup, and it is the smallest normal subgroup H of G such that the quotient G/H is abelian; see Exercise 21.1.1.

Consider the two matrices

$$A_0 = \begin{pmatrix} 1 & 1 \\ 1 & 2 \end{pmatrix}, \quad B_0 = \begin{pmatrix} 2 & 1 \\ 1 & 1 \end{pmatrix}$$

Theorem 21.1.1 *The commutator subgroup of* $SL_2(\mathbb{Z})$ *is a free group on the two generators* A_0 *and* B_0.

Lemma 21.1.2 *The matrices* A_0 *and* B_0 *generate freely a subgroup of* $SL_2(\mathbb{Z})$.

Proof (following Sc1, proof of Theorem 4.2.10). Consider $\overline{\mathbb{R}} = \mathbb{R} \cup \infty$, the compactification of $\mathbb{R}$. We use the left action of matrices on $\overline{\mathbb{R}}$; see Exercise 6.1.4.

We define *intervals* in $\overline{\mathbb{R}}$. Let $x, y \in \overline{\mathbb{R}}$. If $x = y$, $[x, y]$ is the singleton $\{x\}$. Suppose that $x \neq y$. If $x, y \in \mathbb{R}$, then either $x < y$ and then $[x, y]$ is the usual interval in $\mathbb{R}$, or $y > x$ and then it is $[y, +\infty) \cup \{\infty\} \cup (-\infty, x]$. If $x = \infty, y \in \mathbb{R}$, $[x, y]$ denotes the subset $\{\infty\} \cup (-\infty, y]$ of $\overline{\mathbb{R}}$. If $x \in \mathbb{R}, y = \infty$, it denotes the subset $[x, +\infty) \cup \infty$.

In other words, intervals are cyclic intervals if $\overline{\mathbb{R}}$ is seen as an oriented circle; see Figure 21.1.

From Christoffel Words to Markoff Numbers. Second Edition. Christophe Reutenauer, Oxford University Press.
 DOI: 10.1093/9780197907535.003.0021

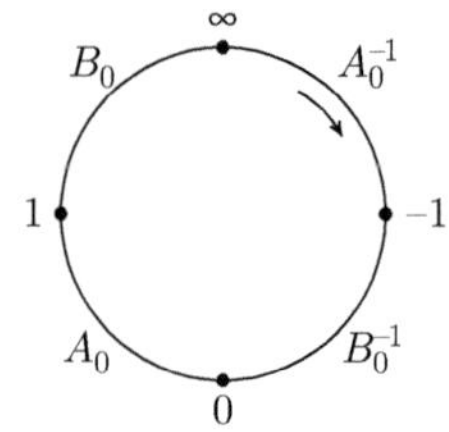

Figure 21.1 Intervals in the circle $\overline{\mathbb{R}}$

We define four special intervals:

$$J_{A_0^{-1}} = [\infty, -1], J_{B_0^{-1}} = [-1, 0], J_{A_0} = [0, 1], J_{B_0} = [1, \infty]$$

Moreover, four other intervals, each obtained by the union of three of the previous ones:

$$I_{A_0^{-1}} = J_{B_0} \cup J_{A_0^{-1}} \cup J_{B_0^{-1}} = [1, 0], I_{B_0^{-1}} = J_{A_0^{-1}} \cup J_{B_0^{-1}} \cup J_{A_0} = [\infty, 1]$$

$$I_{A_0} = J_{B_0^{-1}} \cup J_{A_0} \cup J_{B_0} = [-1, \infty], I_{B_0} = J_{A_0} \cup J_{B_0} \cup J_{A_0^{-1}} = [0, -1]$$

We claim that for $X = A_0, B_0, A_0^{-1}, B_0^{-1}$, acting on $\overline{\mathbb{R}}$, X sends the interval I_X onto J_X. Admitting the claim for the moment, we deduce by induction on $n \geq 1$ that for each reduced product $M = X_n \cdots X_1$ with each X_i in $\{A_0, B_0, A_0^{-1}, B_0^{-1}\}$, M sends I_{X_1} into J_{X_n} (by "reduced", we mean that $X_i \neq X_{i+1}^{-1}$, for $i = 1, \ldots, n-1$, in accordance with "reduced" in Section 20.1); indeed, this follows by induction on n, since $J_X \subset I_Y$ if $Y \neq X^{-1}$, as one sees by inspection of the definitions of the intervals.

Note that if the identity of $\overline{\mathbb{R}}$ sends a subset I of $\overline{\mathbb{R}}$ into a subset J, then necessarily $I \subset J$. Moreover, observe that for $X, Y \in \{A_0, B_0, A_0^{-1}, B_0^{-1}\}$, we have not $I_X \subset J_Y$. This implies that M is not the identity matrix because the latter acts as the identity of $\overline{\mathbb{R}}$. It follows that no reduced product of A_0, B_0, A_0^{-1}, and B_0^{-1} is equal to the identity matrix, and therefore A_0 and B_0 generate freely a subgroup of $SL_2(\mathbb{Z})$.

We now prove the claim. We have $B_0 \cdot z = \frac{2z+1}{z+1}$, hence $B_0 \cdot 0 = 1$, $B_0 \cdot (-1) = \infty$, and by Exercise 21.1.2, B_0 sends $I_{B_0} = [0, -1]$ onto $[1, \infty] = J_{B_0}$. The three other cases are similar. □

The previous proof is an example of a technique initiated by Felix Klein; see the so-called *ping pong lemma*, Section 24 in [dlH].

Proof of Theorem 21.1.1. By Lemma 21.1.2, it is enough to show that the subgroup H generated by A_0 and B_0 is equal to $SL_2(\mathbb{Z})'$. We have $H \subset SL_2(\mathbb{Z})'$ since $A_0 = RL^{-1}R^{-1}L$ and $B_0 = LR^{-1}L^{-1}R$, where R and L are given in (1.1). We know that L and R generate $SL_2(\mathbb{Z})$ (Section 21.1). Therefore, the equalities

$$RB_0R^{-1} = B_0A_0^{-1}, RA_0R^{-1} = B_0, LB_0L^{-1} = A_0, LA_0L^{-1} = A_0B_0^{-1}$$

imply that H is a normal subgroup of $SL_2(\mathbb{Z})$. Finally, the equality $B_0 = LR^{-1}L^{-1}R$ implies that in the quotient $SL_2(\mathbb{Z})/H$, the images of L and R^{-1} commute; since they generate this quotient, it is commutative and therefore $SL_2(\mathbb{Z})' \subset H$ (Exercise 21.1.1). □

21.1.1 Exercises

Exercise 21.1.1 *Show that the commutator subgroup G' of a group G is a normal subgroup of G. Show that if a H is a normal subgroup of G such that G/H is abelian, then $G' \subset H$.*

Exercise 21.1.2 *Let* $M = \begin{pmatrix} a & b \\ c & d \end{pmatrix} \in \mathrm{SL}_2(\mathbb{Z})$. *Show that M sends each interval $[x, y]$ of $\overline{\mathbb{R}}$ onto the interval $[M \cdot x, M \cdot y]$. Hint: use the graph of the real function $\frac{ax+b}{cx+d}$, which is an increasing continuous bijective function from $\mathbb{R} \setminus \frac{-d}{c}$ onto $\mathbb{R} \setminus \frac{a}{c}$ (the case $c = 0$ must be treated separately).*

Exercise 21.1.3 *Show that the homomorphism μ defined in Section 3.1 is a group isomorphism $F(a, b) \to \mathrm{SL}_2(\mathbb{Z})'$. Hint: show first that $\mu(b) = B_0 A_0^{-1} B_0$.*

Exercise 21.1.4 *Show that the matrix $-I_2$ is not in $SL_2(\mathbb{Z})'$. Show that if $X \in SL_2(\mathbb{Z})'$, then $-X$ is not in $SL_2(\mathbb{Z})'$. Hint: the proof of Lemma 21.1.2 shows that no non-empty reduced product of the matrices in Theorem 21.1.1 and of their inverses can act as the identity in $\overline{\mathbb{R}}$, and $-I_2$ acts as the identity.*

21.2 3-matrices and 3-pairs

Following the terminology of [Sc1, Sc2] (after a symmetry), we say that a matrix M in $\mathrm{SL}_2(\mathbb{Z})'$ is a 3-*matrix* if $\mathrm{tr}(M) = 3M_{12}$. Examples of such matrices are the matrices A_0 and B_0 of Section 21.1 and also any matrix of the form $\mu(w)$ for some lower Christoffel word w; see Section 3.1.

Theorem 21.2.1 *(Schmutz-Schaller Sc2) Let $A \in SL_2(\mathbb{Z})'$. Suppose that A is symmetric or a 3-matrix. Then $A \in \mathrm{SL}_2(\mathbb{Z})'$ if and only* $\mathrm{tr}(A) > 0$.

Proof. Let

$$A = \begin{pmatrix} a & b \\ c & d \end{pmatrix} \tag{21.1}$$

(i) Assume that A is symmetric (so that $c = b$) and that $\mathrm{tr}(A) > 0$. We prove that $A \in \mathrm{SL}_2(\mathbb{Z})'$ by induction on $|b|$. We have $ad = 1 + b^2$. If $b = 0$, then $ad = 1$, and since $\mathrm{tr}(A) > 0$, $A = I_2 \in \mathrm{SL}_2(Z)'$. If $b = 1$, then $ad = 2$ so that (since the trace is positive) $A = A_0$ or B_0 is in $SL_2(\mathbb{Z})'$. If $b = -1$, the argument is similar and we find that $A = A_0^{-1}$ or B_0^{-1} is in $\mathrm{SL}_2(\mathbb{Z})'$.

Assume now that $|b| \geq 2$. We have

$$RAL = \begin{pmatrix} a + 2b + d & b + d \\ b + d & d \end{pmatrix}, \; R^{-1}AL^{-1} = \begin{pmatrix} a - 2b + d & b - d \\ b - d & d \end{pmatrix} \tag{21.2}$$

These matrices are in $SL_2(\mathbb{Z})$; using twice Exercise 21.2.1, we see that $d > 0$, and then that the traces of these matrices are positive. By Exercise 21.2.2, we have either $a < |b|$ or $d < |b|$. By Exercise 21.2.3, we may assume that $d < |b|$. Then either $|b + d| < |b|$ or $|b - d| < |b|$. Then we deduce by induction on $|b|$ that $M = RAL$ or $R^{-1}AL^{-1}$ is in $SL_2(\mathbb{Z})'$. If $M = RAL$, then $A = R^{-1}ML^{-1} = LB_0^{-1}ML^{-1}$ (since $B_0 = RL$), hence by Exercise 21.2.3, A is in $SL_2(\mathbb{Z})'$; since so is B_0. If $M = R^{-1}AL^{-1}$, the argument is similar.

Suppose now that A is symmetric and $\mathrm{tr}(A) \leq 0$. Since $ad = 1 + b^2$, the trace does not vanish, hence $\mathrm{tr}(A) < 0$. Then $\mathrm{tr}(-A) > 0$ and by the first part, $-A$ is in $SL_2(\mathbb{Z})'$. Thus A is not in $SL_2(\mathbb{Z})'$ by Exercise 21.1.4.

(ii) We assume now that A is a 3-matrix. Thus $A = \begin{pmatrix} a & b \\ c & 3b - a \end{pmatrix}$. Define

$$B := \begin{pmatrix} \alpha & \beta \\ \gamma & \delta \end{pmatrix} = A_0^{-1}AB_0^{-1} = \begin{pmatrix} a + b - c & -2b + c \\ -2b + c & -a + 4b - c \end{pmatrix}$$

Then $B \in SL_2(\mathbb{Z})$ and B is symmetric. Note that $\alpha + 2\beta + \delta = b = (1/3)\mathrm{tr}(A)$. Since $A_0, B_0 \in SL_2(\mathbb{Z})'$, it follows from (i) and Exercise 21.2.1 that $A \in SL_2(\mathbb{Z})' \Leftrightarrow B \in SL_2(\mathbb{Z})' \Leftrightarrow \mathrm{tr}(B) > 0 \Leftrightarrow \alpha + 2\beta + \delta > 0 \Leftrightarrow b > 0 \Leftrightarrow \mathrm{tr}(A) > 0$. □

Following [Sc1, Sc2], a 3-*pair* is a pair (A, B) of matrices in $SL_2(\mathbb{Z})$ such that A, B, and AB are all 3-matrices. Examples of 3-pairs are the pairs $(\mu(u), \mu(v))$, where uv is the standard factorization of a lower Christoffel word (see Section 3.1).

Theorem 21.2.2 *(Schmutz-Schaller [Sc2], Proposition 3.7)*

(i) A matrix $A \in SL_2(\mathbb{Z})$ is a 3-matrix if and only if $A = -L^3A^{-1}L^3$, where $L = \begin{pmatrix} 1 & 0 \\ 1 & 1 \end{pmatrix}$.

(ii) Let A and B be 3-matrices in $SL_2(\mathbb{Z})$. Then (A, B) is a 3-pair if and only if $AB^{-1}A^{-1}B = -L^6$.

(iii) If (A, B) is a 3-pair, then (A, AB), (AB, B), $(A, A^{-1}B)$, and (AB^{-1}, B) are 3-pairs.

Proof.

(i) follows from the matrix equality

$$-L^3A^{-1}L^3 = \begin{pmatrix} 3b - d & b \\ -3a + 9b + c - 3d & -a + 3b \end{pmatrix} = A + (3b - a - d)L^3$$

where A is given by (21.1).

(ii) If (A, B) is a 3-pair, then by (i) applied three times, $AB = -L^3B^{-1}A^{-1}L^3 = -L^3B^{-1}L^3L^{-6}L^3A^{-1}L^3 = BL^{-6}(-A)$, hence $-L^{-6} = B^{-1}ABA^{-1}$ and $-L^6 = AB^{-1}A^{-1}B$. Conversely, if the latter equality holds, then $-L^3(AB)^{-1}L^3 = -L^3B^{-1}A^{-1}L^3 = -L^3B^{-1}L^3L^{-6}L^3A^{-1}L^3 = B(-B^{-1}ABA^{-1})(-A) = AB$ and AB is a 3-matrix by (i).

(iii) By definition of 3-pairs, it is enough to show that $AAB, ABB, A^{-1}B$ and AB^{-1} are 3-matrices. For this we use (i) and (ii). We have $-L^3(AAB)^{-1}L^3 = -L^3(AB)^{-1}L^3L^{-6}L^3A^{-1}L^3 = ABL^{-6}(-A) = ABB^{-1}ABA^{-1}A = AAB$, and therefore AAB is a 3-matrix. The proof for ABB is similar. Next we have $-L^3(A^{-1}B)^{-1}L^3 = -L^3B^{-1}AL^3 = -L^3B^{-1}L^3L^{-3}AL^{-3}L^6 = B(-A^{-1})L^6 = BA^{-1}AB^{-1}A^{-1}B = A^{-1}B$ and therefore $A^{-1}B$ is a 3-matrix. The proof for AB^{-1} is similar. □

Corollary 21.2.3 *(Schmutz-Schaller [Sc2]) For each basis (A, B) of the free group $\mathrm{SL}_2(\mathbb{Z})'$, (A, B) or (B, A) is conjugate within $\mathrm{SL}_2(\mathbb{Z})'$ to a 3-pair. In particular, $\mathrm{tr}(A), \mathrm{tr}(B), \mathrm{tr}(AB)$ are positive.*

Proof. By Theorem 21.1.1, we have a group isomorphism $\mathrm{F}(a, b) \to SL_2(\mathbb{Z})'$, with $a \mapsto A_0, b \mapsto B_0$. This isomorphism allows to identify both groups: $\mathrm{F}(a, b) = \mathrm{SL}_2(\mathbb{Z})'$.

Consider the subgroup H of the group $\mathrm{Aut}(\mathrm{F}(a, b))$ of automorphisms of $\mathrm{F}(a, b)$, generated by G and $\tilde{D}$. We claim that for each automorphism $\varphi \in H$, $(u, v) = (\varphi(a), \varphi(b))$ is a 3-pair. Since (A_0, B_0) is a 3-pair, it suffices to show that if this is true for φ, then also for $\psi = \varphi \circ F$ for any $F \in \{G, \tilde{D}, G^{-1}, \tilde{D}^{-1}\}$. For $F = G$, one has $(\psi(a), \psi(b)) = (\varphi \circ G(a), \varphi \circ G(b)) = (\varphi(a), \varphi(ab)) = (u, uv)$, which is a 3-pair by Theorem 21.2.2 (iii). Similarly, if $F = \tilde{D}$, $(\psi(a), \psi(b)) = (uv, v)$ is a 3-pair. Next, if $F = G^{-1}$, which sends a onto a and b onto $a^{-1}b$, then $(\psi(a), \psi(b)) = (u, u^{-1}v)$, which is 3-pair by the same result. Finally, for $F = \tilde{D}$, we have $(\psi(a), \psi(b)) = (uv^{-1}, v)$ which is also a 3-pair.

Observe that the abelianization of G and $\tilde{D}$ are the generators R, L of $SL_2(\mathbb{Z})$ shown in (1.1). Hence the set of abelianizations of all elements of H is $SL_2(Z)$.

Let (A, B) be a basis of $SL_2(\mathbb{Z})'$. We consider the automorphism ψ of this group sending A_0, B_0 respectively onto A, B. Exchanging A and B if necessary, we may assume that the abelianization of ψ is in $SL_2(\mathbb{Z})$. By the previous observation, the abelianization of ψ is equal to the abelianization of some $\varphi \in H$. By Nielsen's theorem (Corollary 20.2.4), φ and ψ are congruent modulo the subgroup of inner automorphisms of $\mathrm{F}(a, b)$. This means that $(\psi(a), \psi(b))$ and $(\varphi(a), \varphi(b))$ are conjugate in $\mathrm{F}(a, b)$. Since the latter is a 3-pair, the first assertion is proved. The second one follows from Theorem 21.2.1 □

21.2.1 Exercises

Exercise 21.2.1 *Show that for a symmetric matrix $A = \begin{pmatrix} a & b \\ b & d \end{pmatrix}$ in $\mathrm{SL}_2(\mathbb{Z})$, the five following conditions are equivalent: $\mathrm{tr}(a) > 0$; $a > 0$; $d > 0$; $a + 2b + d > 0$; $a - 2b + d > 0$.*

Exercise 21.2.2 *Suppose that a, b,* and *d are integers such that $a > 0, d > 0, |b| \geq 2$, and $ad = b^2 + 1$. Show that either $a < |b|$ or $d < |b|$.*

Exercise 21.2.3 *Show that the conjugation by any matrix N in $GL_2(\mathbb{Z})$ induces an automorphism of $\mathrm{SL}_2(\mathbb{Z})'$. In particular, when $N = \begin{pmatrix} 0 & 1 \\ 1 & 0 \end{pmatrix}$, in which case it is the mapping $\begin{pmatrix} a & b \\ c & d \end{pmatrix} \mapsto \begin{pmatrix} d & c \\ b & a \end{pmatrix}$.*

Exercise 21.2.4 *Let (A, B) be a 3-pair. Denote by h the homomorphism from $\{a, b\}^*$ into $SL_2(\mathbb{Z})$ which sends a onto A and b onto B. Using Section 2.5 and Theorem 21.2.2, show that for each lower Christoffel word $w = uv$, with its standard factorization, $(h(u)$ and $h(v))$ is a 3-pair.*

21.3 Bases and Markoff triples

The next result is due to Perrine ([Pe1], [Pe2], Proposition 6.4.1); the positivity of the triple was stated without proof, and a proof was given by Schmutz-Schaller [Sc2].

Theorem 21.3.1 *For each basis X and Y of the free group $SL_2(\mathbb{Z})'$, $\{\frac{1}{3}\mathrm{tr}(X), \frac{1}{3}\mathrm{tr}(Y), \frac{1}{3}\mathrm{tr}(XY)\}$ is a Markoff triple.*

Proof. We use the matrices A_0, B_0 of Section 21.1. Note that

$$A_0B_0A_0^{-1}B_0^{-1} = \begin{pmatrix} 2 & 1 \\ 1 & 1 \end{pmatrix}\begin{pmatrix} 1 & 1 \\ 1 & 2 \end{pmatrix}\begin{pmatrix} 1 & -1 \\ -1 & 2 \end{pmatrix}\begin{pmatrix} 2 & -1 \\ -1 & 1 \end{pmatrix}$$

$$= \begin{pmatrix} 3 & 4 \\ 2 & 3 \end{pmatrix}\begin{pmatrix} 3 & -2 \\ -4 & 3 \end{pmatrix} = \begin{pmatrix} -7 & 6 \\ -6 & 5 \end{pmatrix}$$

and therefore has trace -2.

Since $SL_2(\mathbb{Z})'$ is a free group with basis A_0 and B_0, it follows from Theorem 20.4.1 that for any other basis X and Y, the trace of $XYX^{-1}Y^{-1}$ is also -2. It follows from the last Fricke relation in Lemma 3.1.4 that one has $\mathrm{tr}(X)^2+\mathrm{tr}(Y)^2+\mathrm{tr}(XY)^2 = \mathrm{tr}(X)\mathrm{tr}(Y)\mathrm{tr}(XY)$. Taking the values modulo 3, we see by case checking that the three traces must be divisible by 3. Thus the triple in the theorem satisfies the Markoff equation. The positivity follows from Corollary 21.2.3. □

It is also shown in [Pe1, Pe2], Propositions 6.5.1 and 6.5.3, that for two bases (A, B) and (A', B') of the free group $SL_2(\mathbb{Z})'$, one has $(\frac{1}{3}\mathrm{tr}(A), \frac{1}{3}\mathrm{tr}(B), \frac{1}{3}\mathrm{tr}(AB)) = (\frac{1}{3}\mathrm{tr}(A'), \frac{1}{3}\mathrm{tr}(B'), \frac{1}{3}\mathrm{tr}(A'B'))$ if and only if, for some matrix N in $GL_2(\mathbb{Z})$, one has $(A', B') = (NAN^{-1}, NBN^{-1})$.

Perrine's theorem (Theorem 21.3.1) has the following extension by Schmutz-Schaller [Sc2]; for $X, Y \in SL_2(\mathbb{Z})'$, one has the equivalence of the following conditions:

(i) (X, Y) is a basis of $SL_2(\mathbb{Z})'$;
(ii) (X, Y) or (Y, X) is conjugate within $SL_2(\mathbb{Z})'$ to a 3-pair belonging to this group;
(iii) the trace of the commutator $XY^{-1}X^{-1}Y$ is equal to -2;
(iv) one has the matrix Markoff equation $\mathrm{tr}(X)^2 + \mathrm{tr}(Y)^2 + \mathrm{tr}(XY)^2 = \mathrm{tr}(X)\,\mathrm{tr}(Y)\,\mathrm{tr}(XY)$.

21.3.1 Exercises

Exercise 21.3.1 *Show that all Markoff triples are of the form given in Theorem 21.3.1. Hint: use Theorem 3.1.1.*

Solution to selected exercises

1.0.5 Let $s = (a_n)_{n\in\mathbb{N}}$, where the period of s is p; thus a_n depends only on the class modulo p of n. Assuming that p is the smallest period, we deduce that $m = a_0 \cdots a_{p-1}$ is a radical word. In particular, m has p distinct conjugates. These are all factors of s. Let $m' = a_i \cdots a_{i+p-1}$; then m' is a factor of s and a conjugate of m. Since m has p distinct conjugates, if i and j are non-congruent modulo p, $a_i \cdots a_{i+p-1}$ and $a_j \cdots a_{j+p-1}$ are distinct. It follows that the mapping $m' \mapsto i \mod p$ is a bijection from the set C of conjugates of m onto $\mathbb{Z}/p\mathbb{Z}$.

Now let w be a word of length $\geq p$, and u and v words of equal length k, such that uw and vw are factors of s. Let $u = a_i \cdots a_{i+k-1}, v = a_j \cdots a_{j+k-1}$. By considering the prefix m' of length p of w, and its first letter, we see from what precedes that $i + k \equiv j + k \mod p$. Hence $u = v$ by periodicity.

2.2.4 Take the notations of the beginning of Section 2.2. Let v denote the commutative image of u. Suppose that $u \neq 1$ and $u \neq w$. Then $A + v$ is a point located strictly under the segment AB, since it is on the path determined by w; it follows that the slope of v is smaller than the slope of this segment. Thus, $B - v$ is strictly above this segment, but we have $w = u'u$ so that, denoting by v' the commutative image of u', $B - v = (A + v' + v) - v = A + v'$, which is therefore located under the segment.

2.5.4 For example, $(u_{10}, v_{10}) = (a^2b, a^2bab)$. The hyperbinary representations of 10, lexicographically sorted, are 0101, 012, 2001, 202, 221; hence the word obtained is $babaa$, equal to the reversal of $v_{10} = a^2bab$.

2.6.2 (i) Suppose that $w = G(u)$. Then $|w|_a = |u|_a + |u|_b$ and $|w|_b = |u|_b$, hence the slope of w is $|w|_b/|w|_a = |u|_b/(|u|_a + |u|_b) \leq 1$.

(ii) We may assume that $w \neq a, b, ab$. By Theorem 2.6.1, $w = G(u)$ or $w = \tilde{D}(u)$, where u is a lower Christoffel word having both letters a and b. In the first case, the slope of w is $|w|_b/|w|_a = |u|_b/(|u|_a + |u|_b) < 1$. In the other case, the slope is > 1.

2.6.3 Suppose that $w = G(u)$ is a lower Christoffel word. Then the slope of w is ≤ 1. By Exercise 2.6.2, $w = G(u')$ for some lower Christoffel word u'. Since G is injective, $u = u'$. The case $f = \tilde{D}$ is similar.

2.6.4 If the algorithm outputs a or b, then w is a lower Christoffel word by Corollary 2.6.2. If w is a lower Christoffel word, then by Theorem 2.6.1, $w = f(ab)$, where f is a product of endomorphisms in $\{G, \tilde{D}\}$; hence $w = fG(b)$ and $w = f\tilde{D}(a)$, which implies that the algorithm applied to w outputs a or b.

2.6.8 If $u = a$, take $(u, v) = (a, b)$; otherwise, since $u \neq b$, u is a proper lower Christoffel word, with standard factorization xy, so that (x, y) is a standard pair; then $(u, v) = (xy, y)$ is a standard pair. Now, we have the composition of endomorphisms $(u, v) \circ (a, a^n b) = (u, u^n v)$, so that $u^n v$ is a lower Christoffel word.

From Christoffel Words to Markoff Numbers. Second Edition. Christophe Reutenauer, Oxford University Press.
 DOI: 10.1093/9780197907535.003.0022

4.1.1 Let $q = xp\tilde{x}$ be a palindrome with a central factor p. We have $G(q)a = G(x)G(p)G(\tilde{x})a = G(x)(G(p)a)a^{-1}G(\tilde{x})a$. Note that $G(\tilde{x})$ begins with a if x is nonempty. In any case, $a^{-1}G(\tilde{x})a \in \{a,b\}^*$ and has the same length as $G(x)$. Hence $G(p)a$ is a central factor of $G(q)a$.

4.1.3 One has $f(K,-1) = 2^2 + K + 2 - 5 = K + 1 \geq K$ and, for $N \geq 0$, $K+2f(N,N-1)+1 = K+2(2^{N+2}+N-2N+2-5)+1 = K+2^{N+3}-2N-6+1 = 2^{N+3}+K-2N-5 = f(K,N)$.

6.1.6 We have $\alpha = \frac{-b+\epsilon\sqrt{\Delta}}{2a}$ and $F'(x) = 2ax + b$, where $\epsilon = \pm 1$. Thus $F'(\alpha) = -b + \epsilon\sqrt{\Delta} + b$, which implies the result.

7.2.2 To deduce (7.7) from (7.6), one uses the equation $\alpha = \frac{a+\sqrt{a^2+4}}{2}$, which is a consequence of $\alpha = [a, a, a, \ldots] = a + 1/\alpha$.

7.2.3 It follows from (7.7) that $u_t = 2p_t - aq_t$; thus $u_t = 2q_{t+1} - q_t$. In particular, u_t is an integer. The recursion is thus the same as for q_t: $q_{t+2} = aq_{t+1} + q_t$. One verifies that $u_{-1} = 2, u_0 = a$. The same initial conditions and recursion are satisfied by the sequence $\mathrm{tr}(P(a)^t)$, since the characteristic polynomial of $P(a)$ is $x^2 - ax - 1$.

7.3.1 If one has equality, then $\xi - p/q = \pm 1/\mathbf{L}(\xi)q^2$. This can happen for at most one q, since ξ is irrational. Indeed, if it happens for two values of q, we have $q^2\xi - pq = \pm(q'^2\xi - p'q')$ and $q \neq q'$, and ξ would be rational. Hence there are only finitely many rational numbers p/q such that one has equality.

9.1.2 Observe that if x and y are distinct letters, and if a finite word u has a block x^k, then the word yuy has this block too. Since w is of odd length, it has at least a block of odd length. It is a block of 1s or of 2s, and it also is a block of $\tilde{w}$. By the observation, s_1 has this block too.

12.4.2 The subgroup $SL_2(\mathbb{Z})$ of $GL_2(\mathbb{Z})$ is of index 2. Hence each $GL_2(\mathbb{Z})$-class is either an $SL_2(\mathbb{Z})$-class, or the union of two distinct $SL_2(\mathbb{Z})$-classes. Consider the matrices

$$A = \begin{pmatrix} 0 & -1 \\ 1 & 0 \end{pmatrix}, \quad B = \begin{pmatrix} 0 & 1 \\ 1 & 0 \end{pmatrix}, \quad C = \begin{pmatrix} -1 & 0 \\ 0 & 1 \end{pmatrix}$$

Then $A \in SL_2(\mathbb{Z})$, $B, C \in GL_2(\mathbb{Z}) \setminus SL_2(\mathbb{Z})$. Thus $f \cdot A = f(-y, x)$ is $SL_2(\mathbb{Z})$-equivalent to f, while $f \cdot B = f(y, x)$ and $f \cdot C = f(-x, y)$ are $GL_2(\mathbb{Z})$-equivalent to f. The $SL_2(\mathbb{Z})$-class of f coincides with its $GL_2(\mathbb{Z})$-class if and only if one of the two latter forms is $SL_2(\mathbb{Z})$-equivalent to f, since the index is 2.

12.4.4 We use a variant of Gauss cycles of binary quadratic forms, for which we follow [Re6], where the reader may find the results not proven here. Let $f(x,y) = ax^2 + bxy + cy^2$. Write $f \xrightarrow{r} g$ if $a + b + c < 0$ and $g(x,y) = f(x+y, y)$, that is, $g = f \cdot R$. Write $f \xrightarrow{l} g$ if $a + b + c > 0$ and $g(x,y) = f(x, x+y)$, that is, $g = f \cdot L$. Write $f \to g$ if $f \xrightarrow{r} g$ or $f \xrightarrow{l} g$.

All the following quadratic forms are binary and indefinite. The form f above is denoted $[a,b,c]$. With this notation, one has $[a,b,c] \xrightarrow{r} [a, 2a+b, a+b+c]$ if $a+b+c < 0$ and $[a,b,c] \xrightarrow{l} [a+b+c, b+2c, c]$ if $a+b+c > 0$. Call *good* a quadratic form $ax^2+bxy+cy^2$ such that $a > 0$ and $c < 0$. Suppose that the form f is good and has integer coefficients. Then, by iterating the operation $\to$, starting from f, one obtains a *cycle of forms*: $f = f_1 \to f_2 \to \cdots \to f_n = f$. Two integral good forms are $SL_2(\mathbb{Z})$-equivalent if and only if they are on the same cycle.

Let $f = 5x^2 - 9xy - 7y^2$, which is good. The cycle containing f is $f = [5,-9,-7] \xrightarrow{r} [5,1,-11] \xrightarrow{r} [5,11,-5] \xrightarrow{l} [11,1,-5] \xrightarrow{l} [7,-9,-5] \xrightarrow{r} [7,5,-7] \xrightarrow{l} f$. This cycle does not contain $f(-x,y) = [5,9,-7]$; hence this latter form, which is good, is not $SL_2(\mathbb{Z})$-equivalent to f. One concludes using Exercise 12.4.2.

12.4.5 An example: let $w = aab$, $\tilde{w} = baa$, $\chi(\tilde{w}) = 221111$; thus $u = r^2l^2rlrl$. The form is $f_{aab}(x,y) = 13x^2 - 23xy - 19y^2$. The cycle is

$$[13,-23,-19] \xrightarrow{r} [13,3,-29] \xrightarrow{r} [13,29,-13] \xrightarrow{l} [29,3,-13] \xrightarrow{l} [19,-23,-13]$$

$$\xrightarrow{r} [19,15,-17] \xrightarrow{l} [17,-19,-17] \xrightarrow{r} [17,15,-19] \xrightarrow{l} [13,-23,-19]$$

14.1.1 The mapping $m \mapsto amb$ is a bijection from the set of central words onto the set of proper lower Christoffel words. One concludes using Exercise 2.2.1.

14.1.2 We show that $\mathrm{Pal}(ba^n) = b(ab)^n$ by induction on n. For $n = 0$, it holds by definition of Pal. Suppose it is true for n. We have, by definition of Pal, $\mathrm{Pal}(ba^{n+1}) = (\mathrm{Pal}(ba^n)a)^{(+)} = (b(ab)^n a)^{(+)}$. The longest palindromic suffix of $b(ab)^n a$ is $(ab)^n a$ so that $(b(ab)^n a)^{(+)} = b(ab)^n ab$ by Proposition 14.1.1, as was to be shown.

14.1.3 The word aub is a lower Christoffel word so that $G(aub) = aG(u)ab$ and $\tilde{D}(aub) = ab\tilde{D}(u)b$ too.

14.1.4 By Proposition 14.1.1 and its notations, $(w)^{(+)} = ps\tilde{p}$; hence its length is at most $|w| + |p| \leq 2|w| - 1$, since $|p| \leq |w| - 1$ because the last letter of w is a palindromic word so that s is of length at least 1. It follows that, since $\mathrm{Pal}(wa) = (\mathrm{Pal}(w)a)^+$, the length of $\mathrm{Pal}(wa)$ is at most $2|\mathrm{Pal}(w)a| - 1 = 2|\mathrm{Pal}(w)| + 1$.

14.1.5 The anti-automorphism $\mathcal{R}$ is the reversal mapping $w \mapsto \tilde{w}$. Hence $\mathcal{R} \circ \mathrm{Pal} = \mathrm{Pal}$. Now, for any letter a, R_a and $\mathcal{R} \circ L_a \circ \mathcal{R}$ are both endomorphisms of A^*; the first (resp. the second) sends a onto a and any other letter b onto ba (resp. onto $\mathcal{R} \circ L_a(b) = \mathcal{R}(ab) = ba$); hence they are equal, and we obtain the formula $R_u = \mathcal{R} \circ L_u \circ \mathcal{R}$ when the word u is a letter. For general u, the formula follows since the right-hand side is conjugation by $\mathcal{R}$, the latter being an involution.

We prove the formula $\mathrm{Pal}(uv) = L_u(\mathrm{Pal}(v))\mathrm{Pal}(u)$ by induction on the length of u. If u is the empty word, it is true. Let $u = aw$ for some letter a; then, by using the induction hypothesis for $\mathrm{Pal}(wv)$, and twice Theorem 14.1.3 (noting that $L_a = G$), $\mathrm{Pal}(uv) = \mathrm{Pal}(awv) = G(\mathrm{Pal}(wv))a = G(L_w(\mathrm{Pal}(v))\mathrm{Pal}(w))a = L_a \circ L_w(\mathrm{Pal}(v))G(\mathrm{Pal}(w))a = L_{aw}(\mathrm{Pal}(v))\mathrm{Pal}(aw) = L_u(\mathrm{Pal}(v))\mathrm{Pal}(u)$.

Now, we have $\mathrm{Pal}(uv) = \mathcal{R}(\mathrm{Pal}(uv)) = \mathcal{R}(L_u(\mathrm{Pal}(v))\mathrm{Pal}(u)) = \mathcal{R}(\mathrm{Pal}(u))\mathcal{R} \circ L_u \circ \mathcal{R} \circ \mathrm{Pal}(v) = \mathrm{Pal}(u)R_u(\mathrm{Pal}(v))$.

The formula $\mathrm{Pal}(w)ab = L_w(ab)$ is true for $w = 1$. Suppose that $w = au$. Then, by Theorem 14.1.3 and induction, $\mathrm{Pal}(au)ab = L_a(\mathrm{Pal}(u))aab = L_a(\mathrm{Pal}(u)ab) = L_a(L_u(ab)) = L_w(ab)$. Similarly, if $w = bu$, $\mathrm{Pal}(bu)ab = L_b(\mathrm{Pal}(u))bab = L_b(\mathrm{Pal}(u)ab) = L_b(L_u(ab)) = L_w(ab)$. Thus the formula is true for any word w.

From this formula we deduce $ba\mathrm{Pal}(w) = \mathcal{R}(\mathrm{Pal}(w)ab) = \mathcal{R}(L_w(ab)) = \mathcal{R} \circ L_w \circ \mathcal{R}(ba) = R_w(ba)$.

14.2.1 If u, v are both proper, then by Theorem 14.2.1, $u = amb, v = anb, u' = bmb, v' = ana$. Hence $w = v'u' = anabmb$, which implies that the longest common prefix of w and v is an.

14.3.2 If w is a central word, it is a palindrome; if w is not the power of some letter, then Exercise 14.1.7 implies directly that $w \in PbaP$. Conversely, if w is a palindrome and of the form $pbaq$ for some palindromes p and q, then $bwa = bpbaqa$ is a product of two palindromes; hence $a\tilde{w}b$ is the product of two palindromes, and $\tilde{w}$ is central by Theorem 14.3.10, thus $w = \tilde{w}$ too.

Let w be a central word, $w = pbaq = qabp$, w, p, and q palindromes. Then the lower Christoffel word awb is equal to the product $aqabpb$ of two palindromes; since the factorization into palindromes is unique by Lemma 4.2.2, we deduce the uniqueness of p and q.

14.3.3 The words $amb = (apb)(aqb)$, apb, aqb are Christoffel words so that for the standard factorisation $w = uv$, we have $u = apb, v = aqb$. It follows from Lemma 14.3.8 that $|u|$ and $|v|$ are periods of m. By Corollary 2.4.3, u is a proper prefix of v or v is a proper suffix of u. In the first case, u is a factor of m, and it is therefore a periodic pattern of m. Since u is unbordered by Exercise 2.2.4, it has no non-trivial period; hence $|u|$ is the smallest period of m, and we have $|u| = \min\{|u|, |v|\} = 2+\min\{|p|, |q|\}$. The other case is similar.

14.5.1 Let y' and x' be the palindromic factorization of u. By Exercise 14.2.4, (ama, x') is the left child of (y', x') in the tree of palindromic factorizations. If v is in the left subtree under u in the tree of Christoffel words, then (ama, x') is an ancestor of (v_1, v_2) in the tree of palindromic factorizations, where v_1v_2 is the palindromic factorization of v; by the same exercise, ama is a prefix of v_1, hence ama is a prefix of v.

14.5.2 (i) Suppose that w is a left child of x. Then $w = uv$ (standard factorization), $v = x$. Let $v = v'b$. Then v' is a prefix of w by Exercise 2.5.3. Then $ww = uvv' \cdots$. Since each circular factor of $x = v$ is a factor of vv' and is of length $\leq |v|$, it is a factor of ww of length $\leq |w|$, hence a circular factor of w.

(ii) Let $y = y'b$. Then by Exercise 2.5.3, y' is a prefix of xy. We have $xy^nxy^n = xy^ny' \cdots$, hence each conjugate of y^n is a circular factor of xy^n. Hence each circular factor of y^n is a circular factor of xy^n. It follows by (i) that each circular factor of y^n is a circular factor of w.

14.5.3 Suppose that v is a left child of u. Comparing with the tree of Christoffel pairs, one has the standard factorizations $u = xy, v = x(xy)$. Since $H = G$ or $\tilde{D}$ preserve lower Christoffel words, one has the standard factorizations $H(u) = H(x)H(y)$ and $H(v) = H(x)(H(x)H(y))$. Hence $H(v)$ is the left child of $H(u)$.

14.5.4 Suppose that x is an ancestor of y in the tree of central words; then axb is an ancestor of ayb in the tree of Christoffel words; by Exercise 14.5.3, $G(axb) = aG(x)ab = aG'(x)b$ is an ancestor of $G(ayb) = aG'(y)b$ in this tree; thus $G'(x)$ is an ancestor of $G'(y)$ in the tree of central words.

Suppose now that g is a palindromic factor of the central word w. Except for trivial cases, we have $w = G'(v)$ or $w = D'(v)$ for some central word v, shorter than w, and we consider the first of these two cases. By Exercise 4.1.2 (ii), there is some palindromic factor f of v such that $g = G'(f)$ or $g = a^{-1}G'(f)a^{-1}$. By induction, there exists an ancestor u of v in the tree of central words such that f is a central palindromic factor of u. Then, by Exercise 4.1.1, $G'(f)$ is a central palindromic factor of $G'(u)$; hence $a^{-1}G'(f)a^{-1}$ is a central palindromic factor of $G'(u)$ too; moreover, $G'(u)$ is an ancestor of $G'(v) = w$ in the tree of central words, by what we have proved above.

14.5.5 If w is a leaf of T, then it is not an internal node so that $|w| \geq n+1$; all the ancestors of w have length $\leq n$, hence so have also the two words of the standard factorization of w, by Theorems 2.4.1 and 14.5.1. Conversely, suppose that the lower Christoffel word w of length $\geq n+1$ has the standard factorization $w = uv$, with $|u|, |v| \leq n$; then by Theorem 14.5.1, u or v is the parent of w in the tree of Christoffel trees; hence the parent of w has length $\leq n$, and is therefore in T; since w is not an internal node of T, it is a leaf of T.

14.5.6 Let $(u,v) = \mathcal{N}(m)$ (a node in the tree of Christoffel pairs). By Proposition 14.5.2, $|u|+|v|$ is equal to the sum of the entries of the matrix $\nu(w)$. By Theorem 2.6.1, $\mathcal{N}(m) = \rho(m)$; thus $|u|+|v| = |\rho(m)(ab)|$. By Lemma 14.1.5, $\rho(m)(ab) = a\mathrm{Pal}(m)b$.

14.5.7 (i) This is because $(2,1)$ is the first row of $\mu(a)$ and ${}^t(2,1)$ is the second column of $\mu(b)$.

(ii) Both sides of the equality $\mu(x) = P^{-1}(\omega G(x))P$ are homomorphisms defined on the free monoid. Therefore it is enough to prove it for $x = a$ and $x = b$.

(iii) Likewise, it is enough to prove the homomorphisms equality $\nu\theta = \omega$ on the generators a, b. The equality $S\nu\theta\mathrm{Pal}(av) = S\omega\mathrm{Pal}(av)$ follows. One has $(1,1)\omega G\mathrm{Pal}(v)\,{}^t(3,2) = (1,1)\omega G\mathrm{Pal}(v)\omega(a)\,{}^t(1,1) = S\omega\mathrm{Pal}(av)$, since by Theorem 14.1.3, $G(\mathrm{Pal}(v))a = \mathrm{Pal}(av)$.

(iv) This follows from (ii) and from $(1,1)P = (2,1)$ and $P^{-1}\,{}^t(3,2) = {}^t(2,1)$.

(v) The length of $a(\mathrm{Pal}\,\theta\,\mathrm{Pal}(av))b$ is, by Exercise 14.5.6, equal to $S\nu\theta\mathrm{Pal}(av) = (1,1)\omega G\mathrm{Pal}(v)\,{}^t(3,2) = (2,1)\mu(\mathrm{Pal}(v))\,{}^t(2,1)$, by 3. and 4, and the latter number is equal to $\mu(w)_{12}$ by 1.

(vi) By what precedes, and Section 3.3, the set of proper Markoff numbers is equal to the set of numbers of the form $S\omega(\mathrm{Pal}(av))$, $v \in \{a,b\}^*$. Hence also to the set of $S\omega(\mathrm{Pal}(bv))$, $v \in \{a,b\}^*$ (exchanging a an b); noting that $S\omega(\mathrm{Pal}(1)) = 2$, we obtain that the set of Markoff numbers $\neq 1$ is equal to the set of $S\omega(u)$ for all central words u. Since $\omega Pal = \nu\theta Pal = \nu\mathrm{Antipal}$, the last assertion follows, because the central words are the words $\mathrm{Pal}(v)$, $v \in \{a,b\}^*$.

15.2.1 By Theorem 15.2.1 and its notations, $A(w_i) = A(w_{i-1})+1$, since $w_{i-1} = uabv$, $w_i = ubav$, so that the area under w_i is one more than the area under w_{i-1}. The interval and lexicographic properties follow, and also the fact that the smallest value is attained for w; by Exercise 2.3.1 this value is $\frac{1}{2}(|w|_a - 1)(|w|_b - 1)$.

15.2.2 For each factorization $w = ubv$, the indicated occurrence of b appears in all prefixes of w which end in a letter of bv; there are $|bv|$ such prefixes, hence $\sum_{w=ps}|p|_b$ is equal to $\sum_{w=ubv}|bv| = M(w)$. The equality $M(w) = \binom{w}{b} + \binom{w}{ba} + \binom{w}{bb}$ is seen directly from the definition of $M(w)$. Since $A(w) = \binom{w}{ba}$ by Exercise 2.3.2, one deduces $M(w) = A(w) + \binom{w}{b} + \binom{w}{bb}$. The interval property then follows from Exercise 15.2.1, since the second and the third term of the latter sum are invariant after conjugating w. By this exercise, the smallest value is $\binom{w}{b}+\binom{w}{bb}+\frac{1}{2}(|w|_a-1)(|w|_b-1) = 1+|w|_b-1+\frac{1}{2}|w|_b(|w|_b-1) + \frac{1}{2}(|w|_a-1)(|w|_b-1) = 1+\frac{1}{2}(|w|_b-1)(2+|w|_b+|w|_a-1) = \frac{1}{2}(|w|+1)(|w|_b-1)+1$.

15.2.3 (i) The word w has n circular factors of length n, being radical; it has $k+1$ circular factors of length $k = 0, 1, \ldots, n-1$. Thus $N = n + \sum_{0\leq k\leq n} k+1 = n+\frac{1}{2}n(n+1) = \frac{1}{2}n(n+3)$.

(ii) The word w has 1 circular factor of length 0; since it is the ℓ-th power of some word of length n/ℓ, for the lengths $k = 1, \ldots, n$, it has at most n/ℓ circular factors of length k. Thus $N \leq 1 + n^2/\ell$. The inequality $1 + n^2/\ell < n(n+3)/2$ is equivalent to $2\ell + 2n^2 < \ell n^2 + 3n\ell$, which is true, since $2 \leq \ell$ and $1 \leq n$.

15.2.4 By Exercise 14.5.2, first part, the set of circular factors of x of length k is contained in the set of circular factors of w of length k. By Lemma 15.2.2, both sets have the same cardinality $k+1$; hence they are equal.

15.2.6 The set $\{|u|_b, u \in L_k\}$ is an interval of $\mathbb{N}$; this is because, for two consecutive circular factors u and v of the same length of w, one has $|v|_b = |u|_b + \epsilon$, $\epsilon = -1, 0$ or 1. Moreover, for words of a given length in $\{a,b\}^*$, their commutative image is completely determined by their number of b's. From these two observations, the equivalence of the three conditions follows.

15.2.7 (i) Let u and v be two factors of equal length of ${}^{\infty}w^{\infty}$. Then they are of the form $w_1^k x, w_2^k y$, where w_1 and w_1 are conjugates of w and x and y are circular factors of w of equal length. Thus $||u|_a - |v|_a| = ||x|_a - |y|_a| \leq 1$.

(ii) Let $s = \tilde{u}xyv$, with $\{x,y\} = \{a,b\}$. Since s is periodic, we may consider the periodic pattern of s which ends at x and begins at y; thus $\tilde{u}x$ and yv may be written $\tilde{u}x = {}^{\infty}(ymx)ymx, ymx(ymx)^{\infty} = yv$, where m is a finite word. Let p be the longest prefix of m such that $\tilde{p}$ is a suffix of m. Then $|p| \leq N =$ period of s minus 2. If $p = m$, we have $m = \tilde{p}$, hence $m = \tilde{m}$, $u = myu', v = mxv'$, and we are done. Otherwise, $m = px'f = gy'\tilde{p}$, $x', y' \in A$. We must have $x' \neq y'$ by maximality of p. If we have $x' = y, y' = x$, then ypy and $x\tilde{p}x$ are factors of s, contradicting the balance property of s. Thus $x' = x, y' = y$, and $u = pyu', v = pxv'$, and we are done, too.

(iii) Follows from (i) and (ii), using Corollary 15.2.3.

15.3.2 Consider the circular word (w). If u and v are two consecutive factors of length k of (w), then $u = am, v = mb$ for some letters a and b; it follows that $a = b$, since am and mb have the same commutative image. Thus (w) has the period k; in other words $C^k(w) = w$. Since $k < |w|$, it follows, by Exercise 1.0.2, that $C^d(w) = w$ for some proper divisor d of $|w|$, and w is not radical.

15.3.3 Consider the lower Christoffel word of slope 7/3, hence of length 10, and let $m = 4$ and $C = (1,1,2)$.

15.3.5 By Lemma 15.3.4 and Theorem 15.3.3, $N > 1 + 2 + \cdots + n + n = n(n+3)/2$. The remaining assertions follow from Exercise 15.2.3.

15.3.6 Let $w = amb$ be a lower Christoffel word of length $N > n$, and let F be the set of its circular factors of length n. With the notations of Lemma 15.2.2, where we replace n by N and k by n, $F = \{p_0 \leq p_1 \leq \ldots \leq p_{N-1}\}$. Since $w_0 = amb, w_{N-1} = bma$, one has $p_0 = a\ell(F), p_{N-1} = b\ell(F)$, where $\ell(F)$ is the prefix of length $n-1$ of m. Since $\ell(F)$ is left special with respect to the set of circular factors of w, it is unique by Corollary 15.3.5.

15.3.8 We may assume that w is of length at least 2, and therefore contains both letters. Let x and y be two conjugates of w, consecutive in the lexicographic order. If they begin with the same letter, then $C(x) < C(y)$; hence x is not a descent of C. Thus the only possible descent is when x begins with a and y with b. Then x is the largest conjugate beginning with a. We cannot have $x = aax'$, since then $x <_{lex} ax'a$, and x and y are not consecutive. Similarly, we cannot have $y = bby'$. Thus $x = ab\cdots$, $y = ba\cdots$, $C(x)$ begins with b and $C(y)$ begins with a; therefore $C(x) > C(y)$ and x is a descent of C.

15.3.9 Suppose that (ii) holds. Let q be the inverse of m modulo n. Then $C(w_i) = w_{i+q}$, where the indices are modulo n; indeed, since C^n is the identity on words of length n, $w_{i+q} = C^{(i+q)m}(w_0) = C^{im+1}(w_0) = C(w_i)$. We may identify the set L of conjugates of w with $\{0, 1, \ldots, n-1\} = \mathbb{Z}/n\mathbb{Z}$ through the bijection $i \mapsto w_i$; in this way C becomes a permutation σ which is the translation $x \mapsto x + q$ modulo n. Moreover, σ, as a word, is $\sigma(0)\sigma(1)\cdots\sigma(n-1)$, equal to $q(q+1)\cdots(n-1)01\cdots(q-1)$; hence its unique descent is $p-1$, where $p+q=n$. Thus, by Exercise 15.3.8, $|w|_a = p, |w|_b = q$. By Exercise 1.0.4, w_0 is obtained from the cycle form of $(0, \ldots)$ of σ by replacing each i by a if $i \in \{0, 1, \ldots, p-1\}$ and by b if $i \in \{p, p+1, \ldots, n-1\}$. By Exercise 15.1.2, it then follows that w_0 is a lower Christoffel word.

15.3.10 It follows from Corollary 15.3.6 that a Christoffel word w of length n has $\{n, 1\}$ as a circular attractor. By Theorem 14.2.1, there are two Christoffel words in the conjugation class of w, namely $w = v'u'$ and $u'v'$; we thus obtain that $\{|v'|, |v'|+1\}$ is also a circular attractor for w.

Conversely, if w has an attractor of the form $\{k, k+1\}$ or $\{n, 1\}$, then the number of circular factors of a given length $\ell < n$ is at most $\ell+1$, since there are $\ell+1$ occurrences of circular factors intersecting the attractor. By Lemma 15.3.4, there must be at least $\ell+1$ circular factors of length ℓ of w, since the latter is radical. Hence there are exactly $\ell+1$, and w is conjugate to a Christoffel word by Theorem 15.3.3.

15.4.1 Consider the associated circular word, as in the proof of Lemma 14.2.2. Since there are two palindromic conjugates, there are two axes of symmetry. Each of them passes through a unique letter of the word, since its length is odd. Hence the two axes are distinct. It follows, as in the mentioned proof, that the word is not radical.

15.5.1 Let $u \in \{a, b\}^*$ and $v = G'(u) = G(u)a$. By Exercise 15.4.3, u is full if and only if for each letter x, $\mathrm{pal}_x(u) = |u|_x$. In this case, by Exercise 4.1.2, $\mathrm{pal}_a(v) = \mathrm{pal}_a(u) + \mathrm{pal}_b(u) + 1$ and $\mathrm{pal}_b(v) = \mathrm{pal}_b(u)$. Thus $\mathrm{pal}_a(v) = |u|_a + |u|_b + 1 = |v|_a$ and $\mathrm{pal}_b(v) = |u|_b = |v|_b$. Therefore v is full. Since $G(u)$ is a prefix of v, $G(u)$ is full by Exercise 15.4.3.

17.2.1 One has $\ell_1 < u \leq h_{j+1} \leq h_1 \leq \ell_1$ (the first inequality by Theorem 17.2.1, and the last since h_1 is a prefix of ℓ_1).

17.2.2 If s, p are non-empty, then they are different from x and w, since $|x| \leq |w|$. Then, by Theorem 17.2.1, $s < x < p < w < s$, a contradiction.

17.4.1 Assume the contrary. Then x has the non-trivial factorization $x = m_1m_2$, where m_1 is a suffix of u and m_2 a prefix of v; we have $m_1 \neq u$ and $m_2 \neq v$, by Corollary 17.4.4: indeed, if $m_1 = u$, then x is a Lyndon word prefix of w longer than u; if $m_2 = v$, x is a Lyndon word suffix of w longer than v. By Lemma 14.3.5, the word w' obtained from w by suppressing its last letter has the periodic pattern u. Now, x is a factor of w', hence x has the period $|u|$. Since it is a Lyndon word, it has no non-trivial period, hence $|x| \leq |u|$. Thus $|m_2| < |u|$. Since $w = uv$, w' is a prefix of u^∞, and since m_2 is a proper prefix of v, m_2 is a proper prefix of u. Thus $m_2 <_{lex} u \leq_{lex} m_1 <_{lex} x$ (the second inequality because u is a Lyndon word with non-trivial suffix m_1). Hence $m_2 <_{lex} x$, a contradiction, since m_2 is a non-trivial suffix of the Lyndon word x.

17.4.2 (ii) Arguing as in Exercise 17.2.2, we see that we cannot have $x = sp$, for s (resp. p) a non-trivial proper suffix (resp. prefix) of w; it follows that, if x^n is a circular factor of w, but not a factor, then x is a proper prefix and a proper suffix of w, contradicting the fact that w is a Lyndon word, and thus is unbordered.

Thus x^n is a factor of w. Arguing as in Exercise 17.4.1, we see that we have, for some i, j, $n = i+j$, x^i is a suffix of f and x^j is a prefix of g, where fg is the standard factorization of w. Assume $i, j \geq 1$. If x is a proper prefix of g, and a proper suffix of f, then by Exercise 2.5.3, x is a prefix and a suffix of w, contradicting the fact that w is unbordered. Hence either $f = x$ or $g = x$. We consider only the case $g = x$, the other being similar. Then g is a suffix of f and, by Corollary 2.4.3, we have $f = f'g$ (standard factorization), with g^{n-1} a suffix of f. Applying this corollary iteratively, we see that $f = f''g^{n-1}$, with $f''g$ being a lower Christoffel word in standard factorization. Thus $w = uv^n$, $u = f''$, $v = g$, with uv being a lower Christoffel word in standard factorization.

Assume now that $i = 0$ or $j = 0$. Then x^n is a factor of f or of g, and we argue by induction to conclude.

For the final assertion, one uses Exercise 14.5.2.

18.1.7 By Theorem 18.1.4, m has a factor x which is a Christoffel word, and m has $|x|$ as a least non-zero period. Then each factor of m of length $> |x|$ has a border; since x is unbordered by Exercise 2.2.4, we deduce that $n = |x|$. Since $n = |f|$ is a period of m, m is a factor of f^∞; consequently, each factor of length n of m is a conjugate of f.

18.1.9 Since w is Sturmian, F is a subset of the set of circular factors of some Christoffel word. It follows from Corollary 15.3.5 that F contains at most one right special word (with respect to F) of each length; since each suffix of a right special word is right special, and since there is no right special word of length $\geq R$, there is one right special word for each length $0, 1, \ldots, R-1$, and no one for length $\geq R$. Moreover, a word in F which is not a proper prefix of another word in F must be a suffix of w; hence there is at most one such word of each length; it follows that the words in F of length $0, 1, \ldots, K-1$ have all a right extension in F, and those of length $K, \ldots, |w|$, except one of them, have all a right extension.

It follows from these two observations that if $x < m$, then each word of length x in F has a right extension to a word of length $x+1$ in F, and that there is one which has two such extensions; hence $f(x+1) = f(x)$. Moreover, if $m \leq x < M$, then $f(x+1) = f(x)$ (one has to consider the two cases $m = K$ and $m = R$). And for $x \geq M$, there is, for each length x, exactly one word which has no right extension, and no one which has two; thus $f(x+1) = f(x) - 1$.

One has $f(1) = 1$, hence $f(m) = m+1 = f(M)$; moreover, $f(|w|) = 1$, hence $f(M) = 1 + (|w| - M)$, and finally $m + 1 = 1 + |w| - M$, and $|w| = m + M = K + R$. Next, $|F| = \sum f(x) = 1 + 2 + \cdots + (m+1) + (M-m)(m+1) + m + \cdots + 2 + 1 = m(m+1) + m + 1 + (M-m)(m+1) = (m+1)(M+1) = (K+1)(R+1)$.

18.1.10 Let $w = v'u'$, palindromic factorization. Then $|u'| = |u, |v'| = |v$. Suppose that u, v are proper Christoffel words. Then $u = amb, v = apb$, $w = ambapb = apabmb, u' = bmb, v' = ama$. By the geometric interpretation of the palindromic factorization, u' (resp. v') is unioccurrent in F, the set of factors of w because the corresponding path links the two parallel lines in Figure 8.2. Hence $u' = bmb$ has no right extension in F. But the first shorter suffix of w, namely mb, has another occurrence in $w = a\underline{mb}apb$. Hence $K = |u'|$.

Now, a right special word with respect to F is necessarily right special with respect to the set of circular factors of w, hence a suffix of $mbap$ (Corollary 15.3.5). The suffix bap is not right special with respect to F: indeed $bapa = bv'$ is not in F, since v' is unioccurrent. But ap is right special, since $apa, apb \in F$. Hence $R = |bap| = |v'|$.

18.1.12 (ii) Let $1 \neq w = a^{i_0}ba^{i_1}\cdots a^{i_{n-1}}ba^{i_n}$, for some $n \geq 0$ and exponents $i_0, i_n \geq 0$, $i_1, \ldots, i_{n-1} \geq 1$. Let $w' = a^{i_0-1}ba^{i_1-1}\cdots a^{i_{n-1}-1}ba^{i_n-1}$, where a^{i_0-1} (resp. a^{i_n-1}) is removed if $i_0 = 0$ (resp. $i_n = 0$).

If w' is Sturmian, then w' is a factor of some lower Christoffel word u. By Lemma 18.1.3, we may assume that $u = xw'a^hby$, $G(u) = G(x)G(w')a^{h+1}bG(y) = G(x)a^{i_0}ba^{i_1}\cdots a^{i_{n-1}}ba^{i_n-1}a^{h+1}bG(y) = G(x)wa^hbG(y)$, hence w is a factor of $G(u)$, and w is Sturmian.

If w is Sturmian, then it is a factor of some lower Christoffel word v. Suppose that there is some exponent $i_j \geq 2$. Then aa is a factor of v, hence bb is not a factor, and the slope of v is ≤ 1 (Exercise 2.2.2). It follows from Exercise 2.6.2 that $v = G(u)$ for some lower Christoffel word u. Then v is in the submonoid generated by $a = G(a)$ and $ab = G(b)$, and it follows by inspection that w' is a factor of u, hence is Sturmian. If all the exponents i_j are ≤ 1, then $w' = b^n$ is Sturmian.

18.2.2 (i) Imagine u drawn on a circle. Each letter b on this circle appears in j circular factors of length j of u. Hence the sum $S(u,n,j) = \sum_k |u_k|_b$ is equal to $j|u|_b$.

Let $j = 1, \ldots, n$. Denote by $v_1, \ldots, v_n$ (resp. $w_1, \ldots, w_n$) the n circular factors of length j of v (resp. w), with multiplicities, ordered in increasing value of their b-degree.

(ii) By contradiction, suppose that there is some i and j such that $(*)$ $S(v,i,j) > S(w,i,j)$, and choose i minimum. Then $S(v,i-1,j) \leq S(w,i-1,j)$. Since $S(v,i,j) = S(v,i-1,j)+|v_i|_b$, and similarly for w, we deduce that $|v_i|_b > |w_i|_b$; hence $|v_i|_b \geq |w_i|_b+1$. But the set of circular factors of w is balanced (Corollary 15.2.3), hence for any k, $|w_k|_b \leq |w_i|_b+1$. By definition of the order of the v_k, we have $|v_i|_b \leq |v_k|_b$, for any $k \geq i$; consequently, $|w_k|_b \leq |w_i|_b+1 \leq |v_i|_b \leq |v_k|_b$. Since $S(v,n,j) = S(v,i,j)+\sum_{k>i}|v_k|_b$, and similarly for w, we deduce from $(*)$ and the latter inequalities that $S(w,n,j) < S(v,n,j)$. This contradicts (i).

(iii) By contradiction, suppose that $S(w,i,j) \leq S(v,i,j)$ for any i,j. By (ii), we deduce that: $(**)$ $S(w,i,j) = S(v,i,j)$ for any i,j. The set of circular factors of v is not balanced; otherwise, by Theorem 18.1.4, all conjugates of v are Sturmian, hence v is, by Exercise 18.2.1, conjugate to a power of a Christoffel word, and then to a Christoffel word, since $|v|_a, |v|_b$ are relatively prime because so are $|w|_a = |v|_a, |w|_b = |w|_b$.

Thus we may choose $j = 1, \ldots, n$, and $h < i$ such that $|v_h|_b + 2 \leq |v_i|_b$. Since w is balanced, $|w_k|_b = N$ or $N+1$ for any k. Thus $|v_h|_b < |w_h|_b$ or $|v_i|_b > |w_i|_b$. But $(**)$ implies that $|w_k|_b = |v_k|_b$ for any k, a contradiction.

18.2.3 Assume that F is the set of circular factors of length n of some Christoffel word w of length $> n$. By Theorem 14.3.1, F contains a Christoffel word amb, hence it contains its conjugation class C, by Exercise 17.4.3. Since amb is radical, C has n elements, and by Theorem 15.3.3, F has $n+1$ elements. Thus F is obtained from C by adding one word. By Theorem 8.2.3, the minimum and the maximum of F, for the lexicographical order, are amb and bma. Likewise, the minimum and the maximum of F are $a\ell(F)$ and $b\ell(F)$, by Exercise 15.3.6. It follows that either $amb = a\ell(F)$ and $F = C \cup \{b\ell(F)\}$, or $bma = b\ell(F)$ and $F = C \cup \{a\ell(F)\}$. In the first case, $b\ell(F) = bmb$ and $F = C \cup \{bmb\}$; similarly, in the second case, $F = C \cup \{ama\}$.

By Exercise 17.4.2, amb is an ancestor of w in the tree of Christoffel word. By Exercise 14.5.1, if w is in the left (resp. right) subtree under w. then ama (resp. bmb) is a factor of w, hence is in F; these two cases are mutually exclusive, since F is balanced.

18.5.4 (i) By Theorem 8.2.3, $w = ah^{(+)}b$ is the smallest element of its conjugation class; hence its prefix of length n is the smallest element of the set of prefixes of length n of its conjugates; this set is the set of circular factors of length n of w. Now, h is a prefix of $h^{(+)}$; hence ah is the prefix of length n of w. The other assertions follow.

(ii) By Exercise 15.3.6, $\ell(F)$ is a prefix of m. It then follows from Proposition 14.1.6 that $\ell(F)^+$ is a prefix of m. By Corollary 14.1.7, $\ell(F)^+$ is an ancestor of m in the tree of central words; by the construction of the latter, amb is an ancestor of $a\ell(F)^+b$ in the tree of Christoffel words. By Exercise 15.2.4, the words amb and $a\ell(F)^+b$ have the same circular factors of length n; hence $F = \mathcal{F}(\ell(F))$, and $\mathcal{F}$ is surjective.

(iii) By (i), the smallest element of $\mathcal{F}(h)$ is ah; hence, by Exercise 15.3.6, $ah = a\ell(\mathcal{F}(h))$, $h = \ell(\mathcal{F}(h))$. We deduce by (ii) that $\mathcal{F}$ and ℓ are inverse of each other. By Exercise 18.5.2, the number of left special words of length $n-1$ is $\varphi(1)+\cdots+\varphi(n)$, and thus so is the number of factors sets of length n.

(iv) If p is a prefix of m, of length $n-1$, then by Corollary 15.3.5, p is left special with respect to the set of factors of amb, hence also with respect to the set of Sturmian words; thus $\mathcal{F}(p)$ is a factor set; by (ii), it is contained in the set of circular factors of amb.

Conversely, it F is a factor set contained in the set of circular factors of amb, then by Theorem 15.3.3, it is of length $n < |amb|$. Now, F contains the word ap, where p is the prefix of length n of m. By Corollary 15.3.5, $ap, bp \in F$, and by Exercise 15.3.6, $p = \ell(F)$; hence $F = \mathcal{F}(\ell(F)) = \mathcal{F}(p)$.

18.5.6 If m is a Markoff word, then xmy and $x\tilde{m}y$ are factors of some bi-infinite word ${}^{\infty}w^{\infty}$ for some Christoffel word w and letters x, y with $\{x, y\} = \{a, b\}$, by Theorem 4.2.1. Hence amb and bma are Sturmian by Lemma 18.1.3 and Corollary 18.1.2, and finally m is a central word by Theorem 18.5.1.

Conversely, if m is a central word, we see that m is a Markoff word by considering the bi-infinite word ${}^{\infty}(amb)^{\infty}$, which has the Markoff property by the converse of Theorem 4.2.1; see Exercise 15.2.7.

19.1.2 We have, by Theorem, 19.1.1 $w(\xi) = aw(\eta)$. By Corollary 19.1.4, the semi-convergents of ξ are the nodes in the Stern-Brocot tree coded by the prefixes of $w(\xi)$, and the semi-convergent 1 is coded by the empty word; similarly, the semi-convergents of η are the nodes in the Stern–Brocot tree coded by the prefixes of $w(\eta)$. Thus we have a bijection from the set of semi-convergents of ξ, except 1, onto the set of semi-convergents of η, sending the semi-convergent $\frac{s}{r}$ of ξ coded by ap, p prefix of $w(\eta)$, onto the semi-convergent of η coded by p. Note that $s < r$, since this node is in the left subtree. Hence, by Theorem 19.1.1, $ap = w(\frac{s}{r}) = aw(\frac{s}{r-s})$, and it follows that the latter semi-convergent is $\frac{s}{r-s}$.

20.3.1 We consider only the case where that x and y are both non-empty.

(i) If $g = 1$, then $x, y \in \{a, b\}^*$, and clearly (x, y) and (u, v) are conjugate in $\{a, b\}^*$. If $g \neq 1$, there must be some simplification in the product gug^{-1}, since u is cyclically reduced. By symmetry, we may thus assume that $u = u'c$, $c = a$ or b, $g^{-1} = c^{-1}g'^{-1}$. Then $g = g'c$. Similarly, there is a simplification in the product gvg^{-1}. Since $c, v \in \{a, b\}^*$, this simplification cannot occur between g and v, so it occurs between v and g^{-1}. Therefore, $v = v'c$ and we deduce that $(x, y) = g(u, v)g^{-1} = g'c(u'c, v'c)c^{-1}g'^{-1} = g'(cu', cv')g'^{-1}$ and we conclude by induction on the length of g that $x, y \in \{a, b\}^*$, and that (x, y) and (u, v) are conjugate in $\{a, b\}^*$.

(ii) It follows from the definition of $\to$ that there is a unique maximal chain containing (u, v). If the chain is infinite, then $u^\infty = v^\infty$ so that u and v are powers of the same word. If the latter condition holds, then the chain is clearly infinite.

20.4.1 Justin's formula for letter a is given in Theorem 14.1.3; for letter b, it is $\mathrm{Pal}(bv) = D(\mathrm{Pal}(v))b$. One also uses the identities $G(K) = aKa^{-1}$ and $D(K) = bKb^{-1}$. The formulas for av and bv then easily follows from the formulas for v.

21.2.1 We have $ad = b^2 + 1$; hence $a, d \neq 0$, and they have the same sign. Thus $a > 0 \Leftrightarrow d > 0$. Moreover, if $a + d > 0$, then $a > 0$. Conversely, if $a > 0$, then $d > 0$ and $a + d > 0$.

One has $(a + d)^2 - 4b^2 = a^2 + 2ad + d^2 - 4(ad - 1) = (a - d)^2 + 4 > 0$. Suppose that $a + d > 0$; since $(a + d)^2 > (2b)^2$, we deduce that $a + d > 2b$. Conversely, suppose that $a + d - 2b > 0$; since $(a + d - 2b)(a + d + 2b) > 0$, we have $a + d > -2b$, thus $a + d > |2b| \geq 0$.

21.2.2 Suppose that $a \geq |b|, d \geq |b|$. If both inequalities are actually equalities, then $b^2 + 1 = ad = b^2$, a contradiction. Hence we have, for example, $a \geq |b|, d \geq |b| + 1$; then $b^2 + 1 = ad \geq b^2 + |b| \geq b^2 + 2$, a contradiction.

Bibliography

[A] M. Aigner: *Markov's theorem and 100 years of the uniqueness conjecture*, Springer Verlag, 2013. xii, 29, 37, 41, 59, 80, 91, 92, 98, 219

[AB] P. Alessandri, V. Berthé: Three distance theorem and combinatorics on words, *L'Enseignement Mathématique* **44**, 1998, 103–132. 133

[AC] M.-C. Anisiu, J. Cassaigne: Properties of the complexity function for finite words, *Revue d'analyse numérique et de théorie de l'approximation* **33**, 2004, 123–139. 175

[Ba] P. Bachmann: *Die Arithmetik der quadratischen Formen*, Teubner, Leipzig, Berlin, 1923. 97

[Bf] I. Banfield: Christoffel words and the strong Fox conjecture for two-bridge knots, *arXiv*, 2023. 106

[BDJR] N. Bédaride, E. Domenjoud, D. Jamet, J.-L. Rémy: On the number of balanced words of given length and height over a two-letter alphabet, *Discrete Mathematics and Theoretical Computer Science* **12:3**, 2010, 41–62. 176, 183

[BBH] A. Beineke, T. Brüstle, L. Hille: Cluster-cylic quivers with three vertices and the Markov equation (with an appendix of Kerner O.), *Algebra and representation theory* **14**, 2011, 97–112. 27

[BPR] E.A. Bender, O. Patashnik, H. Rumsey Jr.: Pizza slicing, Phi and the Riemann's Hypothesis, *The American Mathematical Monthly* **101**, 1994, 307–317. 183

[BM] F. Bergeron, and M. Mazin: Combinatorics of triangular partitions, *Enumerative Combinatorics and Applications* **3**, 2023, Article S2R1. 183

[BeRe] F. Bergeron, and C. Reutenauer: Golden ratio and phyllotaxis, *Journal of Mathematical Biology* **78**, 2019, 1–19. 91

[B0] J. Berstel: Mots de Fibonacci, *Séminaire d'Informatique théorique*, Institut de Programmation, Paris, 1980. xiii

[B1] J. Berstel: Fibonacci words, a survey, in: G. Rozenberg and A. Salomaa, editors, *The Book of L*, Springer-Verlag, 1985, 11–25. xiii

[B2] J. Berstel: Tracé de droites, fractions continues et distances discrètes, in: M. Lothaire, *Mots, Mélanges offerts à Schützenberger*, Hermès, Paris, 1990, 298–309. 7, 97

[BDFPRR] J. Berstel, C. De Felice, D. Perrin, C. Reutenauer, G. Rindone: Bifix codes and Sturmian words, *Journal of Algebra* **369**, 2012, 146–202. 138

[BLRS] J. Berstel, A. Lauve, C. Reutenauer, and F. Saliola: *Combinatorics on words: Christoffel words and repetitions in words*, CRM Monograph series, American Mathematical Society, 2008. 12

[BdL] J. Berstel, A de Luca: Sturmian words, Lyndon words and trees, *Theoretical Computer Science* **178**, 1997, 171–203. 7, 16, 119, 149, 171, 176, 194, 195

[BP] J. Berstel, M. Pocchiola: A geometric proof of the enumeration formula for Sturmian words, *International Journal of Algebra and Computation* **3**, 1993, 349–355. 183

[BS] J. Berstel, P. Séébold: Morphismes de Sturm, *Bulletin of the Belgian Mathematical Society - Simon Stevin* **1**, 1996, 175–189. 18, 145

[Be] V. Berthé: Fréquence des facteurs des suites sturmiennes, *Theoretical Computer Science* **165**, 1996, 295–309. 133

[BdLR] V. Berthé, A de Luca, C. Reutenauer: On an involution of Christoffel words and Sturmian morphisms, *European Journal of Combinatorics* **29**, 2008, 535–553. 143, 145, 189, 195

[BeR] V. Berthé, C. Reutenauer: On the three-distance theorem, *The Mathematical Intelligencer* **46**, 2024, 183–188. 125, 137

[BBEHCS] A. Blumer, J. Blumer, A. Ehrenfeucht, D. Haussler, and M.T. Chen, The smallest automaton recognizing the subwords of a text, *Theoretical Computer Science* **40**, 1985, 31–55. 156

[B] E. Bombieri: Continued fractions and the Markoff tree, *Expositiones Mathematicae* **25**, 2007, 187–213. xi, 23, 37, 54, 59, 70, 98

[Bn] R. Bondil: Géométrie des fractions continues, *Revue de la filière mathématique* **124-1**, 2013. 200

[Bl] E. Borel: Contributions à l'analyse arithmétique du continu, *Journal de mathématiques pures et appliquées* **9**, 1903, 329–375. 80

[Bo] J.-P. Borel: Image par homographie de mots de Christoffel, *Bulletin of the Belgian Mathematical Society - Simon Stevin* **8**, 2001, 241–255. 120

[BL] J.-P. Borel, F. Laubie: Quelques mots sur la droite projective réelle, *Journal de Théorie des Nombres de Bordeaux* **5**, 1993, 23–51. 13, 14, 16, 163, 166

[BR] J.-P. Borel, C. Reutenauer: On Christoffel classes, *RAIRO-Informatique Théorique et Appliquée* **40**, 2006, 15–27. 107, 126, 135

[BF] M. Boshernitzan, A.S. Fraenkel: A linear algorithm for nonhomogeneous spectra of numbers, *Journal of Algorithms* **5**, 1984, 187–198. 176

[BBGL1] S. Brlek, A. Blondin-Massé, and A. Garon. S. Labbé: Christoffel and Fibonacci tiles, *Lecture Notes in Computer Science* **5810**, 2009, 67–78. 13

[BBGL2] S. Brlek, A. Blondin-Massé, A. Garon, and S. Labbé: Two infinite families of polyominoes that tile the plane by translation in two distinct ways, Theoretical Computer Science , 2011, 4778–4786. 13

[BHNR] S. Brlek, S. Hamel, M. Nivat, C. Reutenauer: On the palindromic complexity of infinite words, *Pattern Recognition* **42**, 2004, 2239–2246. 145

[BLPR] S. Brlek, J.-O. Lachaud, X. Provençal, C.Reutenauer: Lyndon+Christoffel = digitally convex, *International Journal of Foundations of Computer Science* **15**, 2009, 293–306. 178

[BDLF] M. Bucci, A. De Luca, G. Fici: Enumeration and structure of trapezoidal words, *Theoretical Computer Science* **468**, 2013, 12–22. 175

[BDLZ] M. Bucci, A. De Luca, L.Q. Zamboni: Reversible Christoffel factorizations, *Theoretical Computer Science* **495**, 2013, 17–24. 133

[BV] J. Buchmann, and U. Vollmer, *Binary quadratic forms: An algorithmic approach*, Springer Verlag, 2007. 47

[Bu] D.A. Buell, *Binary quadratic forms: Classical theory and modern computations*, Springer, 1989. 47

[BRe] Y. Bugeaud, C. Reutenauer: On the conjugates of Christoffel words, *arXiv*, 2023. 153

[BRS] Y. Bugeaud, C. Reutenauer, S. Siksek: A Sturmian sequence related to the uniqueness conjecture for Markoff numbers, *Theoretical Computer Science* **410**, 2009, 2864–2869. 29

[BFPRS] E.B. Burger, A. Folsom, A. Pekker, R. Roengpyta, J. Snyder: On a quantitative refinement of the Lagrange spectrum, *Acta Arithmetica* **102**, 2002, 55–82. 70

[BW] M. Burrows, D.J. Wheeler: A block sorting data compression algorithm, *Technical Report, DIGITAL System Research Center*, 1994. 183

[CW] N. Calkin, H.S. Wilf: Recounting the rationals, *American Mathematical Monthly* **107**, 2000, 360–363. 194, 196

[CC] N. Carey, D. Clampitt: Partitioned factors in Christoffel and Sturmian words, *Theoretical Computer Science* **743**, 2018, 53–63. 130, 131

[CdL] A. Carpi, A de Luca: Harmonic and gold Sturmian words, *European Journal of Combinatorics* **25**, 2004, 685–705. 196

[Ca1] J.W.S. Cassels: The Markoff chain, *Annals of Mathematics* **50**, 1949, 676–685. xi, 98

[Ca2] J.W.S. Cassels: *An introduction to Diophantine approximation*, Cambridge University Press, 1957. xi, 94, 98

[CRS] G. Castiglione, A. Restivo, M. Sciortino: Circular Sturmian words and Hopcroft's algorithm, *Theoretical Computer Science* **410**, 2009, 4372–4381. 178

[CLDR1] M. Castrillón López, and M. Domínguez Romero: A meeting-point between mathematics and the theory of musical scales: Well-formed scales, *La Gaceta de la Real Sociedad Matemática Espanõla* **16**, 2013, 87–106. 152

[CLDR2] M. Castrillón López, M. Domínguez Romero: On the step-patterns of generated scales that are not well-formed, in: Mathematics and Computation in Music, *Lecture Notes in Computer Science* **9110**, 2015, 361–372. 152

[C1] E.B. Christoffel: Observatio arithmetica, *Annali di Matematica Pura ed Applicata* **6**, 1875, 145–152. xi, xiii, 97, 147

[C2] E.B. Christoffel: Lehrsätze über arithmetische Eigenschaften der Irrationalzahlen, *Annali di Matematica Pura ed Applicata* **15**, 1887, 253–276. xiii, 97, 189

[Ch1] W.-F. Chuan: α-words and factors of characteristic sequences, *Discrete Mathematics* **177**, 1997, 33–50. 107, 171

[Ch2] W.-F. Chuan: Unbordered factors of the characteristic sequences of irrational numbers, *Theoretical Computer Science* **205**, 1998, 337–344. 176

[Ch3] W.-F. Chuan: Sturmian morphisms and α-words, *Theoretical Computer Science* **225**, 1999, 129–148. 217

[Ch4] W.-F. Chuan: Characterizations of α-words, moments and determinants, *Fibonacci Quarterly* **41**, 2003, 194–208. 130

[Ch5] W.-F. Chuan: Moments of conjugacy classes of binary words, *Theoretical Computer Science* **310**, 273–285 (2004). 130

[Cl] D. Clampitt: Sensitive interval property for scales as words in the free group F_2, in: Mathematics and Computation in Music, *Lecture Notes in Computer Science* **6726**, 2011, 56–68. 152

[CN] D. Clampitt, T. Noll: Modes, the height-width duality, and Handschin's tone character, *Music Theory Online* **17.1**, 2011. 152

[ClS] D. Clampitt, J. Shafer: Greek ethnic modal names vs. *Alia musica's nomenclature*, in: Mathematics and Computation in Music, *Lecture Notes in Computer Science* **9110**, 385–390 (2015). 152

[Cle] D. Clemens: Von der diophantischen Approximation zu Matchings in Schlangengraphen - Eigenschaften und Anwendungen von Markoffzahlen und Cohnmatrizen, *Diplomarbeit* (Master's thesis), Freie Universität Berlin, 2011. 91

[CM1] J. Clough, G. Myerson: Variety and multiplicity in diatonic systems, *Journal of Music Theory* **29**, 1985, 249–270. 131

[CM2] J. Clough, G. Myerson: Musical scales and the circle of fifths, *American Mathematical Monthly* **93**, 1986, 695–701. 131

[Co1] H. Cohn: Approach to Markoff's minimal forms through modular functions, *Annals of Mathematics* **61**, 1955, 1–12. xi, 23, 219

[Co2] H. Cohn: Markoff's forms and primitive words, *Mathematische Annalen* **61**, 1972, 8–22. xi, 14, 16, 23, 209

[Ch] P.M. Cohn: *Free rings and their relations*, Academic Press, 1985. 212

[CRST] S. Corteel, G. Rémond, G. Schaeffer, H. Thomas: The number of plane corner cuts, *Advances in Applied mathematics* **23**, 1999, 49–53. 183

[CH] E.M. Coven, G.A. Hedlund: Sequences with minimal block growth, *Mathematical Systems Theory* **7**, 1973, 138–153. 139

[CrH] M. Crochemore, and C. Hancart, Automata for matching patterns, in: G. Rozenberg, and A. Salomaa, *Handbook of formal languages, 2, Linear modeling: Background and application*, Springer Verlag, 399–462, 1997. 156

[CS] J.D. Currie, K. Saari: Least periods of factors of infinite words, *RAIRO-Theoretical Informatics* **43**, 2009, 165–178. 131

[CF] T.W. Cusick, M.E. Flahive: *The Markoff and Lagrange spectra*, American Mathematical Society, 1989. 89, 94, 98

[D'A] F. D'Alessandro: A combinatorial problem on trapezoidal words, *Theoretical Computer Science* **273**, 2002, 11–33. 175

[DL] D. Damanik, D. Lenz: Powers in Sturmian sequences, *European Journal of Combinatorics* **24**, 2003, 377–390. 194

[D'AdLDL] A. D'Aniello, A. de Luca, A. De Luca: On Christoffel and standard words and their derivatives, *Theoretical Computer Science* **658**, 2017, 122–147. 21, 107

[Da] H. Davenport: *The higher arithmetic: An introduction to the theory of numbers*, Hutchinson University Library, London, 1964. 197

[DLR] A. De Luca, C. Reutenauer: Christoffel words and the Calkin-Wilf tree, *The Electronic Journal of Combinatorics* **18**(2), 2011, P22. 18

[DLEZ] A. De Luca, M. Edson, L.Q. Zamboni: Extremal values of semi-regular continuants and codings of interval exchange transformations, *Mathematika* **69**, 2023, 432–457. 69

[DLPPRT] B. Dequesne, M. Lapointe, Y. Palu, P.-G. Plamondon, C. Reutenauer, H. Thomas: A generalization of perfectly clustering words and band bricks for certain gentle algebras, *Combinatorial Theory* **5**, 2025, nr. 6, 41 pp. 183

[Di] W. Dicks: A commutator test for two elements to generate the free algebra of rank two, *Bulletin of the London Mathematical Society* **14**, 1982, 48–51. 212

[D1] L.E. Dickson: *Studies in the theory of numbers*, Chelsea, New York, 1957 (first edition 1930). xi, 94, 97

[D2] L.E. Dickson: *Introduction to the theory of numbers*, Dover, 1957. 47, 48, 49

[DNC] M. Domínguez Romero, T. Noll, M. Castrillón López: An extension of Christoffel duality to a subset of Sturm numbers and their characteristic words, *Theoretical Computer Science* **412**, 2011, 2942–2954. 152

[DP] X. Droubay, G. Pirillo: Palindromes and Sturmian words, *Theoretical Computer Science* **223**, 1999, 73–85. 139, 140

[DJP] X. Droubay, J. Justin, G. Pirillo: Episturmian words and some constructions of de Luca and Rauzy, *Theoretical Computer Science* **255**, 2001, 539–553. 143, 176

[DGB] S. Dulucq and D. Gouyou-Beauchamps: Sur les facteurs des suites de Sturm, *Theoretical Computer Science* **71**, 1990, 381–400. 171, 172

[Du] J.-P. Duval: Relationship between the period of a finite word and the length of its unbordered segments, *Discrete Mathematics* **40**, 1982, 31–44. 172, 174

[ES] A. Ehrenfeucht, D.M. Silberger: Periodicity and unbordered segments of words, *Discrete Mathematics* **26**, 1979, 101–109. 172

[ERSS] J. Ellis, F. Ruskey, J. Sawada and J. Simpson: Euclidean strings, *Theoretical Computer Science* **301**, 2003, 321–340. 129

[EMSV] C. Epifanio, F. Mignosi, J. Shallit, I. Venturini, On Sturmian graphs, *Discrete Applied Mathematics*, **155**, 2007, 1014–1030. 138

[FZ] S. Ferenczi and L.Q. Zamboni: Clustering words and interval exchanges, *Journal of Integer Sequences* **16**, 2013, 13.2.1, 9 p. 125

[Fi1] G. Fici: Special factors and the combinatorics of suffix and factor automata, *Theoretical Computer Science* **412**, 2011, 3604–3615. 156

[Fi2] G. Fici: On the structure of bispecial Sturmian words, *Journal of Computer and System Sciences* **80**, 2014, 711–719. 188

[FiDL] G. Fici, A. De Luca: On the Lie complexity of Sturmian words, *Theoretical Computer Science* **938**, 2022, 81–85. 194

[FLLLMPP-G] G. Fici, A. Langiu, T. Lecroq, A. Lefebvre, F. Mignosi, J. Peltomäki and É. Gaston-Prieur: Abelian powers and repetitions in Sturmian words, *Theoretical Computer Science* **635**, 2016, 16–34. 54

[FW] N.J. Fine and H.S. Wilf: Uniqueness theorem for periodic functions, *Proceedings of the American mathematical Society* **16**, 1965, 109–114. 111

[Fo] D. Fowler: *The mathematics of Plato's Academy*, Clarendon, 1987. 197

[Fr] R. Fricke: Ueber die Theorie der automorphen Modulgruppen, *Nachrichten von der Königlichen Gesellschaft der Wissenschaften zu Göttingen* 1896, 91–101. 24

[FKS] F. Franek, A. Karaman, W.F. Smyth: Repetitions in Sturmian strings, *Theoretical Computer Science* **249**, 2000, 289–303. 176

[Fd] A. E. Frid, Sturmian numeration systems and decompositions to palindromes, *European Journal of Combinatorics* **71**, 2018, 202–212. 153

[F] G. F. Frobenius: Über die Markoffschen Zahlen, *Sitzungsberichte der Königlich Preussischen Akademie der Wissenschaften zu Berlin* **26**, 1913, 458–487. xi, 29, 91, 92, 93, 94, 97

[GK] J. Gilman, L. Keen: Enumerating palindromes and primitives in rank two free groups, *Journal of Algebra* **332**, 2011, 1–13. 200

[GJWZ] A. Glen, J. Justin, S. Widmer, L. Zamboni: Palindromic richness, *European Journal of Combinatorics* **30**, 2009, 510–531. 142, 145

[GLS] A. Glen, A. Lauve, F. Saliola: A note on the Markoff condition and central words, *Information Processing Letters* **105**, 2008, 241–244. 188

[GKP] R.L. Graham, D. Knuth, and O. Patashnik: *Concrete mathematics*, Addison Wesley, 2nd edition, 1994. 189, 190, 191

[Ha1] J. Hančl: Sharpening of theorems of Vahlen and Hurwitz and approximation properties of the golden ratio, *Archiv der Mathematik* **105**, 129–137, 2015. 59

[Ha2] J. Hančl: Second basic theorem of Hurwitz, Lithuanian Mathematical Journal **56**, 72–76, 2016. 59

[HW] G.H. Hardy, E.M. Wright: *An introduction to the theory of numbers*, Oxford University Press, 5th edition, 1979. 37, 41, 43, 54, 59

[HN1] T. Harju, D. Nowotka: Minimal Duval extensions, *International Journal of Foundations of Computer Science* **15**, 2004, 349–354. 176

[HN2] T. Harju, and D. Nowotka: Periodicity and unbordered words: A proof of Duval's conjecture, *Lecture Notes in Computer Science* **2996**, 2004, 294–304. 172

[dlH] P de la Harpe: *Topics in geometric group theory*, Chicago University Press, 2000. 220

[He] P.J. Heawood: The classification of rational approximations, *Proceedings of the London Mathematical Society* **20**, 1922, 233–250. 98

[HeN] L. Hegedüs and B. Nagy: On periodic properties of circular words, *Discrete Mathematics* **339**, 2016, 1189–1197. 131

[HR] C. Hohlweg, C. Reutenauer, Lyndon words, permutations and trees, *Theoretical Computer Science* **307**, 2003, 173–178. 184

[Ho] S. Holub, A proof of the extended Duval's conjecture, Theoretical Computer Science **339**, 2005, 61–67. 172

[H] M.G. Humbert: Sur les fractions continues ordinaires et les formes quadratiques binaires indéfinies, *Journal de Mathématiques Pures et Appliquées 7ème série* **2**, 1916, 104–154. 200

[H1] A. Hurwitz: Ueber die angenäherte Darstellung der Irrationalzahlen durch rationale Brüche, *Mathematische Annalen* **39**, 1891, 279–284. 53, 80, 97

[H2] A. Hurwitz: Ueber die angenäherte Darstellung der Zahlen durch rationale Brüche, *Mathematische Annalen* **44**, 1894, 417–436. 190, 193

[JZ] O. Jenkinson and L.Q. Zamboni: Characterisations of balanced words via orderings, *Theoretical Computer Science* **310**, 2004, 247–271. 126, 138, 177, 178

[J] J. Justin: Episturmian morphisms and a Galois theorem on continued fractions, *Theoretical Informatics and Applications* **39**, 2005, 207–215. 102, 105, 106, 196

[KR] C. Kassel, C. Reutenauer: Sturmian morphisms, the braid group B_4, Christoffel words and bases of F_2, *Annali di Matematica* **186**, 2007, 317–339. xiii, 209, 216

[KT] C. Kassel, V. Turaev: *Braid groups*, Springer Verlag, 2008. 5

[K1] F. Klein: *Ausgewählte Kapitel der Zahlentheorie I*, Göttingen 1896 (available at Cornell University, digital collection of historical mathematical monographs). 197

[K2] F. Klein: Sur une représentation géométrique du développement en fraction continue ordinaire, *Nouvelles Annales de Mathématiques 3ème série* **15**, 1896, 327–331. 197

[KZ] A. Korkine, G. Zolotareff: Sur les formes quadratiques, *Mathematische Annalen* **6**, 1873, 366–389. 87, 88

[LL] S. Labbé, M. Lapointe: The q-analog of the Markoff injectivity conjecture holds over the language of a balanced sequence, *Combinatorial Theory* **2**, 2022, nr.9, 26pp. 136, 138

[LLS] S. Labbé, M. Lapointe, W. Steiner: A q-analog of the Markoff injectivity conjecture holds, *Algebraic Combinatorics* **6**, 2023, 1677–1685. 27, 29

[Lm] P. Lampe: Diophantine equations via cluster transformations, *Journal of Algebra* **462**, 2016, 320–337. 27

[LR] Y. Lanciault, C. Reutenauer: A symmetry property of Christoffel words, *arXiv*, 2024. 139

[La] M. Lapointe: Study of Christoffel classes: normal form and periodicity, *Lecture Notes in Computer Science* **10432**, 2017, 109–120. 131, 180

[LR1] M. Lapointe, C. Reutenauer: On the Frobenius conjecture, *Integers* **21**, 2021, #A67 29

[LR2] M. Lapointe, C. Reutenauer: On perfectly clustering words, to appear in Electronic Journal of Combinatorics **32** (3), 2025, 3. 8

[Ls] A. Lasnier: Christoffel words and Markoff triples: An algebraic approach, *Communications in Algebra* **41**, 2013, 1195–1216. 27

[Lu] E. Laurier: Opérations sur les mots de Christoffel, *Journal de Théorie des Nombres de Bordeaux* **11**, 1999, 111–132. 120

[Le] D. Lee: Primitivity preserving endomorphisms of free groups, *Communications in Algebra* **30**, 2002, 1921–1947. 217

[LLRS] K. Lee, L. Li, M. Rabideau, and R. Schiffler: On the ordering of the Markov numbers, *Advances in Applied Mathematics* **143**, 2023, 102453, 29 pp. 29

[L] A. M. Legendre: *Essai sur la théorie des nombres*, Duprat, Paris, 1797 (an VI). 42

[LS] F. Levé, P. Séébold: Proof of a conjecture on word complexity, *Bulletin of the Belgium Mathematical Society - Simon Stevin* **8**, 2001, 277–291. 175

[Li] F.M. Liang: A short proof of the *3d* theorem, *Discrete Mathematics* **28**, 1979, 325–326. 133

[Lp] E.P. Lipatov: On a classification of binary collections and properties of homogenetity classes, *Problemy Kibernetiki* **39**, 1982, 67–84. 181

[Lo] M. Lothaire: *Algebraic combinatorics on words*, Cambridge University Press 2002. xiii, 18, 51, 137, 140, 143, 149, 183, 186, 187, 214

[dL0] A de Luca: A combinatorial property of the Fibonacci words, *Information and Processing Letters* **12**, 1981, 193–195. xiii

[dL1] A de Luca: Sturmian words: structure, combinatorics, and their arithmetics, *Theoretical Computer Science* **183**, 1997, 45–82. 101, 102, 110, 147, 185, 186

[dL2] A de Luca: On the combinatorics of finite words, *Theoretical Computer Science* **218**, 1999, 13–39. 175

[dLDL1] A de Luca and A. De Luca: Combinatorial properties of Sturmian palindromes, *International Journal of Foundations of Computer Science* **17**, 2006, 557–573. 174, 185

[dLDL2] A de Luca and A. De Luca: Pseudopalindrome closure operators in free monoids, *Theoretical Computer Science* **362**, 2006, 282–300. 121

[dLDL3] A. de Luca, A. De Luca, Some characterizations of finite Sturmian words, *Theoretical Computer Science* **356**, 2006, 118–125. 172, 176

[dLDL4] A de Luca and A. De Luca: Sturmian words and the Stern sequence, *Theoretical Computer Science* **581**, 2015, 26–44. 18

[dLM] A de Luca and F. Mignosi: Some combinatorial properties of Sturmian words, *Theoretical Computer Science* **136**, 1994, 361–385. 107, 108, 110, 112, 113, 147, 150, 171, 183, 185, 187

[MRS] S. Mantaci, A. Restivo, M. Sciortino: Burrows-Wheeler transform and Sturmian words, *Information Processing Letters* **86**, 2003, 241–246. 126, 171, 183

[MRRRS] S. Mantaci, A. Restivo, G. Romana, G. Rosone, M. Sciortino: A combinatorial view on string attractors, *Theoretical Computer Science* **850**, 2021, 236–248. 139, 152

[M1] A.A. Markoff: Sur les formes quadratiques binaires indéfinies, *Mathematische Annalen* **15**, 1879, 381–496. xi, 35, 77, 78, 84, 88, 93, 94, 97

[M2] A.A. Markoff: Sur les formes quadratiques binaires indéfinies (second mémoire), *Mathematische Annalen* **17**, 1880, 379–399. xi, 25, 35, 77, 78, 91, 97

[Mel] G. Melançon: Visualisation de graphes et Combinatoire des mots, *Mémoire d'habilitation à diriger des recherches*, Université de Bordeaux 1, 1999. 167

[Me2] G. Melançon: Lyndon factorizations of Sturmian words, *Discrete Mathematics* **210**, 2000, 137–149. 176, 178

[MR] G. Melançon and C. Reutenauer: On a class of Lyndon words extending Christoffel words and related to a multidimensional continued fraction algorithm, *Journal of integer sequences* **16**, no. 9, Article 13.9.7, 2013, 30 pp. 168, 170

[Mi] F. Mignosi: On the number of Sturmian factors, *Theoretical Computer Science* **82**, 1991, 71–84. 181, 183

[MS] F. Mignosi, P. Séébold: Morphismes sturmiens et règles de Rauzy, *Journal de Théorie des nombres de Bordeaux* **5**, 1993, 221–233. 143

[MZ] F. Mignosi and L. Q. Zamboni: *A note on a conjecture of Duval and Sturmian words*, Theoretical Informatics and Applications **36**, 1–3 (2002). 172

[Mo] R. Morikawa: Disjoint sequences generated by the bracket function, *Bulletin of the Faculty of Liberal Arts. Nagasaki University. Natural Sciences* **26**, 1985, 1–13. 131

[MH] M. Morse, and G.A. Hedlund: Symbolic dynamics II: Sturmian trajectories, *American Journal of Mathematics* **62**, 1940, 1–42. xi

[N] J. Nielsen: Die Isomorphismen der allgemeinen, unendlichen Gruppe mit zwei Erzeugenden, *Mathematische Annalen* **78**, 1918, 385–397. 202, 203, 208, 212

[No] T. Noll: Ionian theorem, *Journal of Mathematics and Music* **3**, 2009, 137–151. 152

[OZ] R.P. Osborne, H. Zieschang: Primitives in the free group on two generators, *Inventiones Mathematicae* **63**, 1981, 17–24. 14, 209

[PR] I. Pak, A. Redlich: Long cycles in *abc*-permutations, *Functional Analysis and Other Mathematics* **2**, 2008, 87–92. 126

[PaR] G. Paquin, C. Reutenauer: On the superimposition of Christoffel words, *Theoretical Computer Science* **412**, 2011, 402–418. 131

[Pa] B. Parvaix: Propriétés d'invariance des mots sturmiens, *Journal de Théorie des nombres de Bordeaux* **9**, 1999, 351–369. 51

[P] A. Pedersen: Solution of problem E 3156, *American Mathematical Monthly* **95**, 1988, 954–955. 147

[PS] J. Peltomäki, A. Saarela: Standard words and solutions of the word equation $X_1{}^2 \cdots X_n{}^2 = (X_1 \cdots X_n)^2$, *Journal of Combinatorial Theory A* **178**, 2021, 105340, 23 pp. 159, 162

[PW] J. Peltomäki, M.A. Whiteland: A square root map on Sturmian words, *The Electronic Journal of Combinatorics* **24(1)**, 2017, P1.54, 50pp. 156

[PeR] D. Perrin, and A. Restivo: A note on Sturmian words, *Theoretical Computer Science* **429**, 265–272 (2012). 138

[Pe1] S. Perrine: L'interprétation matricielle de la théorie de Markoff classique, *International Journal of Mathematics and Mathematical Sciences* **32**, 2002, 193–262. 224

[Pe2] S. Perrine: *La théorie de Markoff et ses développements*, Tessier et Ashpool, 2002. 224

[P1] O. Perron: Über die Approximationen irrationaler Zahlen durch rationale, *Sitzungsbereich der Heidelberger Akademie der Wissenschaften* **4**, 1921, 2–17. xi, 60, 98

[P2] O. Perron: Über die Approximationen irrationaler Zahlen durch rationale II, *Sitzungsbereich der Heidelberger Akademie der Wissenschaften* **8**, 1921, 2–12. xi, 98

[Pi] G. Pirillo: A new characteristic property of the palindrome prefixes of a standard Sturmian word, *Séminaire Lotharingien de Combinatoire* **43**, 1999, B43f. 129, 147

[Pr] X. Provençal: Minimal non-convex words, *Theoretical Computer Science* **412**, 2011, 3002–3009. 178

[RaS] M. Rabideau, R. Schiffler: Continued fractions and orderings on the Markov numbers, *Advances in Mathematics* **370**, 2020, 107231, 18pp. 29

[Rz] G. Rauzy: Mots infinis en arithmétique, in: Automata on infinite words, M. Nivat and D. Perrin (eds.) *Lecture notes in Computer Science* **192**, 1984, 165–171. 149

[R] R. Remak: Über indefinite binäre quadratische Minimalformen, *Mathematische Annalen* **92**, 1924, 155–182. xi, 97

[Re1] C. Reutenauer: Applications of a noncommutative jacobian matrix, *Journal of Pure and Applied Algebra* **77**, 1992, 169–181. 212

[Re2] C. Reutenauer: On Markoff's property and Sturmian words, *Mathematische Annalen* **336**, 2006, 1–12. 131

[Re3] C. Reutenauer: Mots de Lyndon généralisés, *Séminaire Lotharingien de Combinatoire* **54**, B54h, 2006, 16 p. 24

[Re4] C. Reutenauer: Christoffel words and Markoff triples, *Integers* **9**, 2009, 327–332. 23

[Re5] C. Reutenauer: Studies on finite Sturmian words, *Theoretical Computer Science* **591**, 2015, 106–133. 12, 106, 121, 139, 170, 176, 177, 179, 180, 183, 187, 188

[Re6] C. Reutenauer: On the Stern–Brocot expansion of real numbers, *Journal de Théorie des Nombres de Bordeaux* **31**, 2019, 697–722. 226

[Re7] C. Reutenauer: Christoffel words and weak Markoff theory, *Advances in Applied Mathematics* **127**, 2021, 102179, 15pp. 142, 175

[Re8] C. Reutenauer: On quadratic numbers and forms, and Markoff theory, *Journal of Number Theory* **227**, 2021, 265–305 55, 58, 59, 70, 94

[RV] C. Reutenauer and L. Vuillon: Palindromic closures and Thue-Morse substitution for Markoff numbers, *Uniform Distribution Theory* **12**, 2017, 25–35. 121

[Ri1] G. Richomme: Lyndon morphisms, *Bulletin of the Belgium Mathematical Society - Simon Stevin* **10**, 2003, 761–785. 167

[Ri2] G. Richomme: Conjugacy of morphisms and Lyndon decomposition of standard Sturmian words, *Theoretical Computer Science* **380**, 2007, 393–400. 179

[RS] G. Richomme and P. Séébold: On factorially balanced sets of words, *Theoretical Computer Science* **412**, 2011, 5492–5497. 188

[Ro] R.M. Robinson: Problem E 3156, *American Mathematical Monthly* **93**, 1986, 482. 147

[Sa] K. Saari: Everywhere α-repetitive sequences and Sturmian words, *European Journal of Combinatorics* **31**, 2010, 177–192 156

[Sc1] P. Schmutz Schaller: Markoff theory and hyperbolic torus groups, to appear. 21, 219, 221, 222

[Sc2] P. Schmutz Schaller: The Markoff theory and the commutator subgroup SL $(2, \mathbb{Z})'$, *Russian Mathematics*, **66**, 2022, 91–101. 221, 222, 223, 224

[Sc3] P. Schmutz Schaller: Bases of the free group of rank 2, to appear.

[SZ] M. Sciortino, L.Q. Zamboni, Suffix automata and standard Sturmian words, *Lecture Notes in Computer Science* **4588**, 2007, 382–398. 114, 152, 154, 155

[Sb1] P. Séébold: About a family of binary morphisms whose stationary words are Sturmian, *Lecture Notes in Computer Science* xiii

[Sb2] P. Séébold: On the conjugation of standard morphisms, *Theoretical Computer Science* **195**, 1998, 91–109. 214, 217

[Se1] C. Series: The modular surface and continued fractions, *Journal of the London Mathematical Society* **31**, 1985, 69–80. 200

[Se2] C. Series: The geometry of Markoff numbers, *The Mathematical Intelligencer* **7**, 1985, 20–29. xi, 21, 97, 107

[Si] P. Shiu: A footnote to the three gaps theorem, The American Mathematical Monthly **125**, 2018, 264–266. 133

[Sh1] V. Shpilrain: On generators of L/R^2 Lie algebras, *Proceedings of the American Mathematical Society* **119**, 1992, 1039–1043. 212

[Sh2] V. Shpilrain: Recognizing automorphisms of the free groups, *Archiv der Mathematik* **62**, 1994, 385–392. 212

[Sh3] V. Shpilrain: Generalized primitive elements of a free group, *Archiv der Mathematik* **71**, 1998, 270–278. 217

[Sm] J. Simpson: Disjoint Beatty sequences, *Integers* **4**, 2004, A12 ,10 p. 131

[SmP] J. Simpson, and S.J. Puglisi: Words with simple Burrows-Wheeler transforms, *The Electronic Journal of Combinatorics* **15**, 2008, #R83, 17 p. 158, 125, 126,

[SMDS] R. Siromoney, L. Matthew, V.R. Dare, and K.G. Subramanian: Infinite Lyndon words, *Information and Processing Letters* **50**, 1994, 101–104. 179

[S] H.J.S. Smith: Note on continued fractions, *Messenger of Mathematics* **6**, 1876, 1–14. 97, 147, 151, 197

[So] V.T. Sós: On the distribution mod 1 of the sequence $n\,\alpha$, *Annales Universitatis Scientiarum Budapestinensis de Rolando Eötvös Nominatae. Sectio Mathematica* **1**, 1958, 127–134. 137

[St] M.A. Stern: Ueber eine Zahlentheoretische Funktion, *Journal für die reine und angewandte Mathematik* **55**, 1858, 193–220. 18

[Sw] S. Swierczkowski: On successive settings of an arc on the circumference of a circle, *Fundamenta Mathematicae* **46**, 1959, 187–189. 137

[Su] J. Surányi, Über die Anordnung der Vielfachen einer reellen Zahl mod 1, *Annales Universitatis Scientiarum Budapestinensis de Rolando Eötvös Nominatae. Sectio Mathematica* **1**, 1958, 107–111. 137

[T] G. Toussaint: *The geometry of musical rhythms*, CRC Press, Taylor and Francis, 2013. 152

[U] U.U. Umirbaev: Partial derivations and endomorphisms of some relatively free Lie algebras, *Siberian Mathematical Journal* **34**, 1993, 1161–1170. 212

[V] G. Viennot: Algèbres de Lie libres et monoïdes libres, Lecture Notes in Mathematics, **691**, Springer Verlag, 1978. 170

[W] C.M. Weinbaum: Partitioning a primitive word into a generating set, *Mathematische Annalen* **181**, 1969, 157–162. 206

[WW] Z.X. Wen and Z.Y. Wen: Local isomorphisms of invertible substitutions, *Compte Rendu de l'Académie des Sciences, Paris* **318**, Série I, 1994, 299–304. 217

Index